Kew Bulletin Additional Series VII

The FLORA of ALDABRA
and neighbouring islands

F R Fosberg & S A Renvoize

With an account of the mosses by C C Townsend

Illustrations by Mary Grierson & Ann Davies

LONDON: HER MAJESTY'S STATIONERY OFFICE

ISBN 0 11 241156 8*

Preface

This flora has been prepared for the Royal Society by Dr F. R. Fosberg of the National Museum of Natural History, Smithsonian Institution, Washington, DC (recently retired) in collaboration with S. A. Renvoize of the Royal Botanic Gardens, Kew, with a brief account of the mosses by C. C. Townsend (Kew). The illustrations are by Mary Grierson and Ann Davies. The work has been edited for the Kew Bulletin, Additional Series, by Dr G. E. Wickens (Kew).

An identification list of specimens seen and their whereabouts has been prepared by S. A. Renvoize and is obtainable from the Librarian, Royal Botanic Gardens, Kew, Richmond, Surrey, TW9 3AB, UK.

Contents

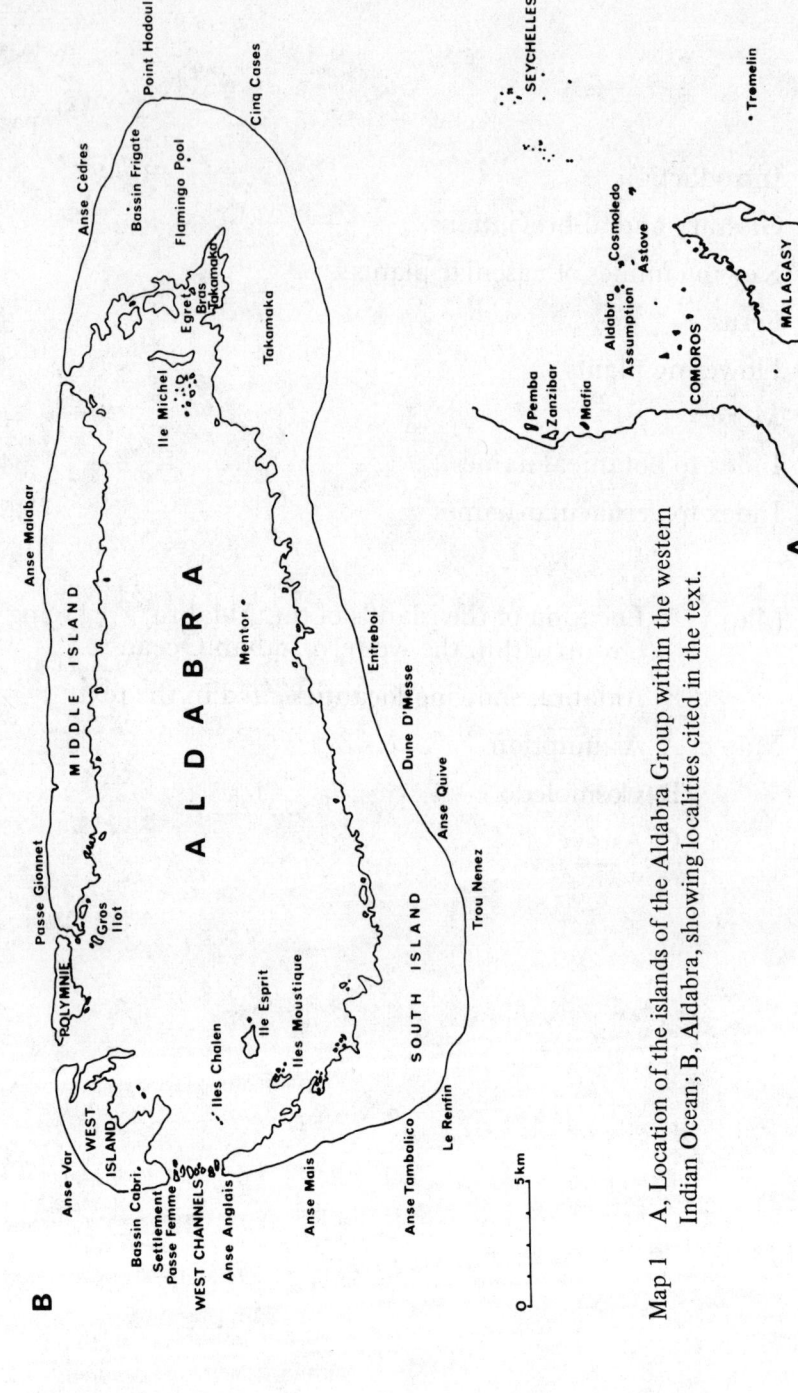

Map 1   A, Location of the islands of the Aldabra Group within the western Indian Ocean; B, Aldabra, showing localities cited in the text.

INTRODUCTION

Aldabra Atoll, Assumption Island, Cosmoledo Atoll, and Astove Atoll form a group of slightly elevated coral islands lying 320-420 km north and slightly west of the northern tip of Madagascar, and about 640-800 km east of the East African coast (Map 1/A).

The surfaces of the islands, especially of Aldabra itself, are mostly 3.5-8 m above mean low-tide level, with a few dunes at least temporarily reaching 10-30 m. The tide has a range of over 3.5 m at springs, and the shores tend to be low undercut cliffs with narrow reef platforms. Some stretches of coast have sandy beaches, usually with broader reefs. Storm-beaches and sand ridges line some of the rock-shores above high tide level, and substantial areas of dunes of coral sand occur on most of the larger islets or land areas of the group.

Aldabra (lat. 9°24' S., long. 46°20' E.), the largest and westernmost of the group, is an elongated ring of land, about 34 x 14 km maximum length and width, with about 97 square km of land area and a large but generally very shallow lagoon with many rocks and islets (Map 1/B).

Assumption (9°46'S., 46°31'E.) lies 32 km to the south of Aldabra. It is much smaller, 6 x 1.6 km maximum length and width, with about 960 ha land area, and no lagoon (Map 2/A).

Cosmoledo (9°41'S., 47°35'E.) is 110 km east of Aldabra. It is a more typical atoll in form, with an irregular oval ring-like reef, about 15 x 11 km in maximum length and width. There are five main islets and a number of smaller ones, totalling about 5 square km in land area: the two principal islets are Menai, on the west, and Wizard, on the east of the reef, both with high dunes (Map 2/B).

Astove (10°06'S., 47°45'E.) is the easternmost of the four islands. It is 5.6 x 4.1 km maximum length and width, with a narrowly horseshoe shaped land area of somewhat over 485 ha with a very shallow lagoon entrance (Map 2/C).

On none of the islands are any rocks exposed other than limestone and phosphate or phosphatic limestone. All except Aldabra have been strip-mined for phosphate rock or "guano", resulting in profound surface alteration, the introduction of exotic species and in the disappearance of some of the native ones, especially certain birds.

The data for the above paragraphs have been mostly adapted or modified from Stoddart, (ed., 1967 & 1970) which may be consulted, along with Stoddart et al. (1971) and Peters & Lionnet (1973) for further geographical and biological information.

The surface of the limestone is coral sand, locally coral gravel, or, mostly, hard limestone. The sand varies from a thin discontinuous veneer to ridges and dunes up to many metres thick. On all islands the sandy or gravelly areas have been largely planted to coconuts. The limestone surface varies from a flat pavement ("platin"), bare or covered by slabs or thin soil, to a moderately rough, pitted surface ("pavé"), to more or less pinnacled micro-karst ("champignon"). This varies from sharp or blunted pinnacle points, to an incredibly intricate and sharply etched fret-work in areas exposed to salt-spray. With the exception of Aldabra the surfaces have been extensively altered by phosphate mining.

These surface types are intimately related to the vegetational features, but the nature of the relationship is not always clear, and the patterns are not yet completely worked out. Geologically the islands, at least Aldabra, are said to have been completely submerged at least three times (Braithwaite el al. 1973). Biologically this seems questionable, because of the extent and differentiation of the flora and fauna, but see Wickens, in press, for further discussion. It is said to have been approximately 80,000 years since the last inundation. During the Pleistocene all the islands must have stood well above the sea-level, and must have also been at least somewhat larger; they probably presented a greater diversity of habitats than now.

Fresh-water ponds are fairly common on Aldabra and small, temporary pools are abundant. Crevices with fresh-water are also common, as well as small underground caverns containing small reservoirs of fresh water.

The climate is strongly seasonal, with a short, hot, usually rainy season from January to March, and a long dry trade-wind season for the rest of the year. Strong winds occur and salt spray is blown inland at times.

There is an abundant bird fauna, which includes sea-birds, shore-birds and wading-birds, as well as seed-eating, fruit-eating, and insectivorous land-birds. Giant tortoises, sea-turtles, and several insectivorous bats and one fruit bat are indigenous. Goats, cats, and rats have been introduced and have become too common. A rich arthropod fauna, of crustaceans, arachnids, and insects is also present.

None of the islands have truly indigenous human populations; all have had intermittent occupation by small settlements of principally Seychellois fisher-men, guano diggers, and other contract labourers. These, of course, have had a considerable influence on the biota and, locally on the geographical features of all the islands. The details have only begun to be elucidated and will not be elaborated on in any detail here apart from noting that phosphate mining and the planting of coconuts and *Casuarina* have had the most conspicuous effects. Since 1969 a research station has been maintained by Aldabra by the Royal Society.

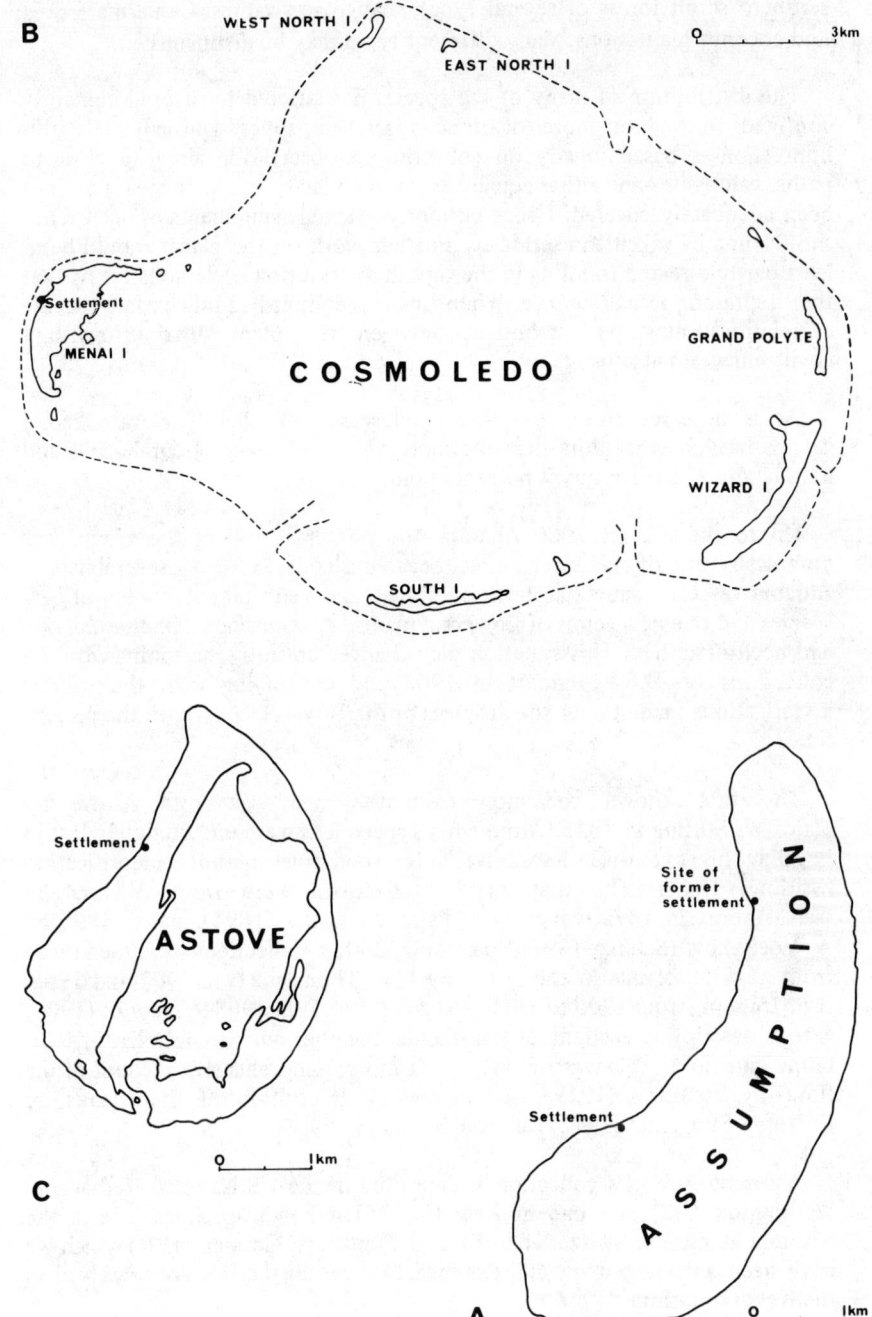

Map 2   A, Assumption; B, Cosmoledo; C, Astove.

The vegetation varies from sea-grass meadows to grasslands, open to closed scrub to scrub forest of several types, mangrove swamps, Casuarina groves, and coconut plantations. Many different types may be distinguished.

The distribution of many of the species is restricted to or predominantly confined to one or more of these vegetation types. Unfortunately this impression is based mostly on collecting or observation along or close to paths, camp sites and other equally accessible places, and large areas have not been adequately covered. Hence currently available statements of occurrence should not be taken too seriously. Further work on the plants should be at least partly directed to filling in the gaps in distribution or demonstrating that they represent actual absence. When this is accomplished much more reliable correlations may be established between the plant distributions and environmental patterns.

As is observed throughout the coral seas, even slightly elevated coral islands have a richer flora than the more numerous sea-level coral atolls and islands; the Aldabra group is no exception.

Up to the end of 1966 Aldabra, the principal atoll of the group, was botanically one of the lesser known oceanic islands. However, several lists of Aldabra vascular plants had been published, and a substantial number of new species and one new genus of flowering plants *(Apteranthera, Amaranthaceae)* had been described. This situation then changed abruptly, beginning with the collections of D.R. Stoddart in 1966 and continuing with the various investigations made under the auspices of the Royal Society until the present time.

The first known collection from the area was from Astove by Major W. Stirling in 1835. Although his specimens were unfortunately lost in Bombay his extensive descriptive notes sometimes permit identification (Stirling, 1843). The first extant collections were from Aldabra by W.L. Abbott in 1892, which were listed by Baker (1894), and in 1895 by A. Voeltzkow (Schinz, 1897; Voeltzkow, 1902). Collections were then made from all four islands in the group by H.P. Thomassett, in 1902 and 1908; R.P. Dupont, from 1906 to 1916, and J.C.F. Fryer in 1908-9. Dupont (1907) gave a descriptive account of the islands, together with a check-list of their fauna and flora. This was followed by a more comprehensive account of the flora by Hemsley (1919). An ecological description of the islands by D. Vesey-Fitzgerald (1942) followed his visit in 1937.

A set of Abbott's collection is deposited in the US National Herbarium, Washington (US) and one in Kew (K). Voeltzkow's specimens are in the herbaria at Zurich, Switzerland (Z), and Frankfurt, Germany (FR) which we have been able to borrow and examine. Almost all the rest are preserved in the Kew Herbarium.

In 1966 and 1967 David Stoddart made a substantial collection on the islands. Charles Rhyne also collected in 1967 in spare time from his algae collecting. Fosberg and Renvoize spent 2 and 3 months respectively, on the island in 1968, with brief visits to Cosmoledo and Astove. M.E. Gwynne and David Wood also made important collections on Cosmoledo and Astove. Stoddart and M.E.D. Poore collected in 1968 on all four of the islands covered by this flora. Francis Merton collected some material on Aldabra in 1973 and 1974 to support his vegetation studies; also in 1976 G.E. Wickens added a few specimens during his vegetation studies. Renvoize visited Aldabra again, briefly, in 1975 and filled in certain gaps in the collections. Jack Frazier over a period of several years made important collections on Aldabra, Assumption and Cosmoledo. Royal Society Research Station personnel, David Wood and Roger and Sarah Hnatiuk also made valuable collections, mostly incidental to other duties during their terms at the station. More recently on Aldabra C.W.D. Gibson has discovered a number of new records from the eastern end of South Island.

All of the above collections have been available for the preparation of the present flora. The best sets of the collections made during the last ten years are at Kew and the US National Herbarium; duplicates have been sent to various other herbaria.

The research leading to the present flora is mostly published in a series of precursory papers in the Kew Bulletin (Renvoize, 1971 b & 1972; Fosberg, 1974, 1977 a-c & 1978 a-e). In these papers certain species are either described for the first time, or are reinterpreted and in some cases have been given a different taxonomic treatment. These papers also contain a more detailed discussion than is appropriate in the flora. These papers are included in the flora citations together with other relevant references including those to original places of publication, basionyms, and occasional other notes necessary to understanding the nomenclature of the species. Usually only synonyms that have been used in literature on Aldabra are listed.

In all there are 274 species and varieties admitted for the angiosperm and pteridophyte flora. Of the angiosperms 185 are indigenous, of which 43 are endemic to the group. The remaining 87 have probably been introduced in historic time, though this may be hard to document. Some of them are probably no longer present.

The number of endemics is high for such low coral islands, especially in view of the suggestion by Braithwaite et al. (1973) that Aldabra, at least has been completely submerged several times in its geological history. In fact, this proportion of endemics suggests that the basis for the claim of several full submergences should be carefully reexamined. The fact that many of the endemics are found on more than one or on all four islands in the group may be significant in this connection. Either these islands have, in their history, been more extensive and intimately connected, or the dispersal mechanisms of the species in the flora have been remarkably effective.

The relationships of the flora have been considered in a preliminary way by Renvoize (1971 & 1975). At that time the flora was imperfectly known and the status of a number of the species had not been established. A brief reconsideration now seems appropriate. The statistics of the flora, as presently understood (Wickens, in ed.), may be summarized as follows:

I.   ANGIOSPERMS   272

   a)   Terrestrial

      1)   Native taxa: 176

            Endemics   42

                  Aldabra   17
                  Aldabra, Assumption   4
                  Aldabra, Assumption, Cosmoledo   2
                  Aldabra, Assumption, Astove   2
                  Aldabra, Assumption, Cosmoledo,
                     Astove   9
                  Aldabra, Cosmoledo   2
                  Aldabra, Cosmoledo, Astove   1
                  Aldabra, Astove   2
                  Assumption   1
                  Cosmoledo   1
                  Cosmoledo, Astove   1
                                                        42

                  Pantropical   30
                  Palaeotropical   14
                  Indo-Pacific   26
                  Indian Ocean   19
                  African — Madagascar   20
                  Africa — Mascarenes   2
                  Africa   3
                  Madagascar   16
                  Asia   3
                  Not known   1
                                                       134

      2)   Introduced   87

                  Weeds   41
                  Cultivated   46
                                                        87

   b)   Marine   9

                  Cosmopolitan   1
                  Indo-Pacific   8
                                                         9

II.   PTERIDOPHYTES   2

            Pantropical   2

   Total Flora   274

The above analysis has been somewhat simplified. Thus plants with an Indian Ocean distribution include a wide range of distribution patterns such as the Aldabra group and the Seychelles, Aldabra group and the Mascarenes as well as more widely ranging patterns that include many of the islands as well as the east coast of Africa. The status of a number of species is questionable. Is the *Casuarina* truly indigenous or could it have been introduced by Arab traders centuries ago or more recently by Europeans? Perhaps we shall never know.

REFERENCES

BAKER, J.G. (1894). Flora of Aldabra Islands. Bull. Misc. Inf. Kew 1894: 146–151.

BRAITHWAITE, C.J.R., TAYLOR, J.D. & KENNEDY, W.J. (1973). The evolution of an atoll, the depositional and erosional history of Aldabra. Phil. Trans. Roy. Soc. Lond. B, 266: 307–340.

DUPONT, R.P. (1907). Report on a visit of investigation to St. Pierre, Astove, Cosmoledo, Assumption and the Aldabra Group of the Seychelles Islands. Seychelles: Government Printer.

FOSBERG, F.R. (1974). Miscellaneous notes on the flora of Aldabra and neighbouring islands: *Flacourtiaceae, Guttiferae, Meliaceae, Anacardiaceae, Rubiaceae, Compositae, Plumbaginaceae, Boraginaceae, Convolvulaceae, Amaranthaceae, Euphorbiaceae, Moraceae.* Kew Bull. 29: 253–266.
.. ..   (1977a). Miscellaneous notes.. :IV   A new *Bulbostylis* and observations on *Cyperus (Cyperaceae).* Kew Bull. 31: 829–835.
.. ..   (1977b). Miscellaneous notes.. :V   *Pandanus tectorius* Parkinson, sensu latissimo *(Pandanaceae).* Kew Bull. 31: 837–840.
.. ..   (1977c). Miscellaneous notes.. :VI   *Portulaca (Portulacaceae)* in the Aldabra Group. Kew Bull. 32: 253–258.
.. ..   (1978a). Miscellaneous notes.. :VII   *Eugenia elliptica* var. *levinervis (Myrtaceae).* Kew Bull. 33: 133–134.
.. ..   (1978b). Miscellaneous notes.. :VIII   *Hedyotis* subgenus *Oldenlandia (Rubiaceae).* Kew Bull. 33: 135–140.
.. ..   (1978c). Miscellaneous notes.. :IX   The native Aldabra *Solanum (Solanaceae).* Kew Bull. 33: 141–142.
.. ..   (1978d). Miscellaneous notes.. :X   The Aldabra *Clerodendrum (Verbenaceae).* Kew Bull. 33: 143–144.
.. ..   (1978e). Miscellaneous notes.. :XI   Critical notes on *Euphorbiaceae.* Kew Bull. 33: 181–190.

FRYER, J.C.F. (1911). The structure and formation of Aldabra and neighbouring islands with notes on their flora and fauna. Trans. Linn. Soc. Lond. II, Zool. 14: 397–442.

HEMSLEY, W.B. (1916). Flora of Seychelles and Aldabra: new phanerogams, chiefly of the Perry Sladen Trust Expedition, with some emendations in synonymy. Journ. Bot. 54, Suppl, 1–4 and continuation in Journ. Bot. 54: 361–363.
.. ..   (1917). Plants of Seychelles and Aldabra. Journ. Bot. 55: 285–288.

HEMSLEY, W.B. (1919). Flora of Aldabra with notes on the flora of the
    neighbouring islands. Bull. Misc. Inf. Kew 1919: 108–153.

PETERS, A.J. & LIONNET, J.F.G. (1973). Central Western Indian Ocean
    Bibliography. Atoll Res. Bull. 165: 1–322.

RENVOIZE, S.A. (1971a). The origin and distribution of the flora of
    Aldabra. Phil. Trans. Roy. Soc. Lond. B, 260: 227–236.
.. ..   (1971b). Miscellaneous notes on the flora of Aldabra and
    neighbouring islands: 1 Five new species of grasses. Kew Bull. 25:
    417–422.
.. ..   (1972). Miscellaneous notes..   :II   A new species of
    Dichrostachys (Leguminosae) from Aldabra. Kew Bull. 26: 433–438.
.. ..   (1975). A floristic analysis of the western Indian Ocean
    coral islands. Kew Bull. 30: 133–152.

SCHINZ, H. (1897). Zur Kenntnis der Flora der Aldabra-Inseln. Abhand.
    Senck. Nat. Ges. 21: 77–91.

STIRLING, W. (1843). Narrative of the wreck of the ship 'Tiger' of Liverpool
    (Captain Edward Searight), on the desert island of Astove, on the morning
    of the 12th of August, 1838; with some particulars regarding the ship, the
    members of the crew on the island, and their subsequent deliverance; by
    Major (late Captain) W. Stirling. Exeter: Stirling, private publication.

STODDART, D.R. (Ed.) (1967). Ecology of the Aldabra Atoll. Atoll Res.
    Bull. 118: 1–141.
.. ..   (Ed.) (1970). Coral islands of the Western Indian Ocean.
    Atoll Res. Bull. 136: 1–122.
.. ..   (Ed.) (1971). A discussion on the results of the Royal
    Society Expedition to Aldabra 1967–68. Phil. Trans. Roy. Soc. Lond. B,
    260: 1–654.

VESEY-FITZGERALD, D. (1942). Further studies of the vegetation on
    islands in the Indian Ocean. Journ. Ecol. 30: 1–16.

VOELTZKOW, A. (1902). Die von Aldabra bis jetzt bekannte Flora und
    Fauna. Abhand. Senck. Nat. Ges. 26: 539–565.

WICKENS, G.E. (in press). Speculations on seed dispersal and the flora of
    the Aldabra archipelago. Phil. Trans. R. Soc. Lond. B.

GLOSSARY OF BOTANICAL TERMS

achene — a small, dry, indehiscent, 1-seeded fruit.

acuminate — tapering to a protracted point.

acute — sharp, regularly pointed to an angle of less than 90°.

aerogenous — tissue containing air-spaces.

alveolate — honey-combed.

androgynophore — a stalk bearing both stamens and ovary above the point of perianth attachment.

anthesis — the period when a flower is expanded and pollination takes place; loosely applied for the flowering period.

anthocarp — a fruit surrounded by the persistent perianth base.

apiculate — ending abruptly in a short point.

apocarpous — carpels free and separate from one another.

appressed — lying close and flat.

areole — the area enclosed in one unit of a reticulate arrangement.

aril — a fleshy outgrowth from the testa of a seed.

arista — a long bristle.

attenuate — gradually tapering.

axile placentation — ovules attached to the axis formed by the union and fusion of the partitions of the ovary.

axillary — arising from the angle between a leaf or bract and a stem.

baccate — berry-like.

bast fibres — fibres composed of long, slender, thick-walled cells with tapering ends, occurring in stem (as opposed to leaves, as in *Agave*).

berry — a juicy, indehiscent fruit with soft pericarp (q.v.) and the seeds immersed in a pulp.

bipinnate — when the primary divisions (pinnae) of a pinnate leaf are themselves once pinnate (q.v.).

bract — a small, often modified, leaf subtending a flower or inflorescence.

bracteole — a small bract on the pedicel, or close under the flower, or a secondary bract associated with a larger one, often paired; perhaps stipular in origin.

capitate — a) like the head of a pin, e.g. the stigma in some flowers;
  b) gathered into compact, head-like clusters, e.g. flowers of *Compositae*.

capitulum — a dense, head-like, inflorescence of usually sessile flowers.

capsicin — an acrid, alkaloid substance found in some species of *Capsicum*.

capsule — a dry fruit composed of 2 or more united cells which, when ripe, split regularly into valves or open by slits or pores.

carpel — an organ containing 1 or more ovules, free or several united to form the ovary.

cartilaginous — cartilage-like, tough but pliable.

caruncle — an outgrowth near the attachment scar of a seed.

caryopsis — a 1-seeded, dry fruit with the thin pericarp usually adherent to the seed.

ciliate — with a fringe of hairs.

circumscissile — opening by a line around the circumference of a fruit or anther etc. (see also calyptra).

cladode — a modified stem having the appearance and function of a leaf.

clavuncle — cup-like or flange-like expansion of the summit of a style supporting a surrounding stigma, at times indistinguishable from the stigma.

cleistogamous — when fertilization occurs within an unopened flower.

coccus — an ill-defined term applied to the seeded segments into which some fruits break up, e.g. *Malvaceae, Euphorbiaceae,* etc.

cochleate — having a coiled appearance like a snail's shell.

compound leaf — blade divided to the rachis (q.v.) or petiole (q.v.) to form separate leaflets (c.f. simple leaf).

cordate — heart-shaped; as in the base of some leaves.

corona — a circle of appendages between the stamens and the corolla.

corymb — a flat-topped, racemose inflorescence in which the branches, or pedicels, start from different points but all reach the same level.

crenate — shallowly notched or scalloped with rounded or blunt teeth (cf. serrate).

cupule — a small cup.

cuspidate — abruptly tipped with a hard rigid point.

cyathium — a cup-shaped structure with stamens and stalked ovary, each stamen and ovary being a separate flower, as in *Euphorbia.*

cyme — an inflorescence in which the central flower opens first, and the first branches are usually forked or opposite.

cymule — a diminative cyme or a portion of one.

cystoliths (crystoliths) — mineral concretions within tissue.

decurrent — when veins or ridges are continued from an organ down into its axis.

deliquescent — readily becoming liquid after maturity.

deltoid — shaped like an equal-sided triangle.

dioecious — with the male and female flowers on separate plants.

disseminule — the organ of dispersal.

distichous — regularly arranged in 2 opposite rows.

domatia — cavities, formed by the plant, often harbouring insects or mites in an apparently symbiotic relationship.

drupe — a fleshy fruit with a hard endocarp (q.v.) enclosing the seed, e.g. cherry, plum, sometimes more than 1 endocarp enclosing seeds.

endocarp — the innermost layer of the pericarp (q.v.).

endosperm — nutritive material within the seed surrounding the embryo.

epicalyx — whorl of bracts immediately below the flower resembling an extra calyx whorl.

epilithic — growing on rocks.

epigynous — with petals and stamens inserted above the ovary, i.e. ovary inferior (cf. hypogynous).

epiphyte — a plant which grows on another plant without deriving nourishment from it.

ericoid — having the habit of *Erica* (heather).

exocarp — the outer layer of the pericarp (q.v.).

fascicle — a cluster of flowers or leaves etc. arising from the same point.

follicle — a fruit formed from a single carpel and opening along the inner suture.

fascicle — the stalk by which the ovule is attached to the placenta.

glaucous — with a pale bloom, or pale bluish-green.

glomerule — a small compact cluster.

gynophore — a stalk supporting the ovary, formed by an elongation of the receptacle (q.v.).

hastate — with 2 diverging, triangular lobes, as in the base of some leaves.

haustorium — specialized organ of a parasitic plant which penetrates the host and withdraws food.

heterostylous — having styles of different lengths.

hirtellous — softly and minutely hairy.

hispid — stiffly hairy.

hyaline — almost transparent.

hypanthium — part of the flower between the base of the calyx lobes and the ovary (calyx tube), in some families used for the part adnate to the ovary.

hypocotyl — the part of the seedling stem below the cotyledons.

hypogynous — with petals and stamens inserted on the receptacle below the ovary, i.e. ovary superior (cf. epigynous).

involucre — a number of bracts surrounding the base of a flower-head.

labellum — the lowest petal of an orchid, usually enlarged and different in shape from the 2 lateral petals.

lanceolate — relatively narrow and narrowed at both ends; often used as narrow but broadest towards the base.

lenticels — corky spots on the bark.

loculicidal — when a ripe capsule splits along the back of the carpels (cf. septicidal).

mericarp — the usually 1-seeded segment of a schizocarp (q.v.).

mesocarp — the middle part of a 3-layered pericarp (q.v.) usually fleshy or pulpy.

monoecious — when male and female flowers are separate but borne on the same plant (cf. dioecious).

monopodial — having 1 continuous main stem or axis.

muticous — blunt.

palmate leaf — a compound leaf (q.v.). where several leaflets diverge from the same point.

palmatifid — with the margins palmately cleft.

panicle — an inflorescence in which the axis is divided into branches bearing several flowers.

papillate — covered with minute protruberances.

pappus — a ring of hairs or scales around the top of a fruit *(Compositae)*.

parietal — when the ovules are attached to the inner surface of the ovary wall.

pedicel — the stalk of an individual flower.

peduncle — the common stalk of several pedicelled or sessile flowers.

peltate — of a leaf with the petiole attached to the undersurface within the periphery of the blade.

pepo — an inferior, 1-celled, many-seeded, pulpy fruit with a rind *(Cucurbitaceae)*.

pericarp — the wall of a ripened ovary.

perigone — a perianth, usually angled, which is not clearly differentiated into calyx and corolla, as in *Nyctaginaceae*.

perigynous — with sepals, petals and stamens inserted on an open receptacle around the ovary but not united to it.

petiolate — leaf with petiole present.

petiole — the stalk by which the leaf blade is attached to the stem.

petiolule — the stalk by which a leaflet in a compound leaf (q.v.) is attached to the rachis (q.v.) or petiole.

phalange — elongate unit of a ripe aggregate or multiple fruit.

phyllode — flattened petiole or leaf-rachis with the form and functions of a leaf.

pilose — hairy, with long, simple hairs.

pinnate leaf — a compound leaf (q.v.) with leaflets arranged along the rachis (q.v.).

pneumatophore — aerial root of trees in wet habitats, e.g. mangroves.

pollinia — clusters of pollen grains.

polygamous — when a species has unisexual and bisexual flowers on the same or different individuals.

prophyll — the first 'leaf' of a shoot.

pubescent — with short, soft hairs.

putamen — the shell of a nut or the bony part of a stone fruit.

pyrene — a nutlet or kernel, the stone of a drupe.

raceme — an inflorescence in which the flowers are borne on pedicels along an unbranched axis.

rachis (rhachis) — the principle axis of an inflorescence or compound leaf.

receptacle — the extremity of the axis on which the flower parts are inserted.

rhaphide (raphide) — a needle-shaped crystal of calcium oxalate.

sagittate — arrow-like, as in the base of some leaves (cf. hastate).

samara — a winged, indehiscent, 1-seeded fruit.

saprophyte — a plant living upon dead or decaying tissues of other plants and animals.

scape — a naked flower stalk arising from a rosette of leaves.

scarious — thin and dry, not green.

schizocarp — a dry compound fruit which splits into usually 1-seeded segments at maturity; each segment is known as a mericarp.

sclerenchymatic — having the nature of rigid, thick-walled, sclerenchyma tissue.

sclerified — hardened, like stone cells.

scorpioid — when the main axis of an inflorescence is coiled in bud, the flowers being usually 2-ranked.

septicidal — when a ripe capsule splits along the lines of junction of the carpels (cf. loculicidal).

septum — a dividing partition.

sericeous — silky, with appressed, soft, straight hairs.

serrate — margin with saw-like teeth (cf. crenate).

sessile — without a stalk.

simple leaf — leaf blade not divided to the midrib (cf. compound leaf).

sinus — recess between teeth or lobes on a margin or leaf-base.

sorus — a cluster of sporangia with or without a cover.

spadix — a flower-spike with thickened and fleshy axis *(Palmae)*.

spathulate — spatula-shaped.

sporangium — the spore-bearing organ (e.g. Ferns)

spore — microscopic, unicellular reproductive disseminule having the diploid number of chromosomes.

staminode — an abortive or vestigial stamen without a perfect anther.

stellate hairs — star-shaped hairs with several arms radiating from one point.

stipule — leaf-like or scale-like appendage of the leaf, usually at the base of the petiole.

striate — marked with parallel, longitudinal lines, grooves or ridges.

strigose — with stiff hairs lying close along the surface.

subrotate — almost wheel-shaped.

suffrutescent — slightly shrubby.

suture — the line of junction or seam (of a carpel).

tepal — a division of a perianth not differentiated into sepals and petals.

testa — the outer coat of a seed.

thyrsoid — having a panicle with the secondary and alternate axes cymose.

tomentose — densely covered with short, felted hairs.

torus — see receptacle.

triseriate — in 3 series or lines.

tuberculate — covered with wart-like protruberances.

umbel — an inflorescence in which the divergent pedicels or rays spring from the same point.

umbo — a small boss or protuberance.

utricle — a thin-walled, usually indehiscent or irregularly rupturing sac surrounding the fruit (e.g. *Cyperaceae, Amaranthaceae*).

whorl — an arrangement of similar parts in a circle at the same level.

ABBREVIATIONS

For the frequently used bibliographical references the following abbreviations are being employed:

A.S.N.G.   Abhandlungen der Senckenbergischen naturforschenden Gesellschaft

B.M.I.K.   Bulletin of Miscellaneous Information, Royal Botanic Gardens, Kew.

F.M.C.   Flore de Madagascar et des Comores (families or groups of families published as separate fascicles)

F.M.S.   Baker, J.G. Flora of Mauritius and the Seychelles.

F.T.E.A.   Flora of Tropical East Africa (families published as separate fascicles)

F.Z.   Flora Zambesiaca

K.B.   Kew Bulletin

P.T.R.S.   Philosophical Transactions of the Royal Society of London.

The usual contractions found in taxonomic literature are used for the remaining references (for Dupont (1967) see Introduction).

For citing the distribution within the Aldabra group of islands the following abbreviations are used:

ALDABRA: W — West Island; P — Polymnie; M — Middle Island; S — South Island. (islets within the lagoon are fully named.)

COSMOLEDO: W — Wizard Island; M — Menai Island; WN — West North Island.

Localities cited in the text are shown in Maps 1 and 2.

KEY TO FAMILIES OF VASCULAR PLANTS

1. Non-flowering plants (ferns), reproducing by means of spores; leafy ferns bearing sporangia on the undersurface of the leaves   **1. Polypodiaceae**

1. Flowering plants, reproducing by means of seeds

   2. Submerged aquatics, mainly marine   **GROUP 1**

   2. Terrestrial plants, if aquatic then emergent above the surface

      3. Plants essentially devoid of functional leaves

         4. Twining, parasitic, wiry plants bearing haustoria
            58. Lauraceae *(Cassytha)*

         4. Plants not as above

            5. Trees, pine-like in appearance; branchlets needle-like, jointed, photo-synthetic, bearing numerous whorls of scale-like, reduced leaves
               65. Casuarinaceae *(Casuarina)*

            5. Shrubs, vines or herbs; reduced leaves, if present, not whorled

               6. Feathery or broom-like, non-fleshy shrubs, milky sap absent

                  7. Feathery, scrambling, much-branched, woody perennial with clusters of linear, green, leaf-like, modified stems; scale leaves papery   **68. Liliaceae** *(Asparagus)*

                  7. Tufted, broom-like shrub, stems usually leafless but with triangular bracts c 1 mm long at the nodes   **40. Plumbaginaceae** *(Plumbago)*

               6. Herbs, scramblers or creepers

                  8. Herbs or somewhat woody below; milky sap absent, plants fleshy, jointed; flowers inconspicuous, in cylindrical spikes
                     57. Chenopodiaceae *(Arthrocnemum)*

                  8. Scrambling or creeping succulent plant; milky sap present, flowers in umbels   **45. Asclepiadaceae** *(Sarcostemma)*

      3. Leaves functional, well-developed

         9. Leaves simple or 1-foliolate, may be deeply divided or variously dissected

            10. Leaves clearly opposite or whorled

               11. Trees, shrubs or woody climbers   **GROUP 2**

               11. Herbs, sometimes woody at the base   **GROUP 3**

    10. Leaves alternate or congested at base of plant or at stem tips so that it is difficult to be sure of the arrangement

        12. Trees, shrubs or woody climbers   GROUP 4

        12. Herbs, sometimes woody at the base   GROUP 5

    9. Leaves compound   GROUP 6

GROUP 1. Submerged vascular aquatics, including an alga with which they could be confused

1. Leaves (or branchlets) opposite or whorled

    2. Leafless, branchlets thread-like, whorled
            Algae *(Chara zeylanica* Kl. ex Willd. var. *diaphana* (Meyer) R.D.W.)

    2. Leaves opposite

        3. Leaves elliptic or oblong, relatively broad
               74. Hydrocharitaceae *(Halophila)*

        3. Leaves very narrowly linear   **72. Najadaceae** *(Najas)*

1. Leaves alternate or 2-ranked and crowded

    4. Stems delicate, thread-like, branched; rhizomes absent
               73. Potamogetonaceae *(Ruppia)*

    4. Rhizomes and erect stems cord-like, coarse

        5. Roots and leafy branches arising from every 4th node
               73. Potamogetonaceae *(Thalassodendron)*

        5. Roots and leafy branches arising from either every node or irregularly distributed

           6. Leaves circular in cross-section   **73. Potamogetonaceae** *(Syringodium)*

           6. Leaves flat, grass-like

               7. Leaves with 3 nerves, lateral nerves faint
               73. Potamogetonaceae *(Halodule)*

              7. Leaves with more than 3 nerves

                 8. Roots and leafy branches at each node; roots without dense white hairs; leaf-sheaths with ligules, main nerves not connected by cross nerves   **73. Potamogetonaceae** *(Cymodocea)*

8. Roots and leafy branches irregularly disposed on rhizome; internodes 2–7 mm long; roots with dense, white hairs on proximal part; leaf-sheaths without ligules, cross nerves abundant
74. Hydrocharitaceae *(Thalassia)*

GROUP 2. Trees, shrubs or woody climbers; leaves opposite or whorled, simple or 1-foliolate, may be deeply divided or variously dissected

1. Stipules present, evident, often falling and leaving scars

   2. Milky sap present; stipules triangular, 1.2 mm long, soon falling; leaves deciduous, spirally arranged at branch tips
62. Euphorbiaceae *(Euphorbia)*

   2. Milky sap absent

      3. Stipules not interpetiolar, a separate pair of stipules or scars for each leaf   **28. Myrtaceae** *(Eugenia)*

      3. Stipules interpetiolar, 1 stipule or scar on each side of stem, connecting leaf bases

         4. Petals united; ovary inferior; calyx lobes 4–6, small, terrestrial trees and shrubs   **37. Rubiaceae**

         4. Petals free; ovary superior *(Cassipourea)* or if apparently inferior, mangroves with aerial roots; calyx lobes 4–15, conspicuous, linear-lanceolate   **26. Rhizophoraceae**

1. Stipules absent or difficult to see

   5. Flowers in heads subtended by 1 or more whorls of bracts

      6. Bracts 4, violet, oblanceolate, 3 x 1.3 cm; woody climber
52. Verbenaceae *(Congea)*

      6. Bracts numerous, up to 1 cm long

         7. Ovary inferior; shrubby herb   **38. Compositae** *(Melanthera)*

         7. Ovary superior; prickly shrub   **52. Verbenaceae** *(Lantana)*

   5. Flowers in looser inflorescences

      8. Ovary completely inferior or apparently so

         9. Parasitic shrubs

            10. Flowers irregular; stamens elongated, 1 cm or more long
60. Loranthaceae *(Bakerella)*

10. Flowers regular; stamens very short   **61. Viscaceae** *(Viscum)*

9. Free-living trees and shrubs

   11. Leaves gland-dotted; flowers solitary, axillary
                                          26. Myrtaceae *(Eugenia)*

   11. Leaves not gland-dotted; flowers in cymose clusters
                                         55. Nyctaginaceae *(Pisonia)*

8. Ovary superior, sometimes closely invested by the united calyx

   12. Milky sap present

      13. Erect trees or shrubs; leaves crowded towards the branch tips

         14. Deciduous; leaves spirally arranged; whorls of cymes subterminal, bearing naked male or female flowers subtended by bracts fused to form a cup-like structure   **62. Euphorbiaceae** *(Euphorbia)*

         14. Leaves opposite, pairs often markedly unequal; flowers large and showy, in subterminal cymes   **44. Apocynaceae**

      13. Twining and scrambling woody climber; cymes much-branched
                                  45. Asclepiadaceae *(Secamone)*

   12. Milky sap absent

      15. Shrub bearing 4 spines up to 3 cm long at each node
                                  43. Salvadoraceae *(Azima)*

      15. Not as above

         16. Flowers irregular (bilaterally symmetrical)

            17. Climbing shrub; cymes congested, subtended by 4 oblanceolate bracts, 3 x 1.3 cm   **52. Verbenaceae** *(Congea)*

            17. Not as above

               18. Corolla blue, trumpet shaped, 2-lipped
                                  51. Acanthaceae *(Barleria)*

               18. Corolla white, yellow or greenish, salver-shaped, not markedly 2-lipped   **52. Verbenaceae**

         16. Flowers regular (radially symmetrical)

            19. Leaves small, oblanceolate to elliptic, less than 2 cm wide

               20. Flowers 4-merous, petals united; fruit a subglobose, 1-seeded drupe   **43. Salvadoraceae** *(Salvadora)*

20. Flowers 6-merous, petals free; fruit a depressed, cylindrical, many-seeded capsule   **29. Lythraceae** *(Pemphis)*

19. Leaves large, more than 2 cm wide

  21. Leaves with c 100 closely parallel veins; flowers in axillary clusters; fruit subglobose to globose, 25 cm in diameter, without attached, enlarged calyx   **8. Guttiferae** *(Calophyllum)*

  21. Leaves with c 10 pairs veins

    22. Flowers large; stamens 12-many; fruit globose, 3–4 cm in diameter, lower part enveloped in the leathery, deeply 8-lobed calyx   **30. Sonneratiaceae** *(Sonneratia)*

    22. Flowers small; stamens 5; fruit usually paired berries or spindle-shaped capsules; calyx 5-lobed, not enveloping the fruit   **44. Apocynaceae**

GROUP 3. Herbs, sometimes somewhat woody at the base; leaves opposite or whorled, simple or 1-foliolate, may be deeply divided or variously dissected

1. Stipules present, evident, often falling and leaving scars

  2. Sap milky; styles 3, ovary 3-celled; stipules persistent
            62. Euphorbiaceae *(Euphorbia)*

  2. Sap not milky

   3. Leaves fleshy

    4. Stipules not interpetiolar, reduced to axillary tufts of hairs; flowers yellow   **7. Portulacaceae** *(Portulaca)*

    4. Stipules interpetiolar, fused to the expanded leaf-base to form a membraneous sheath; flowers white or pink
            35. Aizoaceae *(Trianthema)*

   3. Leaves not noticeably fleshy; stipules interpetiolar, flowers white or pink   **37. Rubiaceae** *(Hedyotis)*

1. Stipules absent or difficult to see

  5. Perianth in 1 series, either all sepaloid or all petaloid

   6. Flowers in terminal spikes   **56. Amaranthaceae** *(Achyranthes)*

   6. Flowers variously arranged, not in spikes

7. Flowers terminal, in paniculate cymes; fruit a 5-ribbed, glandular
   anthocarp   **55. Nyctaginaceae** *(Boerhavia)*

7. Flowers axillary

  8. Plants conspicuously fleshy; flowers solitary
   35. Aizoaceae *(Sesuvium)*

  8. Plants not noticeably fleshy; flowers in axillary cymes or several in
   an axil, less than 4—5 mm long

    9. Leaf-pairs markedly unequal; flowers sessile
   56. Amaranthaceae *(Alternanthera)*

    9. Not as above

      10. Flowers in cymes; perianth united, funnel- to bell-shaped; fruit
       club-shaped, 5-ribbed and sticky, or obovoid to subglobose,
       prominently wrinkled   **55. Nyctaginaceae**

      10. Flowers on axillary pedicels or cymose; perianth segments free;
       fruit a thin-walled capsule   .   **36. Molluginaceae** *(Mollugo)*

5. Perianth in 2 series (calyx reduced to hairs, bristles, scales or spines in
   Compositae)

  11. Plants fleshy, mat-forming; flowers solitary, axillary
    49. Scrophulariaceae

  11. Plants not fleshy

    12. Flowers regular (radially symmetric)

      13. Climbers, scramblers or erect herbs; flowers small, in axillary
     . umbellate clusters   **45. Asclepiadaceae**

    13. Erect or prostrate herbs; flowers not as above

      14. Flowers condensed into heads (capitula) surrounded by 1 or more
       whorls of usually highly modified bracts   **38. Compositae**

      14. Flowers not as above

        15. Flowers solitary or paired or in short, few-flowered cymes
         44. Apocynaceae *(Catharanthus)*

        15. Flowers and fruits in grooves in rachis of a long spike, flowers
         slightly irregular   **52. Verbenaceae** *(Stachytarpheta)*

    12. Flowers irregular (bilaterally symmetric)

    16. Parasitic herb; flowers scarlet   **49. Scrophulariaceae** *(Striga)*

    16. Not as above

      17. Flowers in whorls or pseudo-whorls

18. Flowers orange or white; fruit separating into 4 nutlets
54. Labiatae

18. Flowers blue, mauve or lilac; fruit a capsule
51. Acanthaceae *(Hypoestes)*

17. Flowers not in whorls or pseudo-whorls

19. Flowers and fruits in groves in rachis of a long spike
52. Verbenaceae *(Stachytarpheta)*

19. Not as above

20. Plants glandular, flowers solitary, axillary
53. Dicrastylidaceae *(Nesogenes)*

20. Plants not glandular, flowers in panicles or spikes
51. Acanthaceae

GROUP 4. Leaves alternate or congested at base of plant or at stem tips so
that it is difficult to be sure of the arrangement, simple or
1-foliolate; trees, shrubs or woody climbers

1. Twiners or hooked or tendril climbers

2. Twiners or plants climbing by means of prickles

3. Twiners without prickles

4. Leaves entire; ovary superior; flowers small, petals free, not showy;
fruit not winged   **3. Menispermaceae** *(Cissampelos)*

4. Leaves palmately dissected; ovary inferior; fruit winged
67. Dioscoreaceae *(Dioscorea)*

3. Plants supported by means of axillary prickles (not normally a
climber)   **20. Rhamnaceae** *(Scutia)*

2. Tendril climbers

5. Ovary superior; multi-filament corona present
32. Passifloraceae *(Passiflora)*

5. Ovary inferior; corona absent   **20. Rhamnaceae** *(Gouania)*

1. Plants not climbers

6. Leaves with parallel veins, veins not conspicuously pinnate

7. Leaves very fleshy, in rosettes; plants stemless or with stems
68. Liliaceae

    7. Leaves not very fleshy

       8. Clumps of tall, woody, jointed canes; leaves differentiated into sheath and blade, diffusely arranged on branches in 2 ranks
                                           76. Gramineae *(Bambusa)*

       8. Branched trees or shrubs, leaves crowded at ends of branches

          9. Leaves several to many cm long, spirally arranged in 3 ranks, margins often spiny; flowers dioeceous, without perianth
                                         71. Pandanaceae *(Pandanus)*

          9. Leaves usually not more than 10–25 cm long, not arranged in 3 clear spirals; flowers bisexual, with perianth
                                       68. Liliaceae *(Dracaena)*

   6. Leaves net-veined, or if apparently parallel, the veins clearly pinnately arranged

      10. Leaves peltately attached

        11. Leaves deeply lobed   **62. Euphorbiaceae** *(Ricinus)*

        11. Leaves heart-shaped, margins entire   **59. Hernandiaceae** *(Hernandia)*

      10. Leaves basally attached

        12. Leaf-blades conspicuously jointed to the petiole, gland-dotted
                                       14. Rutaceae *(Citrus)*

        12. Not as above

          13. Tufts of white hairs present in the leaf axils; drupe white or purplish, c 1.3 cm in diameter   **39. Goodeniaceae** *(Scaevola)*

          13. Hairs absent, except as part of general pubescence of plant

            14. Stipules present or represented by scars at bases of leaves

              15. Inflorescences terminal, on short lateral branches; flowers showy, yellow   **16. Ochnaceae** *(Ochna)*

              15. Inflorescences axillary

                16. Leaves with toothed margins

                  17. Stamens more than perianth parts; fruit 4-lobed (1 or 2 lobes sometimes aborted)   **10. Tiliaceae** *(Grewia)*

                  17. Stamens equal in number or half the number of perianth parts

18. Perianth in 1 series; inflorescence and leaves usually confined to the swollen branch tips

63. Urticaceae *(Obetia)*

18. Perianth in 2 series; stamens opposite the petals; leaves scattered along the branches

20. Rhamnaceae *(Colubrina)*

16. Leaves with entire or subentire margins

19. Flowers large, several cm across

20. Flowers 4-merous, solitary in the upper axils; stamens numerous, free; fruit fleshy, borne on a long column

5. Capparidaceae *(Capparis)*

20. Flowers 5-merous, solitary and axillary or in a terminal, few flowered panicle; stamens numerous, fused into a column; fruit a leathery capsule on a whorl of separate segments or lobes, not borne on a column   **9. Malvaceae**

19. Flowers much smaller, usually several; fruit not stalked

21. Leaves and flowers on deciduous branchlets; leaves on fertile branchlets 2-ranked; flowers monoecious; fruit a capsule   **62. Euphorbiaceae** *(Phyllanthus)*

21. Not as above

22. Leaves orbicular; stipules spine-like; stamens opposite the petals   **20. Rhamnaceae** *(Scutia)*

22. Leaves ovate or elliptic to obovate; stipules not spine-like

23. Branchlets tending to be flattened; milky sap absent; leaves obovate

11. Erythroxylaceae *(Erythroxylum)*

23. Branchlets not flattened; milky sap present; leaves not obovate   **64. Moraceae**

14. Stipules absent or very obscure

24. Inflorescence terminal

25. Inflorescence spicate   **56. Amaranthaceae** *(Deeringia)*

25. Inflorescences paniculate, cymose or corymbiform

26. Inflorescence a corymbiform cluster of heads of very small flowers in cup-like involucres of overlapping, scale-like bracts
38. Compositae *(Vernonia)*

26. Inflorescence a paniculate or cymose cluster of single flowers

    27. Petals absent; leaves narrowly elliptic, viscid; fruit winged
    21. Sapindaceae *(Dodonaea)*

    27. Petals present

        28. Petals united; flowers in cymes; fruit a drupe without a red, fleshy basal part   **46. Boraginaceae**

        28. Petals free; flowers in panicles; fruit with a basal, red, fleshy receptacle   **18. Icacinaceae** *(Apodytes)*

24. Inflorescence axillary

    29. Petals absent; flowers few or solitary   **6. Flacourtiaceae**

    29. Petals present

        30. Flowers in spikes or racemes; ovary inferior
        27. Combretaceae

        30. Flowers solitary or in cymes or fascicles

            31. Leaves conspicuously pubescent, pinnately veined, prominently toothed; flowers yellow, solitary in the upper axils   **31. Turneraceae** *(Turnera)*

            31. Not as above

                32. Stems and leaves with small, usually curved prickles
                48. Solanaceae *(Solanum)*

                32. Prickles absent; long spines or spinous branches sometimes present

                    33. Flowers in small cymes

                        34. Stems thick, milky sap present; stamens and ovary in cup-shaped involucres arranged in small cymes
                        62. Euphorbiaceae *(Euphorbia)*

                        34. Stems slender, milky sap absent

                            35. Fruit a fleshy drupe or a dehiscent, ovoid capsule
                            19. Celastraceae

                            35. Fruit a woody, elongated, 5-ribbed, indehiscent capsule   **25. Brexiaceae**

33. Flowers solitary or in fascicles or in very condensed inflorescences

    36. Flowers unisexual   **62. Euphorbiaceae**

    36. Flowers bisexual

       37. Petals united; fruit a 1-seeded, purplish-black, drupe-like berry   **41. Sapotaceae** *(Sideroxylon)*

       37. Petals free; fruit if fleshy more than 1-seeded

          38. Perianth leathery, flowers 3-merous; fruit a large aggregate; seeds many   **2. Annonaceae** *(Annona)*

          38. Flowers 5-merous; petals thin or very small

             39. Leaves linear-lanceolate; flowers solitary, petals bright yellow, thin, c 5–10 mm long; fruit dry, of 5 separate, 1-seeded carpels
                    15. Surianaceae *(Suriana)*

             39. Leaves broad, suborbicular; flowers small, in small clusters, petals dull yellowish or whitish, 1-2 mm long; fruit a berry or drupe
                    6. Flacourtiaceae

GROUP 5. Leaves alternate or congested at base of plant or at stem tips so that it is difficult to be sure of the arrangement, simple or 1-foliolate; herbs, sometimes woody at the base

1. Twiners or tendril climbers

  2. Twiners

    3. Leaves entire, ovary superior

      4. Inflorescences with leafy bracts; flowers small, petals free, not showy
                3. Menispermaceae *(Cissampelos)*

      4. Inflorescences with small, scale-like or awl-shaped bracts; flowers usually large, petals united, showy   **47. Convolvulaceae** *(Ipomoea)*

    3. Leaves palmately dissected; ovary inferior
              67. Dioscoreaceae *(Dioscorea)*

  2. Tendril climbers

    5. Ovary superior; multi-filament corona present, flowers bisexual
              32. Passifloraceae *(Passiflora)*

5. Ovary inferior; corona absent, flowers unisexual

34. Cucurbitaceae

1. Plants not climbers

6. Leaves with parallel veins, veins not conspicuously pinnate

7. Plants scrambling; 'leaves' (cladodes) narrowly linear 1–2 cm long

68. Liliaceae *(Asparagus)*

7. Not as above

8. Grass-like, non-fleshy plants; flowers without obvious perianth, borne in axils of scale-like bracts

9. Leaves arranged in 2-ranks; stems usually hollow, jointed; leaf-sheaths open; floral bracts arranged in 2-ranks, in short spikelets; inflorescences variously arranged; stamens and ovary enclosed between paired bracts   **76. Gramineae**

9. Leaves spirally arranged; stems usually solid; leaf-sheaths closed, leaves mostly basal except for leaf-like bracts usually in 3 ranks beneath inflorescences, floral bracts spirally arranged or in 2-ranks; spikelets single on stems or in capitate or compound clusters; stamens and ovary in axils of single scales or bracts

75. Cyperaceae

8. Plants with obvious perianth; leaves or stem, or both fleshy

10. Leaves thin, scattered along elongated stems; flowers blue, in spathe-like bracts   **69. Commelinaceae** *(Commelina)*

10. Leaves fleshy, crowded; flowers white or cream, very irregular, with conspicuous spur, not enclosed in spathe-like bracts

66. Orchidaceae

6. Leaves net-veined, or if apparently parallel, the veins clearly pinnately arranged

11. Plant usually unbranched, columnar, thick-stemmed; leaves palmately divided   **33. Caricaceae** *(Carica)*

11. Not as above

12. Leaves peltately attached, deeply lobed

62. Euphorbiaceae *(Ricinus)*

12. Leaves basally attached

13. Tufts of hair present in leaf axils; fleshy herbs; flowers usually bright yellow   **7. Portulacaceae** *(Portulaca)*

13. Hairs absent from leaf axils except as general pubescence of plant

  14. Stipules present

    15. Fertile branches slender with very small, 2-ranked leaves; flowers small, in leaf axils   **62. Euphorbiaceae** *(Phyllanthus)*

    15. Leaves alternate or spirally arranged, if 2-ranked, toothed; flowers variously arranged

      16. Leaves alternate; stipules united at base, not persistent, delicate herb   **63. Urticaceae** *(Laportea)*

      16. Not as above

        17. Perianth segments chaff-like, in 1 series   **56. Amaranthaceae**

        17. Perianth segments not chaff-like, in 2 series, petals coloured or white, showy

          18. Petals united; stamens 5; fruit a berry or 2-celled, unappendaged capsule   **48. Solanaceae**

          18. Petals free except at base; stamens more than 5; fruit not a berry, if a capsule with 5 terminal appendages or more than 2-celled

            19. Stamens fused into a column; fruit not appendaged   **9. Malvaceae**

            19. Stamens free; capsule prismatic and appendaged, or sphaeroid and covered by dense hairs and hooked spines   **10. Tiliaceae**

  14. Stipules absent

    20. Flowers aggregated in heads enclosed in cup-like whorls of small bracts   **38. Compositae**

    20. Not as above, inflorescence open

      21. Flowers yellow

        22. Flowers solitary, axillary, their pedicels fused to the adjacent petiole   **31. Turneraceae** *(Turnera)*

        22. Flowers in racemes   **4. Cruciferae** *(Brassica)*

      21. Flowers blue, axillary; herb silky-haired   **47. Convolvulaceae** *(Evolvulus)*

GROUP 6. Leaves compound

1. Twining climbers; leaves alternate

  2. Stipules present; leaves 3-foliolate or pinnately compound; flowers markedly irregular, bisexual   **24. Leguminosae**

  2. Stipules absent; leaves (1-)3—5-foliolate; flowers regular, minute, unisexual, in racemes   **67. Dioscoreaceae** *(Dioscorea)*

1. Not as above

  3. Leaves palmately compound or 3-foliolate

    4. Leaf segments more than 3

      5. Trees; leaves opposite, 3—5-foliolate   **50. Bignoniaceae** *(Tabebuia)*

      5. Herbs; leaves alternate, 3—7-foliolate   **5. Capparidaceae**

    4. Leaf segments 3

      6. Leaflets obcordate   **13. Oxalidaceae** *(Oxalis)*

      6. Leaflets acute to obtuse or rounded at the apex

        7. Ovary and fruit borne on a column above other flower parts; shrub; leaves alternate   **5. Capparidaceae** *(Maerua)*

        7. Ovary and fruit sessile or almost so

          8. Flowers regular, corolla white, funnel-shaped; scrambling shrub; leaves opposite   **42. Oleaceae** *(Jasminum)*

          8. Not as above; leaves alternate

            9. Flowers small, not noticeably irregular; shrubs

              10. Inflorescence cymose; leaflets uniform
                    17. Meliaceae *(Malleastrum)*

              10. Inflorescence racemose; central leaflet larger than laterals
                    21. Sapindaceae *(Allophylus)*

            9. Flowers strongly irregular; fruit a dehiscent, 1-celled, several-seeded pod   **24. Leguminosae**

  3. Leaves 1-pinnate or 2—3-pinnate

    11. Leaves 1-pinnate, alternate or opposite

12. Leaves very large, in an apical cluster; pinnae with parallel veins; palm trees   **70. Palmae**

12. Leaves net-veined

   13. Leaves alternate, pinnae 2–3 pairs; evergreen mangrove tree or shrub   **17. Meliaceae** *(Xylocarpus)*

   13. Leaves alternate or opposite, pinnae more than 3 pairs or if 3, pinnae less than 1 cm long

      14. Leaves with even pairs of leaflets; stipules present

         15. Leaves opposite and markedly unequal; prostrate herb; flowers yellow, regular   **12. Zygophyllaceae** *(Tribulus)*

         15. Leaves alternate not markedly unequal; flowers regular or irregular; trees, shrubs or herbs   **24. Leguminosae**

      14. Leaflets paired, with an odd terminal leaflet

         16. Pinnae more than 6 pairs, glabrous or almost so, apex acute to acuminate; stipules absent; small tree
                                22. Anacardiaceae *(Operculicarya)*

         16. Pinnae less than 6 pairs, or if more, apex obtuse or plant climbing; stipules present   **24. Leguminosae**

11. Leaves 2–3-pinnate, alternate

  17. Leaves 2-pinnate

    18. Inflorescence paniculate; fruit baccate, 1-seeded; small trees
                        21. Sapindaceae *(Macphersonia)*

    18. Inflorescence spicate, racemose or capitate or pinnae 1 pair; trees, shrubs or herbs   **24. Leguminosae**

  17. Leaves irregularly 2–3-pinnate, tree, fruit 3-sided
                      23. Moringaceae *(Moringa)*

FERNS

1. POLYPODIACEAE

Ferns, diverse in habit. Rhizome (stem) prostrate to erect, usually scaly or hairy, with fibrous roots. Stipes (petioles) in some genera articulated to the rhizome; blades entire to several times pinnate or palmate, simple to compound and segments sometimes articulating to the rhachis; fronds (leaves) either all fertile or some sterile, sometimes dimorphic, variously ornamented with scales or hairs. Sporangia usually in clusters (sori), sometimes spreading over lower surface (acrostichoid), and either naked or with a scale-like covering (indusium), often mixed with sterile, hair-like structures (paraphyses); individual sporangia stalked, surrounded by a vertical ring of thick-walled large cells (annulus), splitting at maturity.

A very diverse, cosmopolitan group, divided into several families by various authors but here retained in a traditional, broad sense. There is also little agreement regarding generic limits.

1. Pinnae ascending, leathery, base attenuate, shortly stalked, net veined; fertile pinnae acrostichoid, indusia absent **1. Acrostichum**
1. Pinnae spreading, articulate to rachis, papery, base truncate, subsessile, venation free; fertile pinnae with sporangia in sori, indusia present
2. Nephrolepis

1. ACROSTICHUM L.

Terrestrial or aquatic ferns. Rhizome thick, erect or more rarely prostrate, very scaly, scales thick in centre and at base, thin at margins; roots abundant, thick, woolly. Fronds erect, not articulated, stiff, 1-pinnate, pinnae shortly stalked, lowest pinnae much reduced, net veined, without included veinlets. Fertile fronds resembling sterile but with upper pinnae (or all) densely covered on lower surface by sporangia and capitate paraphyses, indusia absent; spores tetrahedral, colourless.

A pantropical genus of about 5 species, usually found in swampy places, frequently in mangrove swamps.

Acrostichum aureum L., Sp. Pl.: 1069 (1753); F.M.S.: 514 (1877); Dupont, Report: 41(1907); Hemsley in B.M.I.K. 1919: 135 (1919); F.Z. Pterid.: 99, t. 31 (1970); Fosberg in Amer. Fern. Journ. 61: 97 (1971).

Large leathery fern; rhizomes 2 cm or more thick, branched, erect to creeping, thickly covered below by roots and above by stipes and stipe-bases. Fronds erect, up to 1–1.5 m tall; stipe densely scaly at extreme base, up to 40 (–50) cm long, with scattered basal projections of vestigial pinnae; blade ovate-lanceolate in outline, pinnate, up to 20 cm wide; pinnae shortly stalked, linear-lanceolate, up to 20 × 2–2.5 cm, blunt, strongly ascending, leathery, margins stiff, cartilaginous, rolled towards the lower side, prominently net veined. Terminal 1–8 segments may be fertile, the sporangia closely crowded in a dense network over the entire under surface. Fig. 1/1.

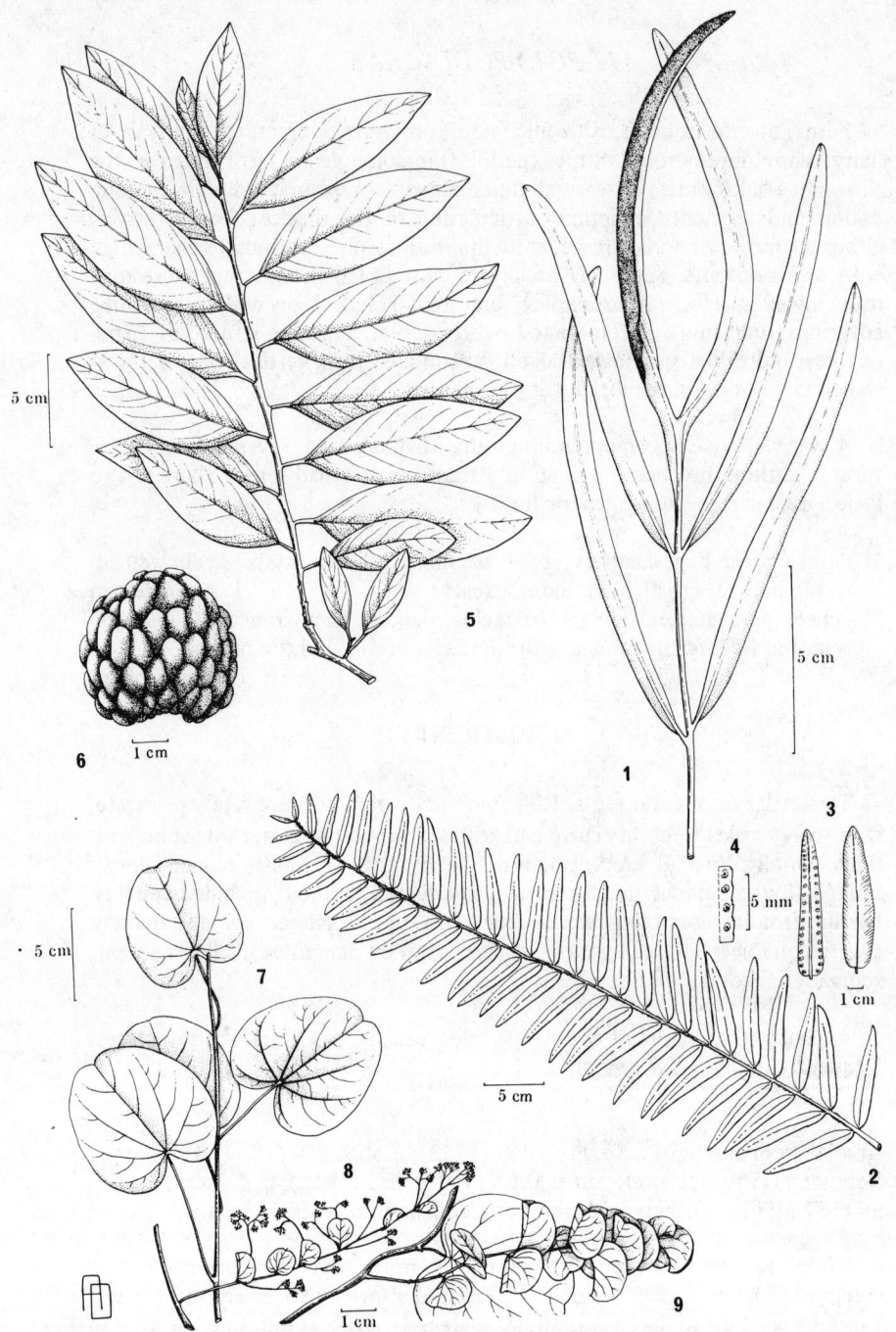

Fig. 1   1, *Acrostichum aureum*, part of frond. 2, *Nephrolepis biserrata*, part of frond; 3, pinnule, lower and upper surface; 4, detail of sori. 5, *Annona squamosa*, leaves; 6, fruit. 7, *Cissampelos pareira* var. *hirsuta*, leaves; 8, male inflorescence; 9, female inflorescence.

Distr. ALDABRA: W, P (isle) M, S; ASSUMPTION; pantropical.

Notes. Very abundant in deep, muddy bottomed pools with vertical walls in platin, and in deep cracks, also locally conspicuous and abundant dwarf forms in pits in the champignon, especially in eastern end of Middle and South Islands; scarce in West Island, also from an islet off Polymnie but not from Polymnie itself. New growth observed during the wet season.

Vernac. 'fougère manglier'.

2. NEPHROLEPIS Schott

Terrestrial or epiphytic ferns. Rhizome erect, bearing several to numerous, slender, often woody stolons, stolons sometimes serving as props, producing roots as well as plantlets or tubers or both; rhizomes and stolons scaly. Fonds erect or pendent, pinnate; stipe and rachis usually scaly; pinnae articulated to rhachis, falling when old, usually asymmetric, sometimes scaly or hairy, at least when young, veins free, forking once to several times, ending short of the margin, with a gland or white incrustation at the end of each on upper surface of the pinna. Sori terminal on some of the veins below, often submarginal, rarely joined into a single, linear sorus near margin, indusium kidney-shaped, round or crescent shaped, variously orientated; spores brown, surface rough.

Pantropical genus of c 35 species, often weedy; several species are widely cultivated.

Nephrolepis biserrata (Sw.) Schott, Gen. Fil. t. 3 (1834); F.Z. Pterid.: 160 (1970); Fosberg in Amer. Fern Journ. 61: 97 (1971).

Aspidium biserratum Sw. in Schrad., J. Bot. 1800, 2: 32 (1801) & Syn. Fil. :242 (1806).
Nephrolepis exaltata Schott var. *biserrata* (Sw.) Baker, F.M.S.:493 (1877).

Rhizome erect, scales lanceolate-attenuate, with the narrow, brown, central portion extending into apex, not black and shining in centre; tubers absent. Fronds narrowly elliptic in outline, up to $100-150 \times 20-30$ cm, tending to be indeterminate in growth; rachis scales peltate, woolly-ciliate on margins; pinnae spreading, linear-lanceolate, apex often prolonged, margins subentire to doubly crenate-toothed, hairy to subglabrous beneath. Sori 1.5–2.5 mm from margin; indusium flat, orbicular-cordate, sinus generally opening toward midrib but orientation somewhat variable. Fig 1/2–4.

Distr. ALDABRA: S; ASSUMPTION; pantropical.

Notes. Only known from one locality on Aldabra, between Anse Takamaka and Cinq Cases, where it occurs in crevices in champignon: one collection only, *Renvoize* 959 (K, US). Found near north point on Assumption: one collection only, *Stoddart* 1097 (K).

Vernac. 'fougère tabac'.

FLOWERING PLANTS

2. ANNONACEAE

Woody plants, usually aromatic, branchlets often zig-zag. Leaves alternate, simple, entire; stipules absent. Flowers usually bisexual, regular. Perianth segments usually in 3 whorls of 3 (calyx rarely 2). Sepals touching but not overlapping, free or basally united. Corolla whorls similar or dissimilar, touching or overlapping, often leathery; receptacle usually somewhat elongate. Stamens numerous, spirally arranged; filaments short and thick; anthers 4-celled, connective prolonged beyond the cells. Carpels several to many, free, or in some genera becoming united in fruit, spirally arranged; ovaries 1-celled, ovules 1-many, placentation usually parietal or sometimes basal; style short or stigmas sessile. Fruit a berry or becoming a massive, united aggregate on a thickened, fleshy receptacle; seeds large, hard.

Medium sized, principally tropical family; several species ornamental and some producing excellent edible fruits.

ANNONA L.

Trees or shrubs. Flowers pedicelled, solitary or in few-flowered cymes, usually bisexual; buds globose, ovoid or elongate, 3-angled. Sepals 3, free, much shorter than the petals. Stamens numerous, connective capitate, dilated or apiculate, prolonged beyond the anther-sacs. Carpels numerous, more or less united, each carpel containing 1 basal ovule. Fruit fleshy, ovoid to globose or cylindrical; seeds numerous.

A genus of 110 species, mainly in the tropics of the New World.

Annona squamosa L., Sp. Pl.: 537 (1753); F.M.S.: 3 (1877); F.M.C. 78: 100, fig. 24/8-9 (1958); F.Z. 1: 145 (1960); F.T.E.A. Annon.: 113 (1971).

Trees or shrubs, 3–6 m tall. Leaves alternate, elliptic-oblong to elliptic-lanceolate, 7–17 × 3–5.5 cm, glabrous or sparsely pubescent; petiole 0.5–1.5 cm long. Flowers solitary. Sepals small, broadly triangular, 2–3 mm long, pubescent. Petals: outer whorl green, purple at the base, long and strap-like, 15–25 × 3–5 mm, pubescent; inner whorl reduced to minute scales or absent. Fruit greenish-yellow, globose or conical, 5–10 cm long, the berry-like carpels loosely cohering to give the surface a chequered appearance; flesh white; seeds dark brown. Fig. 1/5-6.

Distr. ALDABRA: W; native in the West Indies, widely cultivated throughout the tropics.

Notes. Tree in edge of thicket on platin east of Settlement, West Island. Vernac. 'attier' or 'sugar apple'.

3. MENISPERMACEAE

Mostly twining vines. Leaves alternate, usually simple, mostly palmately veined; stipules absent. Flowers unisexual, small, greenish, usually regular. Perianth usually in several series and differentiated into calyx and corolla, each in whorls of 2–3 distinct parts. Male flowers: stamens 6, rarely 3 or many, free or variously united; anthers 4-celled. Female flowers: carpels usually 3–6, separate, rarely 1 or many, free; ovary superior, 1-celled; ovules 2, 1 abortive, placentation parietal; style short or absent, stigma usually curved.

A medium-sized, mostly tropical family, some genera very toxic and of medicinal interest.

CISSAMPELOS L.

Twining lianes with simple, peltate or sub-peltate leaves. Male inflorescence 1–several clusters of corymbose cymules of tiny flowers; sepals 4, obovate, often spreading; petals fused and forming a short, cup-shaped corolla; stamens 4, fused centrally. Female inflorescence of corymbose cymules, small flowers variously arranged in the axils of leaves or smaller bracts; sepals 1, obovate; petals 1; ovary 1-celled. Fruit a 1-seeded, slightly fleshy drupe.

A genus of c 30 species, widespread throughout the tropics.

Cissampelos pareira L., Sp. Pl.: 1031 (1753).
Liane with rather woody stems. Leaves alternate, peltate or subpeltate, orbicular to broadly orbicular, 2–8 × 2–9 cm, apex indented or obtuse, mucronate, base cordate, sparsely hairy above, more densely hairy beneath, rarely completely glabrous, membranous; petiole 2–6 cm long. Male inflorescence axillary with slender, rather spreading branches, up to 15 cm long, often with very small bracts on the main axis which do not conceal the flowers. Male flowers: very small, 0.5–0.75 mm long; sepals not spreading; anthers dehiscing by an outer horizontal slit. Female inflorescence axillary, consisting of a usually unbranched main axis bearing bracts 1–1.5 cm long, similar in shape to the leaves and enclosing small clusters of flowers. Female flowers 1 mm long. Fruit orange or red, obovate, 3.5 mm long, becoming ribbed or wart-covered when dry; seeds 1, depressed globose, c 4 mm in diameter, warty.

A species widespread throughout the tropics.

var. **hirsuta** (Buch. ex DC.) Forman in K.B. 22: 356 (1968).

C. hirsuta Buch. ex DC., Syst. 1: 535 (1817).
C. convolvulacea Willd. var. *hirsuta* (Buch. ex DC.) Hassk., Pl. Jav. Rar.: 171 (1848).
C. pareira var. *orbiculata* (DC.) Miq. in Ann. Mus. Bot. Lugd. Bot. 4: 85 (1868); F.T.E.A. Menisp.: 26, fig. 6 (1956); F.Z. 1: 167, fig. 23 (1960).
C. pareira var. *nephrophylla* (Bojer) Diels in Engler, Pflanzenr. IV, 94: 292 (1910); Hemsley in B.M.I.K. 1919: 115 (1919).

Leaves distinctly peltate with the petiole inserted 1–4 mm from the edge of the blade. Fig. 1/7-9.

Distr. ASSUMPTION; widespread through the Old World tropics.
Notes. Only known from Assumption where it was found growing in guano pits, probably as an introduced weed. Two collections; *Dupont* 104 (male & female plants) & 263 (male plants), not recorded since.
Vernac. 'la liane'.

4. CRUCIFERAE

Herbs or rarely small shrubs, often with a sharp odour and taste. Leaves alternate, simple, entire to deeply and intricately dissected; stipules absent. Inflorescence racemose; flowers bisexual, usually regular. Sepals 4, free. Petals 4, free, rarely less or none. Stamens usually 6, 2 short and 4 long. Ovary superior, 2-celled; style 1 or absent, stigmas 2. Fruit a 2-celled capsule, rarely indehiscent; seeds usually several to many, rarely 1–2 per cell.

A large, mostly north temperate family, a few tropical members; includes many cultivated species, food plants and ornamentals and many widely distributed weeds.

BRASSICA L.

Herbs or small shrubs. Leaves entire or pinnately lobed. Sepals erect, inner larger than the outer. Petals yellow or white, clawed. Ovary sometimes on a short stalk. Fruit elongate, with a long or short beak, valves convex with a prominent median vein; seeds in 1–2 rows in each half.

Brassica nigra (L.) Koch in Röhling, Deutschl. Fl. ed. 3, 4: 713 (1833); F.M.S.: 8 (1877); Schinz in A.S.N.G. 21: 84 (1897); Voeltzkow in A.S.N.G.

26:   550 (1902); Dupont, Report: 34 (1907) Hemsley in B.M.I.K. 1919: 138 (1919).

Sinapis nigra L., Sp. Pl.: 668 (1753).

Annual herbs up to 1 m high branching from near the base or the middle. Lower leaves with 1–3 pairs of lateral lobes and a much larger terminal lobe; upper leaves linear-oblong, margins entire or wavy. Petals yellow, 7–9 mm long. Fruits 10–20 mm long, tapering to a slender beak, on short stalks and held close to the stem; seeds globose, 1.5 mm diameter.

Distr. ALDABRA, ASSUMPTION, COSMOLEDO, ASTOVE; native in Europe, now widely cultivated throughout the world.

Notes.   Recorded by Dupont, Report (1907) for all the islands in the Aldabra group but not recorded in recent years. Only specimen seen is from Aldabra, *Voeltzkow* 58 (Z).

Vernac. 'pissard le chien' or 'black mustard'.

5. CAPPARIDACEAE

Herbs, shrubs, or rarely trees, usually malodorous.   Leaves usually alternate, simple or palmately compound; stipules small or spiny, persistent or falling. Inflorescence racemose, corymbose or flowers solitary or fascicled. Flowers bisexual or rarely unisexual, regular or often irregular; perianth in 2 series, or rarely corolla absent. Sepals 4 or 8, free or united. Petals 4 or more, or absent, often clawed, often unequal. Stamens 4 or more, sometimes numerous. Ovary superior, usually on a prominent column, 1–2-celled placentae parietal; ovules few to many; style 1. Fruit usually a capsule or berry; seeds often kidney-shaped, with curved embryo.

Medium large, widely distributed family, largely tropical, some weedy, a few planted. A species related to the Aldabra *Capparis cartilaginea* produces the edible caper.

1. Fruit an elongated, dry capsule dehiscing by 2 valves; annual or perennial herbs
  2. Stamens 6, all the same length, radiating from an elongate staminal column; ovary also borne on an elongate column   **3. Gynandropsis**
  2. Stamens 2-many, arising directly from the base of the flower; ovary not borne on a conspicuous, elongate column   **2. Cleome**
1. Fruit indehiscent, globose to cylindrical; usually woody plants
  3. Branches spiny   **1. Capparis**
  3. Branches not spiny   **4. Maerua**

1. CAPPARIS L.

Shrubs, small trees or scramblers. Leaves simple and entire; stipules often developed into spines. Flowers bisexual, in terminal or axillary corymbose racemes or solitary. Sepals 4, often unequal and overlapping. Petals 4. Stamens 6—many, free. Ovary globose or cylindrical, 1—several cells, borne on a slender column. Fruit globose or ellipsoid, usually indehiscent; seeds few to many.

A genus of c 250 species, mainly in America but distributed throughout the tropics and subtropics.

Capparis cartilaginea Decne. in Ann. Sci. Nat. II, 2: 273 (1835); F.T.E.A. Capparid.: 59, fig, 10/3-6 (1964).

C. galeata Fresen. in Mus. Senckenb. 2: 111 (1836); Baker in B.M.I.K. 1894: 148 (1894); Schinz in A.S.M.G. 21: 84 (1897); Voeltzkow in A.S.N.G. 26: 550 (1902); Dupont, Report: 34 (1907); Hemsley in B.M.I.K. 1919: 116 (1919).

A spreading, scrambling shrub (up to 1 m high on Aldabra). Young branches white-tomentose, appearing greyish-green. Leaves ovate to orbicular, 2—5.5 × 1.5—5 cm, apex acute, rounded or slightly indented with the midrib ending in a small recurved spine, glabrous; petiole 9—25 mm long; stipules developed into small, recurved spines. Flowers solitary in the axils of the upper leaves; pedicels stout, 3.5—12 cm long. Sepals green or greenish-white, unequal; 3 oblong, 1.5—2 cm long and boatshaped, 1 up to 5.5 cm long and hooded at the apex. Petals white, unequal, one pair broadly ovate, up to 2.5 cm long, the other pair apparently fused, up to 4 cm long and tucked inside the hooded sepal. Stamens numerous, 150—200; filaments up to 5.5 cm long. Ovary and column up to 8 cm in flower, often longer in fruit. Fruit red, ovoid or ellipsoid, up to 5 cm long, ribbed; seeds many, reddish brown, kidney-shaped, c 3 mm long. Fig. 2/1—2.

Distr. ALDABRA: W, P, M, S; ASSUMPTION; ASTOVE; Pakistan and the Middle East southwards to southern Tanzania, Zanzibar and Madagascar.
Notes. Found in open areas of bare limestone rock near the coast; absent from the NE and SW ends of South Island. Flowering mainly during the dry season — flowers only open fully at night and last for less than 24 hours; new leaves are also mostly produced during the dry season.

2. CLEOME L.

Annual or perennial herbs. Leaves alternate, simple or digitately 3—9-foliolate; petiole present. Inflorescence racemose, flowers irregular pedicels with small leaf-like bracts. Sepals 4, free or united at the base. Petals

Fig. 2   1, *Capparis cartilaginea*, habit and flower; 2, fruit. 3, *Cleome strigosa*, habit; 4, flower; 5, fruit. 6, *Gynandropsis gynandra*, habit; 7, flower; 8, fruit. 9, *Maerua triphylla* var. *pubescens*, habit and flowers; 10, fruit.

4. Stamens 2–many, free, filaments equal or unequal. Ovary sessile or on a short slender column, 1-celled; stigma capitate. Fruit an oblong or linear capsule borne on an elongated column, dehiscing by two valves splitting from a central placenta; seeds numerous, kidney-shaped, smooth, ridged or tubercled.

A genus of c 150 species, widespread throughout the tropics and subtropics.

Cleome strigosa (Boj.) Oliv. in Fl. Trop. Afr. 1: 80 (1868); Baker in B.M.I.K. 1894: 146 (1894); Hemsley in B.M.I.K. 1919: 115 (1919); F.T.E.A. Capparid.: 7, figs. 1/1-5 & 2/16 (1964).

Polanisia strigosa Boj. in Ann. Sci. Nat. II, 20: 56 (1843); Voeltzkow in A.S.N.G. 26: 550 (1902) Dupont, Report: 34 (1907).

Annual herb, erect or decumbent, up to 1 m high; stems densely covered with stiff, spreading hairs. Leaves 3–5-foliolate, leaflets obovate, central leaflet up to 5 × 2.4 cm, apex rounded, coarsely hairy. Inflorescence stout, up to 50 cm long. Sepals green, narrowly lanceolate, 5 mm long, united at the base, coarsely hairy. Petals deep purple, upper pair usually yellow at the base, obovate, 10–13 mm long. Stamens 10–15; 7–12 with short filaments 6–9 mm long; 3–5 with long filaments, 10–14 mm long. Ovary and column 5–10 mm long. Capsule 2–3 cm long, valves splitting from the base upwards; Seeds dark brown, kidney-shaped, 1–1.3 mm long, faintly ridged. Fig. 2/4–5.

Distr. ALDABRA: W, P, M, S; Esprit, Michel; ASSUMPTION; COSMOLEDO: W, M; ASTOVE; a seashore species distributed from Somalia southwards to northern Tanzania, Zanzibar, Pemba and Mascarene Is.

Notes. Apparently a fairly recent introduction to the group for Dupont, Report (1907) records it only from Aldabra. Most abundant on disturbed ground near habitation; frequently found growing in sand at the top of beaches, above high water mark. Flowers throughout the year but most abundantly during the mid wet season.

Vernac. 'brède caya'.

3. GYNANDROPSIS DC.

Annual, rarely perennial herbs. Leaves alternate, 3–7-foliolate; petiole present. Inflorescence racemose; flowers irregular, bisexual; pedicels with leaf-like bracts. Sepals 4, free. Petals 4. Stamens 6, radiating from an elongate, slender, staminal column. Ovary oblong, 1-celled, also borne on a slender column arising from the apex of the staminal column. Fruit an elongate capsule dehiscing by 2 valves from a central placenta; seeds numerous, kidney-shaped, ridged.

A genus of 2 species in the Old and New Worlds.

Gynandropsis gynandra (L.) Briq. in Ann. Conserv. Jard. Bot. Geneve 17: 382 (1914); F.Z. 1: 205, fig. 31 (1960); F.T.E.A. Capparid.: 18, fig. 3 (1964); F.M.C. 83: 66, fig. 13 (1965).

Cleome gynandra L., Sp. Pl.: 671 (1753).
Cleome pentaphylla L., Sp. Pl., ed. 2,: 938 (1763), name illegit.
Gynandropsis pentaphylla (L.) DC., Prodr. 1: 238 (1824); F.M.S.: 9 (1877); Hemsley in B.M.I.K. 1919: 116 (1919), name illegit.

Annual herb up to 1 m high; stems glandular hairy. Leaves 3—7-foliolate; leaflets obovate to elliptic, up to 3 cm long; petiole 3—12 cm long. Inflorescence up to 30 cm long. Sepals ovate to lanceolate, up to 8 mm long, glandular hairy. Petals white, pink or lilac, spathulate, 1—2 cm long. Stamens 2.5—4 cm long; staminal column 2 cm long. Ovary and column 0.5—2 cm long from the base of the stamens. Capsule linear or linear-oblong, 3—15 cm long, flattened, glandular hairy to almost glabrous, style persistent; seeds brown, globose-kidney-shaped, 1.5 mm in diameter, transversly striate. Fig. 2/6-8.

Distr. ALDABRA; a pantropical weed.
Notes. Presumably a weed of cultivation; only known from 1 collection, *Dupont* 42 (K).
Vernac. 'brède caya' or 'mosambé'.

4. MAERUA Forssk.

Small trees or shrubs. Leaves alternate or clustered on short side branches, simple or 3—4-foliolate; sessile or petiolate. Flowers bisexual, in terminal or axillary, corymbose racemes, terminal panicles, or solitary. Sepals 4. Petals 0 or 4. Stamens few to many, generally free. Ovary borne on an elongated slender column, globose to cylindric, 1—2-celled; stigma sessile or subsessile. Fruit globose to cylindrical, smooth or covered by warts; seeds subglobose or oblong, smooth or wrinkled.

A genus of c 50 species, mostly in the drier parts of Africa but also extending to Madagascar the Middle East and tropical Asia.

Maerua triphylla A. Rich., Tent. Fl. Abyss. 1: 32, t.7 (1847); F.T.E.A. Capparid.: 43 (1964).

Shrubs or small trees, up to 1 m on Aldabra; branchlets glabrous or pubescent. Leaves, (1—)3(—4)-foliolate, leaflets lanceolate, ovate or elliptic, apex obtuse or acute, central leaflet 2—7 × 1—4 cm; petiole 1.5—4.5 cm long. Flowers in terminal and axillary corymbose racemes although superficially appearing all terminal; pedicels 1—2 cm long. Sepals 4, green, obovate, 6—8 mm long, glabrescent. Petals 4, cream, obovate, 5—7 mm long, glabrous. Stamens 10—28, 1.5—2 cm long, fused for a short distances at the base. Ovary and column 1.5–3 cm long. Fruit globose to cylindrical, woody, often becoming constricted between the seeds, borne on the end of the column.

var. **pubescens** (Klotzsch) De Wolf in K.B. 16: 82 (1962); F.T.E.A. Capparid.: 45, fig. 8 (1964).

Streblocarpus pubescens Klotzsch in Peters, Reise Mossamb. Bot. 1: 166 (1861).
Maerua pubescens (Klotzsch) Gilg in Engl., Bot. Jahrb. 33: 223 (1903); Hemsley in B.M.I.K. 1919: 116 (1919); F.Z. 1: 222 (1961).
M. cylindrocarpa Gilg & Bened. in Engl., Bot. Jahrb. 53: 241 (1915); Hemsley in B.M.I.K. 1919: 116 (1919), name illegit.
M. dupontii Hemsley in J. Bot. 54, Suppl. 2: 2 (1916) & in B.M.I.K. 1919: 116 (1919). Types: Aldabra, *Fryer* 64, *Dupont* 136 & 137 (all K, syn.).

Branchlets sparsely pubescent or glabrescent. Leaves 3(–4)- foliolate. Ovary narrowly cylindrical, 2–3 mm long. Fruit cylindrical, up to 5 cm long when mature, constricted between the seeds, glabrous and granular; seeds ovoid to subglobose, up to 6 mm in diameter, glabrous. Fig. 2/9–10.

Distr. ALDABRA: W, P, M, S, lagoon islets; ASSUMPTION; COSMOLEDO: M; ASTOVE; Uganda and Kenya southwards to Rhodesia and Mozambique, also Mascarene Is.
Notes. In mixed scrub, uncommon. Provided sufficient moisture is available leaf-growth, flowering and fruiting occurs throughout the year.
Vernac. 'bois trois feuilles'.

6. FLACOURTIACEAE

Trees and shrubs. Leaves alternate, usually in 2 ranks, simple; stipules present. Flowers in cymes or racemes, regular; perianth usually in 2 series, sometimes undifferentiated or corolla lacking, 2–15-partite. Petals overlapping. Stamens usually numerous, at least more than sepals or petals, sometimes in fascicles. Ovary usually superior, 1-celled, placentae parietal, ovules usually numerous; styles 1–10. Fruit a capsule or berry; seeds many, often arillate.

A medium-sized tropical family with a few cultivated species.

1. Leaves papery, base obtuse or acute, margins obscurely and minutely crenate; flowers dioecious, in small racemose clusters; fruit with a thin weak rind; without a strong persistent style; seeds more than 2–3
   1. Flacourtia

1. Leaves sub-leathery to leathery, base obtuse, margins subentire; flowers
   solitary or several in a very congested glomerule; fruit with a firm rind and
   a persistent stylar beak; seeds 2–3(–4)   **2. Ludia**

1. FLACOURTIA Comm. ex L'Hér.

Shrubs or small trees, often spiny. Leaves evergreen or deciduous, simple,
tending to be reddish when young. Flowers axillary or terminal, solitary or
usually in very small racemes, dioecious; perianth in a single whorl of 4–6
small, rather scale-like sepals. Male flowers: stamens many, surrounded at
base by a nectariferous disk. Female flowers: ovary 1-celled, on a
nectariferous disk; 4–7 short styles widely separated in a ring, or close
together, rarely coherent in a single column, stigmas small. Fruit usually a
globose berry; seeds many.

A small, Old World genus.

Flacourtia ramontchii L'Hér., Stirp. Nov. 59-62, t. 30 (1785); F.M.S.: 12
(1877); F.M.C. 140: 9 (1946); Fosberg in K.B. 29: 253 (1974).*

Shrub, usually with axillary thorns. Leaves alternate, ovate or oval to
oblong-elliptic, subacute to obtuse, margins obscurely crenulate. Flower
clusters usually terminal. Fruit broadly cylindrical, c 1 cm in diameter; seeds
pear-shaped.

The var. *ramontchii*, of the Malagassic region, does not occur in Aldabra;
no plants were seen that were at all thorny. The Aldabra and Astove plants
form the probably endemic var. *renvoizei*.

var. **renvoizei** Fosberg in K.B. 29: 254 (1974). Type: Aldabra, Cinq Cases,
Fosberg 48856 (US, holo., K, iso.).

F. ramontchii sensu Hemsley in B.M.I.K. 1919: 116 (1919); Fosberg in
   P.T.R.S. B, 260: 218 & 225 (1971).
F. indica sensu Renvoize in P.T.R.S. B, 260: 230 (1971), not (Burm.f.)
   Merrill (1917).

Rather densely branched, unarmed, large shrub or small tree. Leaves
alternate, oblong to obovate, up to 9 × 5.5 cm, apex obtuse or acute, margins
obscurely crenulate, bright glossy green; petiole curved, 5–10 mm long;
stipules absent. Flowers dioecious. Mal flowers: in short, few-flowered
racemes, 1–2 cm long, pedicels short; calyx subrotate, 4–5 mm across,
shallowly and obtusely lobed, minutely glandular puberulent; stamens c 25,

---

*   F.T.E.A. Flacourt.: 57 (1975) includes *F. ramontchii* in the Afro-Asian
    F. indica (Burm.f.) Merrill, but see Fosberg in Kew Bull. loc. cit. for
    further discussion.

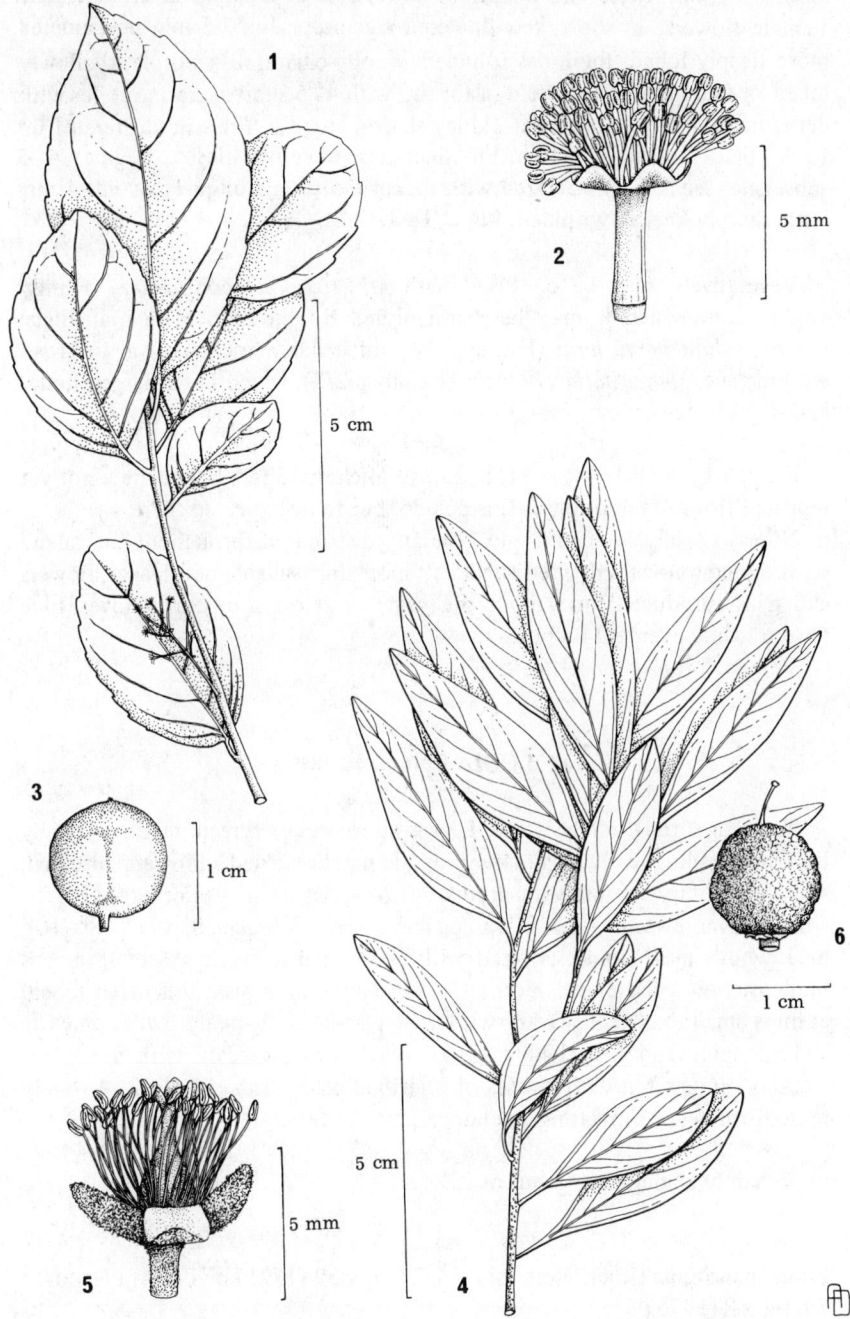

Fig. 3   1, *Flacourtia ramontchii* var. *renvoizei*, habit; 2, male flower;
3, fruit. 4, *Ludia mauritiana*, habit; 5, flower; 6, fruit.

exserted about twice the length of the calyx in a spherical arrangement. Female flowers: in short, few-flowered racemes; calyx 2.5 mm in diameter, more deeply lobed, the lobes rounded, woolly-ciliate; disk purple, shallowly lobed or undulate; ovary ovoid, glabrous, with 4—5 short radiating styles with dark, maroon-black, orbicular-kidney-shaped stigmas. Fruit in clusters of up to 8, black, globose, c 1 cm in diameter, sweetish, fleshy, stigmas 4—5 subsessile, seeds 6—8, brick-red with cream mottling, obliquely ovoid, 4 mm long, sharply keeled, wrinkled. Fig. 3/1—3.

Vegetatively may be confused with *Mystroxylon aethiopicum* (Family 18/2), from which it may be distinguished by the lenticulate branchlets; *Erythroxylum acranthum* (Family 11), which has branchlets oval in cross-section; and *Margaritaria anomala* (Family 62/3), which has enlarged nodes and slightly zig-zag branching.

Distr. ALDABRA: W, P, M, S, Esprit, Michel; ASTOVE; endemic, not yet reported from Assumption or Cosmoledo, but to be expected there.

Notes. Locally abundant and generally distributed throughout the inland scrub communities. Provided sufficient moisture available new leaves, flowers and fruits produced throughout the year. Fruit eaten by turtle doves, blue pigeon, white-eyes and bulbuls.

Vernac. 'prunier' or 'prune'.

2. LUDIA Comm. ex Juss.

Trees or shrubs, rarely spiny. Leaves alternate, evergreen, leathery, rarely thin and falling early, palmately or pinnately veined. Flowers bisexual, axillary, solitary or rarely in groups of 2—3, sessile or pedicellate, pedicels with 2 or more scale-like, overlapping and persistent bracts. Sepals 5—8(—10), in 1 whorl, overlapping, covered with simple hairs. Petals absent. Stamens numerous on a flattended receptacle, subtended by a disk, this often lobed; anthers small, basifixed. Ovary with 2—4 placentas, 2—many ovules on each; style simple, 2—4 branched, stigmas lobed. Fruit baccate with a hard or leathery rind, often with masses of sclerified cells; seeds usually 2—4, rarely more, forming or simulating an endocarp.

A mainly Madagascan genus of 23 species.

Ludia mauritiana Gmel., Syst. Nat. ed. 13, 1: 839 (1791); F.T.E.A. Flacourt.: 53, fig. 18 (1975).

L. sessiliflora Lam., Encycl. Méth. Bot. 3: 613 (1792); F.M.S.: 11 (1877).

Small tree; branchlets light grey-brown, glabrous, lenticels light brown. Leaves alternate, elliptic to obovate or oval, 4—6 × 2—4 cm, apex obtuse to rounded, base wedge-shaped to attenuate, margins entire or obscurely crenate, thinly leathery, glabrous; closely reticulate-venulose, especially beneath; petiole 3–8 mm long. Inflorescence axillary, reduced to a few flowers or a single flower; pedicels stout, 1–1.5 mm long, sometimes bracts closely subtending the calyx. Sepals 5 (rarely 4), orbicular, 1–1.5 mm wide, strongly concave, woolly without, glabrous within, persistent, reflexed in fruit. Petals absent; disk thick, woolly. Stamens up to 40, filaments glabrous, 3 mm long, tapering in distal part, anthers broadly oblong or oval, 0.5 mm long. Ovary broadly ovoid, 1–1.5 mm long, tapering into the style; styles 3, united for 2.25 mm, free for c 0.5 mm. Fruit red, ovoid-subglobose or globose, 7–10 mm long when mature, thinly fleshy, drying to a tessellate surface, style persistent; seeds 2–3, oval to suborbicular, about 1.5–2 mm diameter, compressed, enveloped in a thin, fleshy aril. Fig. 3/4–6.

Rather similar vegetatively to *Flacourtia ramontchii,* from which it may be distinguished by the aereoles on the underside of the leaf. May also be confused with *Erythroxylum acranthum* (Family 11), which has compressed branchlets or *Margaritaria anomala* (Family 62/3) the leaves of which have a distinct cartilaginous margin.

Distr. ALDABRA: S (east); E. Africa, Madagascar, Mauritius and the Seychelles.

Notes. Very common locally in patches of scrub vegetation on rather rough limestone, principally in the Takamaka area.

Vernac. 'prunier marron'.

7. PORTULACACEAE

Usually fleshy herbs or slightly woody sub-shrubs. Leaves alternate to opposite or forming a basal rosette, simple, entire; stipules usually present. Perianth apparently in 2 series. Sepals usually 2, overlapping, free or united. Petals 4–6, rarely 2–3, free, Stamens equalling petals in number and opposite them, or more, to many, rarely fewer. Ovary superior to half inferior, 1-celled, ovules usually several to many, placentation basal (central); style 1, entire or 2–8 stigmatic branches at apex, or stigmas sessile. Fruit a capsule, circumscissile or loculicidal, rarely irregularly rupturing, or an indehiscent nut; seeds somewhat obliquely rounded, with curved embryo.

Pantropical and temperate, rather small family, with a number of ornamental and edible species.

PORTULACA L.

Fleshy herbs, sometimes appearing slightly woody below. Leaves alternate or opposite, axils usually hairy. Flowers usually terminal, solitary or usually in few-flowered clusters. Sepals 2, falling early. Petals brightly coloured, 4 or 5 (–6), membranous, usually lasting only a few hours. Stamens few to many.

Style usually with (2–) 4–8 stigmatic branches or lobes. Ovary 1-celled; with central basal placentation, ovules many. Capsule circumscissile, fused with receptacle in lower part.

A genus of c 200 tropical and subtropical species.

1. Leaves obovate; root not tuberous; capsule lid bell-shaped; sepals keeled
   1. P. oleracea
1. Leaves oblong to elliptic, ovate or oblong-lanceolate; root tuberous; capsule lid lowconic, sepals not keeled, oblong-lanceolate
   2. P. mauritiensis

1. Portulaca oleracea L., Sp. Pl.: 445 (1753); F.M.S.: 125 (1877); F.Z. 1: 363 (1961); Fosberg in K.B. 32: 253 (1977).

Fleshy herb, prostrate to decumbent or ascending; stems reddish, green or brownish green, 10–20 (–40+) cm long; branches and leaves alternate, sub-opposite or opposite, even on same plant. Leaves obovate to spatulate, up to 2 × 4 cm, apex rounded to subtruncate or even very slightly notched, base wedge-shaped; petiole 1–3 mm long; axillary hairs sparse. Flowers sessile, in terminal or rarely axillary or subaxillary heads of 2–6 or more flowers or solitary subtended by a whorl of leaves and small, triangular, scale-like bracts. Sepals united at base, to 4 mm long. Petals yellow, 5, obovate to obovate-oblong, 3–8 mm long, apex indented to 2-fid. Stamens usually 7–12. Ovary half-inferior; style included, stigmatic branches 3–6. Fruit up to 4 × 2–3 mm, circumscissile at about middle, lower half conical, capsule lid bell-shaped, closely invested by the calyx; seeds many, dark brown to black or bluish, kidney-shaped, c 0.7 mm long, somewhat compressed.

1. Many leaf and branch pairs opposite or subopposite; stipular hairs pale reddish brown   **1b. var. delicatula**
1. Leaves and branches predominantly alternate or some subopposite; stipular hairs white
   2. Seeds tuberculate   **1c. var. granulato-stellulata**
   2. Seeds glossy, with very low star-shaped markings which appear chequered at low magnifications   **1a. var. oleracea**

1a. var. oleracea Fosberg in K.B. 32: 254 (1977).

Leaves and branches alternate to subopposite, rarely opposite. Flowers capitate. Seed black or dark brown, glossy, not at all tuberculate, but with very low star-shaped markings elongate in a radial direction under moderate magnification, appearing somewhat chequered. Fig. 4/1.

Distr. ALDABRA: W, P, M, S; ASSUMPTION; pantropical.
Notes. Occurs on the coastal sands.
Vernac. 'pourpier'

1b. var. delicatula Fosberg in K.B. 32: 254 (1977). Type: Aldabra, South Island, Takamaka *Fosberg* 49300 (US, holo., K, iso.).

A slender, brownish, subfleshy herb, very prostrate, flattened to the ground, stems radiating from a slender root, branching and leaf arrangement mostly opposite, occasionally subopposite or even alternate on the same plant. Leaves obovate to broadly spathulate, 2–3 × 1.5–2 mm, thinly fleshy; petioles c 1 mm long, those of involucral leaves somewhat longer; axillary hairs sparse, short, pale reddish brown. Flowers solitary, terminal, surrounded by c 4 involucral leaves, 8–10 mm in diameter. Sepals obtuse, c 2.5 mm long. Petals oblong, deeply 2-fid. Stamens 6–10. Stigmatic branches 2–4. Seeds c 0.7 mm across, glossy, dark brown, with very low star-shaped markings notably elongate in a radial direction on the sides of the seed (similarly to those of var. *oleracea* as interpreted here). Fig. 4/2.

Noted at once in the field as conspicuously different from the other populations of *P. oleracea* seen on Aldabra or elsewhere. Further observations are required to determine with certainty if this variety represents an established endemic population or if it is merely a chance variation that has propagated itself locally.

Distr. ALDABRA: S; endemic.
Notes. Very local, found only near Takamaka Grove on the bottom of shallow pits on rough limestone. Only known from the type collection. Flowers close about 10 a.m.

1c. var. granulato-stellulata v. Poelln. in Occ. Pap. Bishop Mus. 12(9): 5 (1936); Fosberg in K.B. 32: 255 (1977).

Leaves and branches alternate to subopposite, rarely opposite. Flowers capitate. Seeds black, strongly tuberculate with star-shaped markings, tubercles minute but prominent, especially on the periphery of the seed, star-shaped markings with "granules" or tiny prominences around their bases. Fig. 4/3–4.

The tubercles vary somewhat in prominence. In the Aldabra specimens the "granules" are less prominent than in those from Hawaii, the type locality.

Distr. ALDABRA: W, S, lagoon islets; ASSUMPTION; COSMOLEDO: W; ASTOVE; pantropical
Notes. This is the commonest variety in Aldabra.

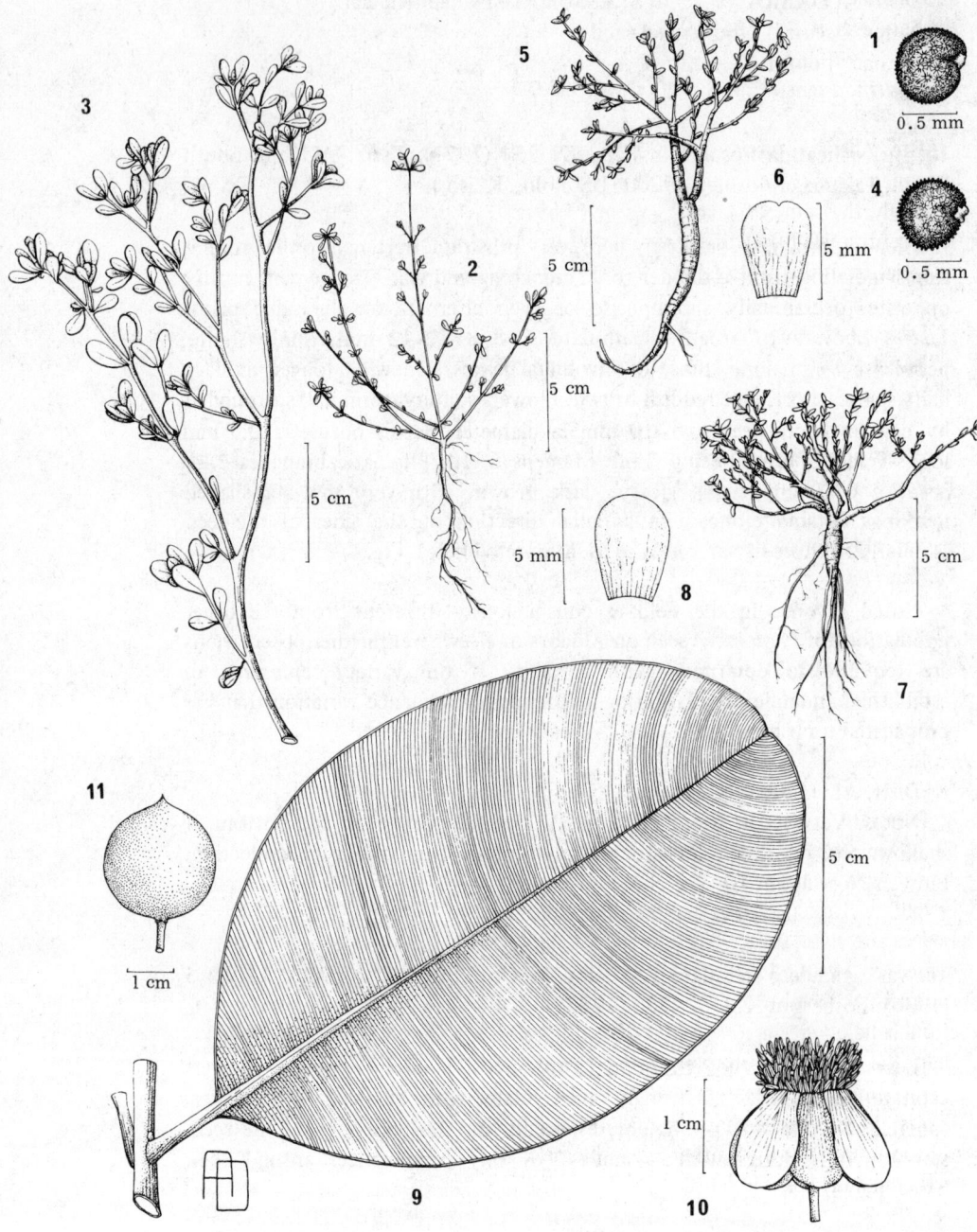

Fig. 4   1, *Portulaca oleracea* var. *oleracea*, seed. 2, *Portulaca oleracea* var.
delicatula, habit. 3, *Portulaca oleracea* var. *granulato-stellulata*,
habit; 4, seed. 5, *Portulaca mauritiensis* var. *aldabrensis*, habit;
6, petal. 7, *Portulaca mauritiensis* var. *grubbii*, habit; 8, petal.
9, *Calophyllum inophyllum* var. *takamaka*, leaf; 10, flower,
11, fruit.

2. Portulaca mauritiensis v. Poeiln. in Fedde, Repert. Sp. Nov. 37: 311(1934); Fosberg in K.B. 32: 255 (1977).

P. quadrifida sensu F.M.S.: 125 (1877), not L. (1767).

The Aldabra plants represent a distinct variety. The one fragmentary specimen from Assumption available for comparison, *Stoddart* 1092, is probably the same but, since the one Cosmoledo plant studied also seems to represent a slightly different entity, it would be unwise to assume that material from the other islands is identical with that from Aldabra.

Distr. (of the species) ALDABRA: ASSUMPTION; COSMOLEDO;? ASTOVE; Mauritius and Diego Garcia Islands.
Vernac. 'pourpier'.

1. Petals obovate, 2-fid or indented at apex   **2a. var. aldabrensis**
1. Petals oblong, not indented   **2b. var. grubbii**

2a. var. aldabrensis Fosberg in K.B. 32: 256 (1977). Type: Aldabra, islet Passe Femme, *Fosberg* 49593 (US, holo., K, iso.).

P. quadrifida sensu Baker in B.M.I.K. 1894: 147 (1894); Schinz in A.S.N.G. 21: 84 (1887); Voeltzkow in A.S.N.G. 26: 550 (1902); Dupont, Report: 34 (1907); Hemsley in B.M.I.K. 1919: 116 (1919), not L. (1767).
Portulaca sp.; Fosberg in P.T.R.S. B, 260: 225 (1971); Renvoize in P.T.R.S. B, 260: 231 (1971).

Prostrate, brownish or purplish, fleshy herb, several to 10 or more stems, 5–20 cm long, radiating from a root crown, grey-corky when old, leaves and stems with scattered crystalline crystoliths; tap-root tuberous thickened, simple or forked, giving rise to elongate, slender, sparsely branched roots. Leaves alternate, somewhat appressed to stem, oblong, elliptic or, when small, ovate, 3–10 × 1.5–2 mm, dorsiventrally compressed, apex obtuse to rounded, base more or less abruptly contracted; petiole 0.5–1.0 mm long; stipular hairs not conspicuous, in small fan-like axillary clusters, much shorter than the leaves. Flowers solitary, terminal, surrounded by 4–6 involucral leaves similar to the stem leaves, and 1 or more, tiny, triangular, scale-like bracts; buds spindle-shaped. Sepals oblong lanceolate, c 4 mm long, apex subacute, apparently neither winged nor keeled. Petals 5(–6), deep yellow, c 1 cm long, apex indented or 2-fid, with a small point in notch. Stamens 12–20. Style branches 4–5, linear. Fruit 3 × 3 mm, lid glossy, firm, broadly conical, blunt; seeds blackish to iridescent blue, very asymmetrical, 0.7–0.8 mm across, plump, sides with prominent, interlocking, star-shaped markings, edges with small, less regular, but concentrically arranged prominences. Fig. 4/5–6.

Distr. ALDABRA: W, M, S, Esprit, Moustique and lagoon islets; ASSUMPTION; ?COSMOLEDO: W; endemic.

Notes. Widely distributed in both coastal and inland areas. Plant eaten and seeds locally dispersed by tortoises.

Vernac. 'pourpier'.

2b. var. grubbii Fosberg in K.B. 32: 258 (1977). Type: Cosmoledo, Wizard Is., *Fosberg & Grubb* 49823 (US, holo., K. iso.).

Portulaca cf. *australis* sensu Fosberg & Renvoize in Atoll Res. Bull. 136: 59, 103 (1970), not Endl. (1834).

Differing from var. *aldabrensis* in the entire, rather than indented or 2-fid petals. Fig. 4/7–8.

Distr. COSMOLEDO; endemic.

Notes. Only known from the type collection.

8. *GUTTIFERAE*

Trees, shrubs or herbs; often with milky sap. Leaves usually opposite, entire; stipules absent. Flowers regular, perianth differentiated into calyx and corolla, each series 2–12-merous. Stamens few to many, free or united in phalanges. Ovary superior, 1–many-celled, placentation parietal, basal or axile, ovules 1–many in a cell; styles free or usually united, stigmas or style branches 1–many. Fruit a capsule, berry or drupe; seeds 1–many, fleshy or appendaged.

A medium-sized tropical family (except for many temperate members of *Hypericum*), with several cultivated members.

CALOPHYLLUM L.

Trees; latex yellow. Leaves leathery, veins pinnate, parallel, closely spaced, no cross veins; petiole bases not noticeably expanded. Flowers in axillary or terminal racemes, these sometimes paniculate. Sepals 4, in 2 dissimilar pairs. Petals 1–4 or absent. Stamens many, free. Ovary 1-celled; style 1, stigma peltate. Fruit a drupe, mesocarp thin, endocarp thin, hard; seed 1, large, basally attached.

A pan-tropical genus centred in the Indo-Malaysian region; 1 very widespread lowland species, *C. inophyllum*, is represented in the western Indian Ocean islands, including Aldabra, by var. *takamaka*.

Calophyllum inophyllum L., Sp. Pl.: 513 (1753); F.M.S.: 16 (1877); Hemsley in B.M.I.K. 1919: 116 (1919); F.M.C. 136: 6 (1951); F.Z. 1: 394, fig. 76 (1971); Fosberg in K.B. 29: 255 (1974).

var. **takamaka** Fosberg in K.B. 29: 255 (1974). Type: Aldabra, South Is., Takamaka Grove, *Fosberg* 49272 (US, holo; K, iso.).

Tree to 15 m tall; bark fissured. Leaves broadly elliptic to oblong or oval, up to 15 × 10 cm, apex and base obtuse to rounded, usually a little more than twice as long as wide, leathery; petiole c 2 cm long. Racemes loose, open, up to 12 cm long; pedicels to 4 cm long, elongating somewhat in fruit; flowers fragrant. Sepals concave, broadly oval, to 8 mm long. Petals white, oval, broadly obovate or narrowly oblong, 10–15 mm long, apex rounded. Style bent in an S-shaped curve near apex; stigma 4-lobed. Fruit subglobose to globose c 25 mm in transverse diameter. Fig. 4/9–11.

Distr. ALDABRA: S; ASSUMPTION; COSMOLEDO: W, M; ASTOVE; Agalega, Mafia Island and Pemba Island.

Notes. A typical strand species but on Aldabra known only from Takamaka Grove. Appears to be able to flower throughout the year.

Vernac. 'bois takamaka' or 'takamaka'.

9. MALVACEAE

Trees, shrubs and herbs, often with stellate pubescence and tough bast-fibre in stems. Leaves alternate, simple, though occasionally lobed and rarely deeply palmately cut, venation frequently palmate; stipules present. Flowers bisexual, rarely unisexual, in some genera subtended by an involucre or epicalyx. Sepals 5, united, margins meeting but not overlapping. Petals 5, united at base with each other and with the filament tube. Stamens many, usually unequal in length, filaments united for most of their length in a tube, free at apex; anthers circular to kidney-shaped, 1-celled or obscurely 2-celled. Ovary superior, (2–)5–many-celled, ovules 1–several per cell, placentation axile; style slender, stigmas usually exserted from staminal tube and equal in number to the cells. Fruit a capsule, schizocarp or berry; seeds 5 (1 per cell) to many.

Large cosmopolitan but principally tropical family, including a number of economic and many ornamental plants.

1. Involucre (epicalyx) present; fruit a capsule
  2. Calyx not gland-dotted   **3. Hibiscus**
  2. Calyx and other flower parts dotted with dark glands
    3. Involucre a whorl of 3, lacerated, leaf-like, persistent bracts; seeds with long wool; sap not yellow   **2. Gossypium**

   3. Involucre reduced to 3, small, ovate or awl-shaped, spirally arranged
      bracts, falling; seeds with short hairs, not conspicuously woolly; sap
      yellow   **5. Thespesia**
1. Involucre absent; fruit a schizocarp.
   4. Flowers less than 20 mm across; mericarps indurated; seeds 1   **4. Sida**
   4. Flowers 20 mm or more across; mericarps parchment-like or firm-
      cartilaginous; seeds several   **1. Abutilon**

1. ABUTILON Mill.

Somewhat woody herbs and shrubs; at least partly stellate-pubescent.
Leaves tending to be cordate, usually palmately veined; stipules various.
Flowers axillary or more rarely in terminal, cylindrical panicles; pedicels
jointed; involucre absent. Sepals united at base or for part of their length.
Corolla frequently orange. Staminal column short, included in corolla unless
corolla becomes reflexed or, rarely, is absent. Carpels 5–many; cells as many
as the carpels, ovules 2–several per cell; style branches 1 per cell. Fruit a
schizocarp, with a number of lobes, splitting into 5–many, usually parchment-
like or cartilaginous, dehiscent mericarps; seeds 2–several per mericarp.

A pan-tropical genus of many species.

1. Style branches and mericarps less than 10   **2. A. fruticosum**
1. Style branches and mericarps more than 10.
  2. Stem branches circular in cross section   **3. A. pannosum**
  2. Stem branches somewhat angled in cross section   **1. A. angulatum**

In our area *A. angulatum* and *A. pannosum* seem scarcely to differ from
each other, though generally separated in African literature. Perhaps they
hybridize on Aldabra.

1. Abutilon angulatum (Guill. & Perr.) Masters in Fl. Trop. Afr. 1: 183
(1868); F.M.C. 129: 138 (1955); F.Z. 1: 488, fig. 93/1 (1961).

Bastardia angulata Guill. & Perr., Fl. Seneg. Tent. 1: 65 (1831).
Abutilon indicum sensu Baker in B.M.I.K. 1894: 147 (1894); Schinz in
    A.S.N.G. 21: 87 (1897); Voeltzkow in A.S.N.G. 26: 551 (1902);
    Dupont, Report: 34 (1907), not *Sida indica* L. (1763).
Abutilon asiaticum sensu Schinz in A.S.N.G. 21: 87 (1897); Voeltzkow in
    A.S.N.G. 26: 551 (1902); Dupont, Report: 34 (1907), not *Sida asiatica*
    L. (1763).
Abutilon "sp. an *A. angulosum* Boj.", Hemsley in J. Bot. 54, Suppl. 2: 4
    (1916) & in B.M.I.K. 1919: 117 (1919), in part.

Shrubs, stems angular in cross section. Leaves broadly ovate-cordate, up
to 11 × 12.5 cm, apex acuminate, margins very shallowly toothed, stellate-

pubescent beneath, thinly so above; petioles longer than blades; stipules small, awl-shaped, thickened at base. Flowers in leafy panicles; pedicels 1–2 cm long. Calyx lobes acute to somewhat acuminate; united part enclosing ovary and enlarging to partly enclose fruit when mature, stellate-pubescent. Corolla orange, maroon at base, 3–4 cm across, reflexed at maturity. Fruit strongly depressed-globose, 1.5 cm across, stellate-pubescent, mericarps c 25; seeds kidney-shaped, 2.5 × 2 mm, surface nodular or smooth. Fig. 5/1-2.

Distr. ALDABRA: W; ASTOVE; tropical Africa, Madagascar.

Notes. Waste ground around habitation sites. Flowers from mid wet season to mid dry season, but most prolifically during the wet season.

Vernac. 'mauve' or 'mauve bâtard'.

2. Abutilon fruticosum Guill. & Perr., Fl. Seneg. Tent. 1: 70 (1831); F.Z. 1: 491, fig. 93/5 (1961)

Slender shrub, finely greyish-tomentose. Leaves ovate to elliptic, up to 5 cm long, apex acute, base strongly cordate, margins subentire to slightly serrate-toothed, densely and finely stellate-pubescent, blade at about right angles to petioles; petioles slender. Pedicels axillary, to 2.5 cm long, jointed and sometimes detaching just below summit. Calyx hemispherical in outline, 5 × 8 mm, divided about half-way into triangular-acute lobes, densely hairy, enlarging slightly when mature. Corolla c 2 cm across when spread out. Staminal column with free filaments at and near summit, base enlarged, stellate-pubescent; style branches 7–8, carpels, cells and mericarps the same number. Schizocarp depressed cylindrical, 7 × 10 mm, valves firm-papery or cartilaginous; seeds kidney-shaped, c 1.5 × 1.5 mm, greyish brown, wrinkled. Fig. 7/1.

Distr. ASSUMPTION; tropical Africa, Arabia and northwest India.

Notes. Waste ground around habitation.

3. Abutilon pannosum (Forst. f.) Schlecht. in Bot. Zeit. 9: 828 (1851).

Sida pannosa Forst. f. in Comm. Soc. Reg. Goetting. 9: 62 (1789).

Abutilon asiaticum sensu Schinz in A.S.N.G. 21: 87 (1897); Voeltzkow in A.S.N.G. 26: 551 (1902) not *Sida asiatica* L. (1763).

Abutilon sp. "an *A. angulosum* Boj.", Hemsley in J. Bot. 54: 4 (1916) & in B.M.I.K. 1919: 117 (1919) in part.

Shrub to 2–3 m tall, stems finely and closely stellate-pubescent. Leaves large, orbicular-cordate, up to 12(–15) cm wide, apex acuminate, margins subentire to obscurely crenate, thinly and very finely stellate beneath, sparsely so above; petioles up to 12(–15) cm long, strong, ascending; stipules small, linear, falling early. Flowers axillary and terminal in long, cylindrical, open panicles; pedicels c 2 cm long. Calyx hemispherical, to 10 × 10 mm, densely stellate-pubescent, divided to below middle into acuminate lobes, enlarging somewhat when mature, enclosing schizocarp. Corolla orange-yellow, c 15 mm long, becoming reflexed. Filaments free at apex of staminal column, column enlarged at base, strongly and stiffly stellate-pubescent. Style branches c 16–20. Schizocarp depressed cylindrical, somewhat contracted apically, rather truncate, c 10 × 15 mm, strongly and densely hairy; mericarps c 16–20, obliquely oblong-elliptic, ends obtuse; seeds kidney-funnel-shaped, 3 × 2 mm, hairy.

Distr. ALDABRA: W, lagoon islets; COSMOLEDO: W; ASTOVE; tropical Africa and tropical Asia.

Vernac. 'mauve' or 'mauve bâtard'.

2. GOSSYPIUM L.

Herbs, shrubs or small trees; usually partly dotted with small, black glands. Leaves alternate, usually palmately lobed, palmately nerved; stipules diminutive, leaf-like. Flowers large, solitary, axilllary, or in few-flowered terminal clusters, enclosed in an involucre formed by a whorl of 3, large, leaf-like bracts. Calyx cup-like. Corolla of 5, large, showy petals, united at base and fused with base of staminal column. Staminal column shorter than corolla, with anthers its full length. Ovary 3–5-celled, 2 or more ovules per cell; style short, unbranched, stigma club-shaped, 5-grooved or slightly lobed. Fruit a globose to ovoid or spindle-shaped capsule, splitting into hard, leathery or stiff, papery, spiny-tipped valves; seeds 2-many, hard, pubescent, usually with long tangled wool, often with an additional finer, short pubescence.

A pan-tropical genus of 20 to many species, depending on taxonomic viewpoint; several species of cotton are of great economic importance and widely cultivated and naturalized, with many forms probably of cultural origin. The taxonomy of these is subject to much discussion and difference of opinion

Gossypium hirsutum L., Sp. Pl. ed.2 :975 (1762); Schinz in A.S.N.G. 21:87 (1897); Renvoize in P.T.R.S. B,260:228 (1971) [as *G. hirtum*].

G. purpurescens sensu Dupont, Report:34 (1907), not Poir. (1811).
G.barbatum sensu Voeltzkow in A.S.N.G.26:551 (1902),not *G.herbaceum* L.
    var. *barbatum* Raf. (1838).

Herbs, usually perennial, woody-based, sometimes annual or tangled shrubs, to 3 m tall; much branched. Leaves as wide or wider than long, up to 12 × 15 cm, 3(–5)-lobed, lobes short-ovoid or deltoid, apices acute to acuminate, base cordate, entire; petioles as long or longer than leaves; stipules ovate to lanceolate, up to 13 mm long, falling. Pedicels mostly 1–2.5(–3) cm long, stiff; involucral bracts roughly elliptic or orbicular, cordate, to 3 cm long, apex divided into 5–9 linear-lanceolate acuminate lobes up to 1.5 or 2 cm long. Calyx up to 1 cm long, lobes obtuse or subacute. Corolla creamy

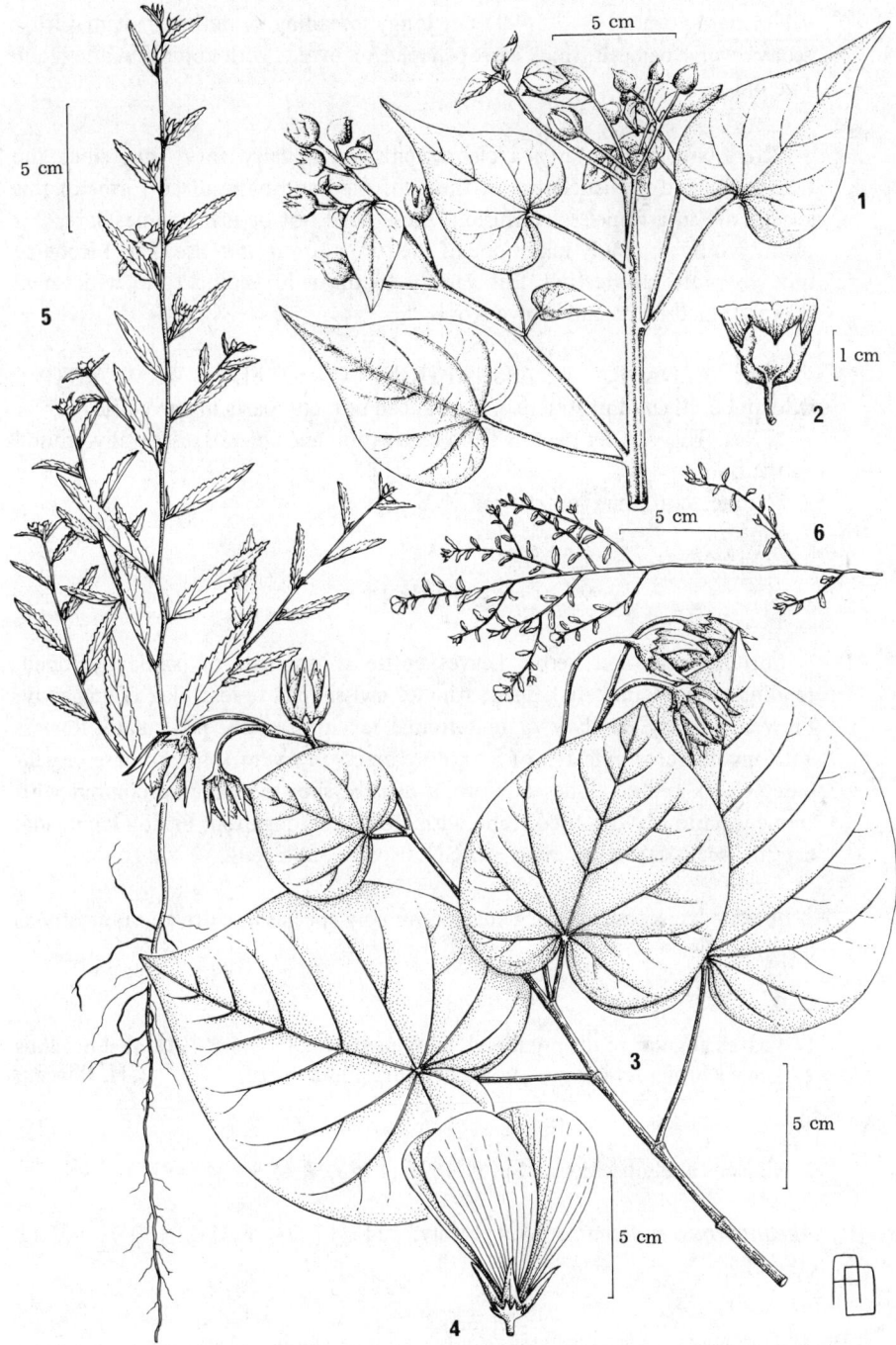

Fig. 5   1, *Abutilon angulatum*, habit; 2, flower. 3, *Hibiscus tiliaceus*,
habit; 4, flower. 5, *Sida acuta*, habit. 6, *Sida parvifolia*, habit.

turning pink, up to 5 cm long, lobes obvate. Capsule subglobose, beaked, valves hard, oval, 1.5—2.5(—3) cm long, spreading or reflexed at maturity; seeds several per cell, thick-carrot-shaped or ovoid, with copious white wool. Fig. 6/1—2.

There are in our area a glabrescent and a hairy form, but since the taxonomy and nomenclature of the cultivated and naturalized forms of this species are in a hopeless confusion, these need not be given names here. Our plants would probably fall in one of the forms of var. *taitense* (Parl.) Roberty but we prefer to restrict this variety to the wild Pacific form with small flowers, small fruits and brown wool.

Distr. ALDABRA: W; ASSUMPTION; COSMOLEDO: W, M; ASTOVE; wild and cultivated in America, introduced in many parts of the world.

Notes. This species persists from cultivation and spreads, especially around cultivation.

Vernac. 'cotonnier' or 'cotton'.

3. HIBISCUS L.

Shrubs, trees and herbs. Leaves entire to toothed or palmately lobed, tending to be palmately veined; stipules awl-shaped to leaf-like, falling early. Flowers solitary, axillary or in terminal racemes, rarely paniculate; flowers with involucre or epicalyx of 5 or more bracts in a whorl, free to base or near base. Calyx united, lobes 5. Corolla usually showy. Staminal column with stamens variously disposed, tube with sterile prolongation. Fruit a loculicidal capsule; seeds several to many per cell, usually pubescent.

A very large pantropical genus, with many species in cultivation, mostly as ornamentals.

1. Leaves angular to deeply lobed, margins toothed   **1. H. abelmoschus**
1. Leaves orbicular-cordate, not lobed, margins subentire   **2. H. tiliaceus**

1. Hibiscus abelmoschus L., Sp. Pl. :696 (1753); F.M.S. :24 (1877).*

Abelmoschus moschatus Med., Malv. :24 (1787); F.M.C. 129:7, fig. 11 (1955).

Coarsely hairy herb or small shrub. Leaves angular to deeply 3—7-palmately lobed, base of blade truncate to cordate, hastate or with an abrupt, narrow, basal sinus, apices of lobes acuminate, margins prominently bluntly serrate,

---

*In most recent floras Sect. *Abelmoschus* is treated as a distinct genus, this species would be known as *A. moschatus* Med. See van Borssum Waalkes in Blumea 14: 89 (1966).

stipules linear, hairy. Flowers solitary, axillary; pedicels straight, hairs coarse, backwardly directed; involucral bracts 6–10, linear-lanceolate, c 1 cm long, covered with sharp appressed, rigid hairs. Calyx 2–3.5 cm long, stellate-pubescent, splitting down one side and falling early with corolla. Corolla yellow with dark purple centre, 4–8 cm long. Staminal column to 2 cm long, glabrous; filaments short. Capsule ellipsoid, to 8 cm long, beaked, appressed-hairy, sutures forming hairy lines, valves elliptic, papery smooth or strongly shaggy-haired within; seeds kidney-shaped, 3 × 2.5 mm, concentrically ribbed and wrinkled. Fig. 7/2.

Distr. ASTOVE; pantropical, from south or southeast Asia, now widely cultivated and naturalized.

Notes. Two collections only from Astove, *Ridgway* 64 (US) and *Veevers-Carter* 64 (EA)

Vernac. 'ambrette'

2. Hibiscus tiliaceus L., Sp. Pl.: 694 (1753); F.M.S.: 24 (1877); Dupont, Report:34 (1907); Hemsley in B.M.I.K. 1919: 143 (1919); F.M.C. 129: 16, fig. 4/3–4 (1955); F.Z. 1: 435, fig. 89/8 (1961).

Tangled large shrub or small tree; young growth stellate-pubescent. Leaves orbicular-cordate, up to 15(–20) cm wide, basal sinus deep, apex shortly acuminate, margin subentire, dark green, minutely pubescent especially in vein axils above, white-pubescent beneath; petiole shorter than to almost as long as blade; stipules oblong to oblong-ovate, enclosing terminal bud, falling early and leaving an encircling scar. Inflorescence a very sparse, few-flowered, terminal panicle, enlarging and becoming more open in fruit; involucral bracts 8–11, short, united c ½-way, lobes triangular, thinly stellate-pubescent, enlarging somewhat in fruit. Calyx united in lower third or half, pubescent, lobes triangular-acuminate, keeled, enlarging slightly, persistent. Corolla bell-shaped, spreading as anthesis proceeds, lemon-yellow with blackish-maroon centre, yellow part turning dull red in late afternoon before dropping. Petals obovate, c 5 cm long, rounded. Staminal column c 2 cm long, with stamens almost the full length; free filaments c 1 mm long. Style and stigma purple. Capsule globose to ovoid or short-cylindric, c 2 × 1.5 cm, beaked, densely tomentose, splitting to base into 5 elliptic or elliptic-oblong, stiff, beaked valves; seeds 5–7 per cell, curved, humped and subtruncate distally, compressed, 4.5 × 3 mm, base broadly stalked, surface dull, glabrous or thinly stellate. Fig. 5/3–4.

An extremely variable species, responding to environmental differences as well as being genetically variable.

Distr. ALDABRA: W; ASTOVE; pantropical.

Notes. Only known on Aldabra from the coastal champignon at Anse Var, where it appears to consist of a single tree surrounded by numerous stems resulting from sucker growth from the pendent branches. Flowers in mid to late wet season.

Vernac. 'varre' or 'var'.

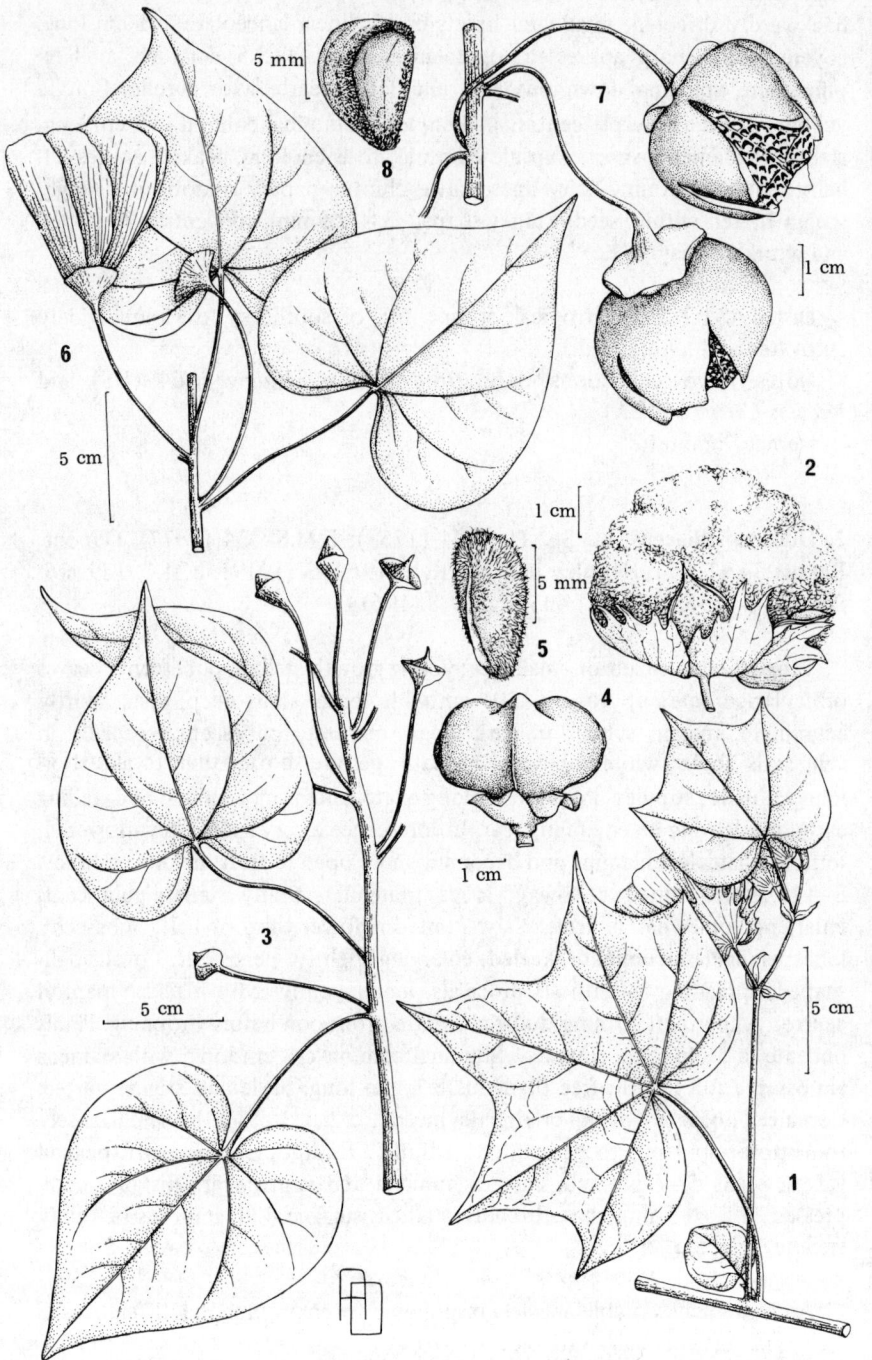

Fig. 6　1, *Gossypium hirsutum*, habit; 2, fruit. 3, *Thespesia populnea*, habit; 4, fruit; 5, seed. 6, *Thespesia populneoides*, habit and flower; 7, fruit; 8, seed.

4. SIDA L.

Herbs or somewhat woody. Leaves usually with toothed or crenate margins, more or less stellate-pubescent; petiolate. Flowers axillary or terminally paniculate, usually without an involucre. Calyx of 5 sepals, united below, united part angular or with midribs and sutures often keeled. Petals usually orange or white, rarely pink, united below, corolla closing toward evening. Staminal column usually shorter than petals, antheriferous near and at apex. Carpels 5–14, ovules 1 per cell; style branches 5–14. Fruit a schizocarp; mericarps with or without 2 small points or awns at the apex, indehiscent or splitting at apex only after falling, leaving a slender column; seeds 1 per mericarp.

A pantropical genus of c 200 species, mainly in the New World.

1. Leaves green beneath, 2 cm or more long; pedicels much shorter than leaves   **1. S. acuta**
1. Leaves greyish or whitish pubescent beneath; pedicels generally at least half as long as leaves
   2. Leaves not much more than 1–1.5 cm long, ovate to elliptic; pedicels mostly shorter than leaves   **2. S. parviflora**
   2. Leaves usually 2 cm or more long; plant greyish; pedicels usually as long or longer than leaves   **3. S. rhombifolia**

1. Sida acuta Burm. f., Fl. Ind.: 147 (1768); F.M.C. 129: 145 (1955); F.Z. 1: 477 (1961); Renvoize in P.T.R.S. B, 260: 229 (1971).

Sida carpinifolia L. f., Suppl.: 307 (1781); F.M.S.: 20 (1877); Hemsley in B.M.I.K. 1919: 116 (1919).

Ascending, somewhat woody herb, to 1 m tall; stems tending to be zigzag, thinly stellate-pubescent. Leaves 2-ranked, ovate to lanceolate or linear, margins serrate, teeth often ending in a short bristle; petiole short, 3–6 mm; stipules lanceolate to linear-lanceolate, tending to be unequal. Flowers axillary; pedicels 2–5 mm long, jointed at or just below the middle. Calyx about as broad as high, 5 mm across, united for about two-thirds, 10-ribbed, lobes triangular-acuminate, ciliate. Corolla yellow-orange, 10–15 mm across. Fruit depressed-globose with a low, strongly furrowed conical projection; mericarps 6–10, obliquely tetrahedral-kidney-shaped, netted-wrinkled, dehiscent above, with 2, somewhat spreading, strong, short, spinous processes; seed triangular, 1 × 1 mm, smooth. Fig. 5/5.

Distr. ALDABRA: W; ASSUMPTION; COSMOLEDO: M; pantropical.
Notes. A variable weedy plant, naturalized in gardens and around settlements. Flowers during the wet season.
Vernac. 'la bolzé'

2. Sida parvifolia DC., Prodr. 1: 461 (1824); Fosberg in P.T.R.S. B, 260: 223-225 (1971); Renvoize in P.T.R.S. B, 260 260: 230 (1971).

Sida spinosa sensu Schinz in A.S.N.G. 21: 87 (1897); Voeltzkow in A.S.N.G. 26: 551 (1902); Dupont, Report: 34 (1907); Hemsley in B.M.I.K. 1919: 116 (1919), not L. (1753).
    Sida spinosa var. *pusilla* sensu Baker in B.M.I.K. 1894: 147 (1894); not Cav. (1781).

Stems spreading from a root-crown, sparsely branched, loosely stellate-hairy. Leaves oval to orbicular, apex mostly obtuse, base rounded to sub-cordate, margins toothed to subserrate, hairy above, stellate beneath; petioles slender, shorter than blades; stipules awl-shaped. Flowers axillary, solitary; pedicels inconspicuously jointed. Calyx 10-ribbed below, 5-lobed, lobes tri-angular, somewhat shorter than united portion, densely to thinly stellate or stellate-hairy. Petals obovate, c 1 cm long. Staminal column antheriferous at apex. Schizocarp hemispherical, 3 mm across, mericarps 5, 2 mm long, dorsally shield-shaped with 2 blunt, triangular beaks above with dehiscence between them, beaks especially minutely stellate-pubescent, on dorsal sides; seeds dark brown, obliquely pear shaped, 2 × 1.5 mm, minutely wrinkled. Fig. 5/6.

Plants with the upper surfaces of the leaves so densely stellate-pubescent that they scarcely look green have been determined as *S. vescoana* Baill., but seem to be no more than extremes of *S. parvifolia*. *S. vescoana* from Madagascar may well be nothing more than this, judging from Baillon's description in Bull. Mens. Soc. Linn. Paris 1: 504 (1885).

Distr. ALDABRA: W, P, M, S; Michel; ASSUMPTION; COSMOLEDO: W, M; ASTOVE; islands of the Indian Ocean.
    Notes. Frequent on coastal sands, champignon and pavé, in sunny areas where there is little competition from other species. Flowers and fruits throughout the year if sufficient moisture is available.
    Vernac. 'herbe dure'.

3. Sida rhombifolia L., Sp. Pl.: 684 (1753); F.M.S.: 20 (1877); F.M.C. 129; 146 (1955); F.Z. 1: 480, fig. 92/A (1961).

Wiry shrub or somewhat woody herb reaching 50–60 cm tall; stems ascending, much-branched, thinly greyish-pubescent when young, dark reddish or maroon when older. Leaves elliptic, oblong, or obovate, up to several cm long, margins finely serrulate in distal two-thirds, often more coarsely toothed apically, greyish green to dull green above, grey beneath; petiole short; stipules slender or awl-shaped, pubescent. Flowers axillary; pedicels fine, ascending, shorter than or exceeding leaves, jointed c third to quarter from apex. Calyx broadly bell-shaped, 5–7 mm or more wide, 5–6 mm high, united about half-way; lobes 5, triangular, acuminate. Corolla dull orange, petals obliquely broadly obovate, apex somewhat shallowly notched.

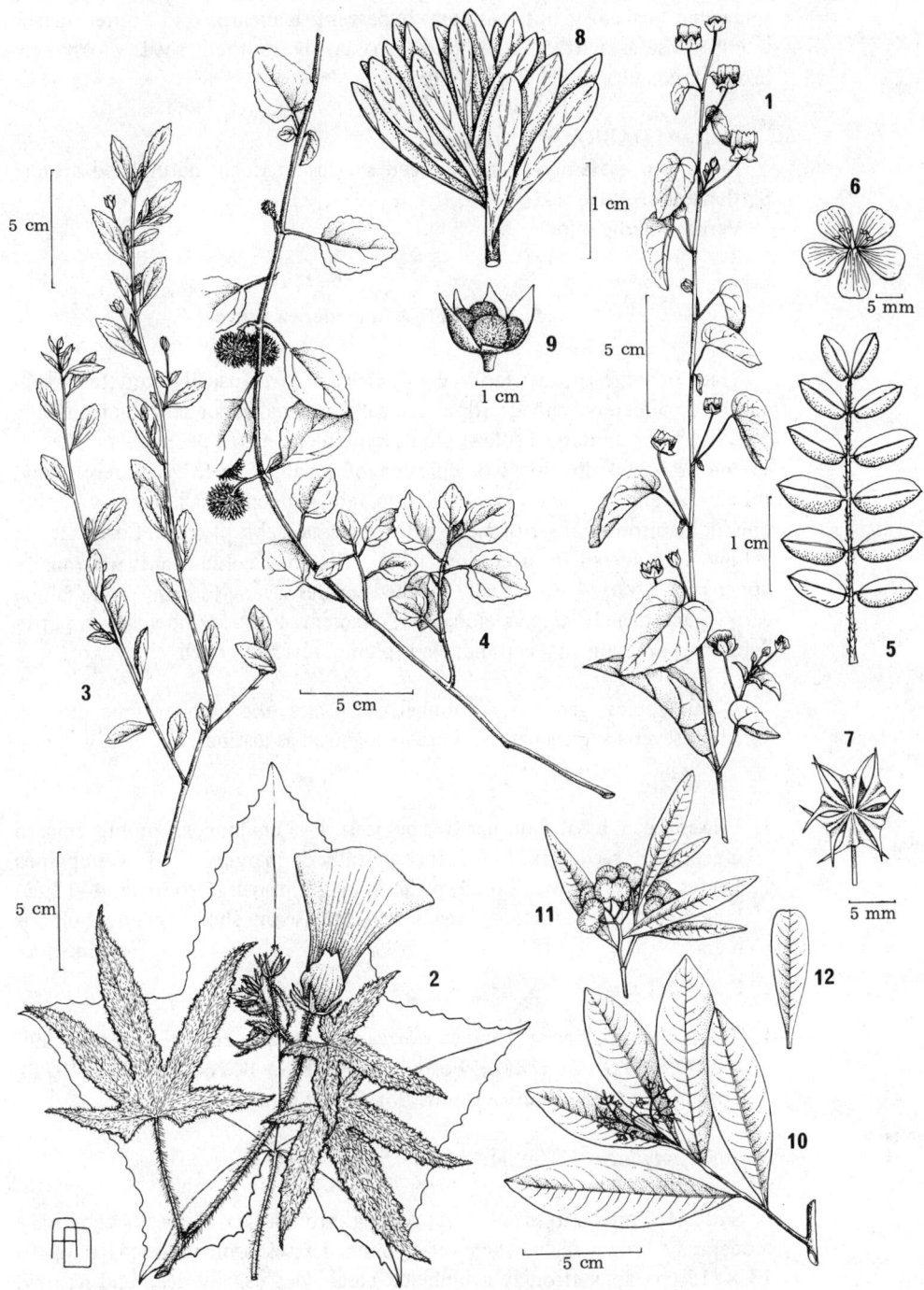

Fig. 7   1, *Abutilon fruticosum*, habit. 2, *Hibiscus abelmoschus*, habit, leaf
and flower. 3, *Sida rhombifolia*, habit. 4, *Triumfetta procumbens*,
habit and fruit. 5, *Tribulus cistoides*, leaf; 6, flower; 7, fruit.
8, *Suriana maritima*, branch tip; 9, fruit. 10, *Dodonaea viscosa*,
habit and flowers; 11, habit and fruit; 12, leaf variation.

Receptacle deeply 10-furrowed. Styles 8—12. Schizocarp depressed-globose, somewhat beaked or not, minutely pubescent; mericarps 8—12, outer surface wrinkled, sides netted, apex blunt to shortly toothed; seeds glossy red-brown, triangular, 2.5 × 2 mm. Fig 7/3.

Distr. ALDABRA: W; pantropical.

Notes. An extremely variable weed species sparingly naturalized around Settlement in Aldabra.

Vernac. 'herbe dure'.

5. THESPESIA Solander ex Correa

Trees or large shrubs; sap yellow, sticky. Leaves usually cordate or sub-cordate, palmately veined; stipules small, awl-shaped or lanceolate, falling early. Flowers axillary; pedicels stout, jointed either to a peduncle or directly to the branchlet; involucre or epicalyx of usually few (3) to rarely many, spirally arranged, usually reduced, separate segments, falling early. Calyx united, hemispherical, subtruncate with very small, abrupt teeth. Corolla large, yellow, turning red in afternoon, falling. Staminal column antheriferous in upper part. Ovary 4—6-, usually 5-celled, ovules several to many; style falling early, unbranched, stigmas elongated, coherent. Fruit an indehiscent, partly dehiscent, or loculicidal capsule; seeds plump, several per cell.

A pan-tropical genus of a number of species, the number depending on which of several segregate genera are recognized as distinct.

1. Leaves green, basal sinus narrow; pedicels 1—5 cm long, ascending; epicarp indehiscent; seeds with long hairs, especially on margins **1. T. populnea**
1. Leaves bronzed, sinus broad; pedicels longer, usually drooping, 4—12 cm long; epicarp dehiscent; seeds covered with short, erect, bulbous hairs **2. T. populneoides**

1. **Thespesia populnea** (L.) Sol. ex Correa in Ann. Mus. Hist. Nat. Paris 9: 290 (1807); F.M.S.: 25 (1877); Fosberg in P.T.R.S. B, 260: 219,225 (1971); Fosberg & Sachet, Smiths. Contr. Bot. 7: 8, figs. 1, 2 & 5 (1972).

Hibiscus populneus L., Sp. Pl.: 694 (1753).

Small to moderate-sized tree; young growth and leaves green, only moderately brown scaly when very young. Leaves orbicular-cordate, up to 13 × 15 cm, apex strongly acuminate, basal sinus usually deep and narrow, greenish, even when young; stipules linear-lanceolate, to 1 cm long, falling early. Pedicels usually 1—5 cm long, erect or ascending, jointed, with 2 scale-like bracts near base, very rarely with 2 joints; flowers erect, not drooping, involucral bracts 3, lanceolate, to 1 cm long; buds and young fruit exuding yellow gum when cut. Calyx hemispherical-bell-shaped, c 5–6 cm long, subtruncate, densely appressed-hairy within. Corolla bright light yellow, turning reddish in afternoon, centre red to dark maroon; broadly bell-shaped, c 5–6 cm long. Staminal column included in corolla. Mature fruit depressed-globose, indehiscent, irregularly crumbling with age, 4–5 celled; seeds several per cell, broadly obovoid, 15 × 9 mm, slightly angled, covered by closely matted silky hair. Fig. 6/3–5.

Distr. ALDABRA: W (north), P, M, S; pantropical.

Notes. Mainly on coastal dunes and adjacent scrub. Flowers mainly during the late wet to early dry season but seasonal rains can produce flowering late in the dry season. Seeds eaten and locally dispersed by tortoises. Viable seeds found along strand line.

Vernac. 'bois de rose'.

2. Thespesia populneoides (Roxb.) Kostel. in Allg. Med. Pharm. Fl. 5: 1861 (1836); Fosberg in P.T.R.S. B, 260, 219, 225 (1971); Fosberg & Sachet, Smiths. Contr. Bot. 7: 10, Figs. 3, 4 & 6 (1972).

Hibiscus populneoides Roxb., Fl. Ind. ed. Carey 3: 191 (1832).
Thespesia populnea sensu Dupont, Report: 34 (1907); Hemsley in B.M.I.K. 1919: 43 (1919); F.M.C. 129: 124, fig. 30 (1955); F.Z. 1: 421 (1961), not (L.) Sol. ex Correa (1807).

Small tree; young growth and leaves with persistent and dense brown scales giving a bronzed appearance. Leaves deltoid to subcordate, or cordate, up to 15 × 12.5 cm, apex tending to be long-acuminate, base with a very shallow broad sinus, usually with prominent domatia in axils of main nerves; stipules awl-shaped to lanceolate, soon falling. Pedicels up to 10–12 cm long, usually curved downward so flowers are drooping, bracts absent; involucral bracts 3, reduced, soon falling, triangular-ovate; young fruit and buds exuding yellow gum when cut. Calyx 8–10 mm long, truncate or with minute teeth. Corolla yellow, centre dull reddish to dark maroon, bell-shaped, 5–6 cm long. Staminal column included. Style exserted from staminal column but included in corolla. Mature fruit with 2 very distinct layers, a smooth exocarp separated from a hard, tough, fluted endocarp by a loose, fibrous-spongy mesocarp which partially disintegrates at maturity, the exocarp then dehiscing into (4–)5(–6) valves; seeds several per cell, broadly obovoid, 13 × 9 mm, covered by a dense, short pubescence of erect, bulbous hairs. Fig. 6/6–8.

Distr. ALDABRA: S (east); ASSUMPTION; COSMOLEDO: W, M; ASTOVE; coasts of the Indian Ocean and its islands, Australia, Malaysia, Indo-China to Hainan; very sparingly and probably introduced in West Africa, cultivated in Brazil and British Guiana.

Notes. Inland platin, locally dominant in low-lying somewhat saline areas, especially around pools and behind mangroves. Flowering during the mid wet season.

Vernac. 'bois de rose'.

10. TILIACEAE

Trees or shrubs, rarely herbs; often with tough bast fibres. Leaves simple, alternate or occasionally opposite, base often oblique, usually 3 or more strong nerves from base; stipules present. Flowers in cymes, panicles or racemes; regular, usually bisexual. Sepals and petals usually 5, rarely 3 or 4, free, or petals lacking. Stamens many, hypogynous, free or rarely in fascicles. Ovary superior, usually 2–10-celled, placentation usually axile, ovules 1–several per cell; style 1, stigmas several. Fruit various.

A medium-sized family, pantroptical with some extension into the temperate zone.

1. Shrubs or small trees; flowers pink white or yellow; fruit indehiscent, ovoid, lobed or entire   **2. Grewia**
1. Herbs or small shrubs; flowers yellow
  2. Fruit dehiscent, elongated and capsular   **1. Corchorus**
  2. Fruit sphaeroid, densely spiny   **3. Triumfetta**

1. CORCHORUS L.

Herbs or small shrublets. Leaves alternate, margins serrate to lobed; stipules slender. Flowers bisexual, leaf-opposed or axillary. Sepals 4–5, narrowly oblong. Petals 4–5, yellow, obovate to linear. Stamens 7–many, free. Ovary 2–5-celled; ovules 2–many per cell; style glabrous, stigma lobed to frilled. Fruit an elongated or ovoid capsule; seeds 2–many.

A genus of c 100 species in tropics and subtropics.

Corchorus aestuans L., Syst. Nat. ed. 10, 2: 1079 (1759); F.Z. 2: 87, fig. 8/E (1963).

Decumbent or erect, branched annual or short lived perennial, up to 0.5 m high; branches shaggy-haired, sometimes tinged with purple, otherwise green. Leaves broadly to narrowly ovate, 2–4.5 × 1–2.5 cm, apex acute, base rounded, margins serrate, petiole up to 10 mm long; stipules up to 5 mm long. Inflorescence 1–2(–3)-flowered, cymes opposite the upper leaves. Sepals brownish-yellow, oblong, 4 mm long. Petals yellow, oblong, 4 mm long. Fruit a 3–5-sided, erect, dehiscent, winged capsule, (10–)25 × 3(–6) mm, glabrous, terminating in 3–5 short, spreading horns; seeds numerous, brown, shortly cylindrical, c 0.8 × 0.8 mm. Fig. 8/1–2.

Distr. ALDABRA: W; ASSUMPTION; COSMOLEDO: W; ASTOVE; pantropical.

Notes. Found in waste places around settlement sites, also occasionally from lagoon islets. Flowers mainly during the wet season.

2. GREWIA L.

Shrubs or small trees. Leaves alternate, margins usually serrate; petiole present. Flowers bisexual, in terminal, axillary or leaf-opposed panicles or cymes. Sepals 5, coloured inside like the petals, linear-oblong, hairy outside. Petals 5, yellow, pale purplish-pink or white, shorter than or as long as the sepals. Stamens numerous, free. Ovary 2–4-celled, ovules 2–many per cell. Fruit a 1–4-lobed drupe or berry; seeds 1 per lobe.

A large genus of c 400 species, mostly in the tropics and subtropics.

1. Small slender trees or shrubs; flowers yellow; fruits small, becoming glabrous with age; leaves thin, 7 cm or less long   **1. G. aldabrensis**
1. Medium-sized to small trees; flowers purplish pink, fruits large, hairy; leaves firm, mostly 12–18 cm long   **2. G. salicifolia**

1. Grewia aldabrensis Baker in B.M.I.K. 1894: 147 (1894); Schinz in A.S.N.G. 21: 87 (1897); Voeltzkow in A.S.N.G. 26: 551 (1902); Dupont, Report: 35 (1907); Hemsley in B.M.I.K. 1919: 117 (1919); Type: Aldabra, *Abbott* s.n. (K, holo.).

Small slender tree, up to 3.5 m, branches dark brown, glabrous; branchlets light brown, subglabrous. Leaves ovate, (2.5–)3.5–7 × (1.5–)2–3.5 cm, apex acuminate, base rounded, margins serrate, glabrous, primary nerves 3; petioles up to 12 mm long; stipules linear, up to 5 mm long. Sepals narrowly obovate, up to 6 mm long, pubescent and green outside, glabrous and yellow inside. Petals yellow, narrowly obovate, up to 6 mm long, glabrous. Ovary sparsely pubescent; style up to 5 mm long. Fruit woody, (2–)4-lobed, lobes obovate, 4–5 mm long, becoming glabrous with age. Fig. 8/3–4.

Distr. ALDABRA: W, S (east), Michel; endemic.

Notes. An occasional constituent of mixed scrub. Flowers during the mid wet season

2. Grewia salicifolia Schinz in A.S.N.G. 21: 87 (1897); Voeltzkow in A.S.N.G. 26: 551 (1902); Dupont, Report, 35 (1907); Hemsley in B.M.I.K. 1919: 117 (1919). Type: Aldabra, *Voeltzkow* 43 (FR, holo.; Z, iso.).

Tree up to 7 m high; bark grey, rough, flaking; branches dark brown, glabrous; branchlets light brown, pubescent. Leaves lanceolate,

13–18 × 2–4 cm, apex acuminate, base rounded, margins minutely serrate, upper surface glabrous, lower surface very shortly pubescent; petioles 6–10 mm long, pubescent; stipules linear, up to 10 mm long. Inflorescence (2–) 3(–4)-flowered, flowers axillary. Sepals pink and pubescent inside, grey and pubescent outside, linear oblong, up to 1 cm long, spreading. Petals pink, ovate, 4–5 mm long. Stamens 4–6 mm long, filaments pink. Ovary pubescent; style up to 5 mm long. Fruit woody, ovoid, up to 2 cm in diameter, 4-lobed, densely pubescent, hairs golden. Fig. 8/5–6.

Distr. ALDABRA: W, M (west), S (east); COSMOLEDO: M; endemic.
Notes. An occasional constituent of mixed scrub. Flowers during the wet season.
Vernac. 'mabolo'.

3. TRIUMFETTA L.

Annual or perennial herbs, shrublets or shrubs. Leaves alternate, simple or digitate. Flowers regular, in terminal or axillary cymes. Sepals 5, usually narrow. Petals 5, yellow or orange, narrow or broad towards the apex. Stamens and ovary raised on a short, glabrous column (androgynophore). Stamens 4–40, inserted on the inner margin, near the apex of the column. Ovary 2–5-celled, often tubercled, with a tuft of bristles from the apex of each tubercle. Fruit a capsule splitting by 3–5 valves; seeds 1–2 seeds per cell.

c 150 tropical species.

Triumfetta procumbens Forst. f., Prodr.: 35 (1786).

Densely hairy, perennial herb; stems prostrate. Leaves ovate or 3-lobed, 1.5–5 × 1.5–5 cm, apex obtuse, margins crenate or crenate-serrate; petioles 1.5–4 cm long. Flowers yellow, in terminal and axillary clusters. Sepals 1 cm long. Petals glabrous, 1 cm long, broad at the apex, narrowing to the base. Stamens numerous, as long as the petals. Ovary hairy and tubercled, each tubercle bearing a stout spine. Fruit spheroid, 1 cm in diameter, densely hairy and spiny, the spines hooked. Fig. 7/4.

Distr. ASTOVE; Indo-Pacific region, mainly oceanic islands.
Notes. Only known from one collection on Astove, *Ridgway* 7 (US). Fruit readily dispersed by becoming attached to the plumage of birds.

11. ERYTHROXYLACEAE

Shrubs or small trees. Leaves usually alternate simple; stipules ± united. Flowers axillary, solitary or in thyrsoid fascicles, regular. Sepals usually 5,

Fig. 8   1, *Corchorus aestuans*, habit; 2, fruit. 3, *Grewia aldabrensis*, habit
and fruit; 4, flower. 5, *Grewia salicifolia*, habit and fruit; 6, flower.
7, *Erythroxylum acranthum*, habit.

free. Petals usually 5, free, appendaged on inner side, alternate with sepals. Stamens 10, in 2 series, basally united. Ovary superior, 2−3-celled, placentation axile; ovules 1−2 per cell. Fruit a berry or drupe.

Pan-tropical family, largely New World.

ERYTHROXYLUM Browne

Trees or shrubs; young shoots conspicuously laterally compressed. Leaves alternate, simple, entire; petiolate. Inflorescence of few-flowered, axillary fascicles or flowers solitary. Sepals 5, triangular, united at the base. Petals 5, free, clawed. Stamens 10, united at the base. Ovary usually 3-celled, ovules 1 per cell; styles 3, free, divergent, stigmas capitate. Fruit a 1-seeded drupe, slightly fleshy.

A genus of c 200 species, widespread throughout the tropics and subtropics.

Erythroxylum acranthum Hemsley in J. Bot. 54, Suppl. 2: 5 (1916) & in B.M.I.K. 1919: 117 (1919). Types: Aldabra, *Dupont* 103, *Thomasset* 233, *Fryer* 35 & 87 (all K, syn.).

Shrubs or small trees, 1−5 m high; branches glabrous, branchlets oval in cross-section, annual rings conspicuous. Leaves obovate, 2.5−5 × 1.5−3 cm, apex rounded to indented, with or without a very small projection, glabrous, leathery, central vein conspicuous; petiole 3−5 mm long. Inflorescence of 3−6 flowers, 4 mm in diameter; pedicels 4−7 mm long. Sepals green, sometimes tinged with orange or pink, 1.5−2.5 mm long, acuminate, glabrous. Petals cream, slightly longer than the sepals, 2.5−3 mm long, glabrous. Ovary ovoid, glabrous; heterostylous. Fruit oblong, 3−4 × 2 mm (immature), red; seeds not seen. Fig. 8/7.

The compressed branchlets clearly distinguish this species from *Flacourtia ramontchii* and *Ludia mauritiana* (Family 6), and *Margaritaria anomala* (Family 62/3) with which it may be vegetatively confused.

Distr. ALDABRA: W, P, M, S, Esprit, Michel; ASSUMPTION; COSMOLEDO: M; endemic.
Notes. A frequent constituent of the inland mixed scrub. Provided sufficient moisture available can flower throughout the year; responds rapidly to unseasonal rainfall.
Vernac. 'sandol' or bois sandol'.

12. ZYGOPHYLLACEAE

Herbs, shrubs or small trees. Leaves usually opposite, pinnately compound, rarely simple; stipules present. Flowers axillary or in cymes; regular, usually bisexual. Sepals 5, rarely 4, free or almost so. Petals 5, rarely 4, free, alternate with sepals. Stamens in 1—several whorls, outer whorl usually opposite the petals. Ovary superior, 4—5-celled; ovules usually several on axile placentae. Fruit usually capsular or a schizocarp, often variously lobed or winged.

Pantropical and subtropical family, best developed in arid regions.

TRIBULUS L.

Annual or perennial herbs; branches prostrate or ascending. Leaves pinnately compound, opposite, one of each pair usually longer than the other; leaflets opposite, entire, sessile or petiole very short; stipules herbaceous. Flowers solitary, axillary. Sepals 5, soon falling or persistent. Petals 5, spreading. Stamens 10. Ovary 5-lobed, consisting of 5 carpels, each containing 3—5 ovules, densely covered with stiff, erect hairs; stigma 5-angled, almost sessile on the ovary or with a distinct style. Fruit 5-partite, readily breaking up when mature, each segment armed with several spines.

A genus of about 25 species extending throughout the tropical and warm parts of the world.

Tribulus cistoides L., Sp. Pl.: 387 (1753); Schinz in A.S.N.G. 21: 85 (1897); Voeltzkow in A.S.N.G. 26: 550 (1902); Dupont, Report: 35 (1907); Hemsley in B.M.I.K. 1919: 117 (1919); F.M.C. 103: 4, fig. 1 (1952); F.Z. 2: 130 (1963).

T. terrestris sensu Baker in B.M.I.K. 1894: 147 (1894); Schinz in A.S.N.G. 21: 85 (1897); Voeltzkow in A.S.N.G. 22: 550 (1902). Dupont, Report: 35 (1907); Hemsley in B.M.I.K. 1919: 117 (1919), not L. (1753).

Annual or perennial herbs; stems long, prostrate, sparsely hairy or glabrescent. Leaves of each pair unequal in length, the larger up to 7 cm long with 4—7 pairs of leaflets, the smaller up to 2.5 cm long with 4 pairs of leaflets; leaflets obliquely-oblong, 5—17 × 2—8 mm, apex obtusely acuminate, pubescent or glabrous above, silky pubescent beneath; stipules soon falling. Pedicel as long as or longer than the subtending leaf. Sepals lanceolate, 8—10 mm long, apex acute, pubescent outside. Petals yellow, broadly wedge-shaped, 7 × 10 mm. Ovary with stiff, bristle-like hairs; style cylindrical, up to 4 mm long. Fruit breaking up into 5 cocci, each coccus armed with 4 stout spines. Fig. 7/5—7.

This is a variable species and some specimens may approach *T. terrestris L.* in general appearance but differ in having a distinct style and larger flowers.

Distr. ALDABRA: W; ASSUMPTION; COSMOLEDO: W, M; ASTOVE; pantropical.

Notes. Occurs around areas of habitation, such as coconut plantations. Flowers during the early dry season (? wet season). Unpleasant underfoot. Vernac. 'pagode'.

13. OXALIDACEAE

Usually herbs, rarely trees or shrubs; often with fleshy tubers or rhizomes. Leaves alternate, compound, frequently palmate, (or 1-foliolate), often sensitive, folding at night; stipules present or reduced or rudimentary. Flowers in cymes, racemes, or solitary; bisexual, open or cleistogamous, regular, 5-merous. Calyx united below. Petals distinct or coherent at base. Stamens 10, or 5 plus 5 staminodes, united at base; anthers dehiscing longitudinally. Ovary superior, 5-celled, ovules 1 or more per cell; placentation axile; styles 5, stigmas capitate or truncate. Fruit a capsule or berry.

A cosmopolitan, largely tropical family, with few genera but many species.

OXALIS L.

Herbs of very diverse habit; sap acid; rhizomes or bulbs underground, or simple root system. Leaves palmately or pinnately 3-foliolate, folding at night or in bad weather. Inflorescences cymose, basal or axillary; flowers variously coloured. Fruit an ovoid to spindle-shaped or prismatic, thin-walled capsule; seeds numerous, often dispersed explosively.

A cosmopolitan genus of a great many species.

Oxalis sp. near **O. bakerana** Exell (= *O. villosa* Baker); Hemsley in B.M.I.K. 1919: 143 (1919).

The collection by *Thomasset* from Wizard Island, Cosmoledo has not been traced. No further collections have been made. The identity of this plant is therefore uncertain.

14. RUTACEAE

Herbs, shrubs or trees; usually glandular and aromatic; often spiny. Leaves alternate or opposite, simple or compound; stipules absent. Flowers commonly bisexual, rarely unisexual and dioecious. Sepals 3–5, free except sometimes at base. Petals 3–5, free or rarely united, rarely absent. Stamens 3–10, in 2 whorls, outermost opposite petals, attached to or at base of a disk, unequal, rarely united or adherent. Ovary superior, entire to deeply lobed, usually 4–5-celled, placentation axile, rarely parietal, ovules 1 or more per cell; styles 4–5, free or united into one. Fruit various.

Large family with tropical and temperate distribution; the genus *Citrus* of considerable economic importance for its edible fruit.

CITRUS L.

Shrubs and small trees; stems dark green; plants aromatic when broken. Leaves alternate, 1-foliolate, simple, dark green, finely gland-dotted; glands translucent; petiole usually winged, obviously articulate to blade; stipules absent, but frequently a spine at one side of axillary bud. Flowers usually bisexual, very fragrant, axillary, solitary, fascicled or in small cymes or panicles. Sepals and petals 5. Stamens 15 or more, bases coherent in bundles. Ovary 8–15-celled; style conspicuous, falling early. Fruit a berry with a tough, glandular rind, the cells filled with large vescicles or "juice sacs", seeds several per cell, attached along axis among these vescicles.

A southeast Asian and Malesian genus of few or many species, very difficult taxonomically, many of the "species" domesticated and perhaps originating in cultivation.

Citrus aurantifolia (Christm.) Swingle in Journ. Wash. Acad. Sci. 3: 465 (1913).

Limonia aurantifolia Christm. in Houtt., Nat. Hist. 2: 618 (1777).

Small tree, glabrous, spiny. Leaves elliptic, 5–6 cm long, margins obscurely crenulate; petiole c 1 cm, very narrowly winged. Flowers several in an axil, c 1 cm long, white. Stamens 20 or more. Fruit green, broadly ellipsoid to globose, 10-celled, very sour but with an agreeable flavour.

Distr. ALDABRA: W; cultivated throughout the tropics.
Notes. One fair-sized tree planted near the settlement on Aldabra succumbed to the drought in 1968. It is not known if others persist elsewhere.
Vernac. 'lime'.

15. SURIANACEAE*

Shrub. Leaves alternate simple; stipules absent. Flowers in few-flowered, axillary inflorescences, regular, bisexual, 5-merous. Sepals 5. Petals 5, yellow. Stamens in 2 whorls of 5, inner whorl fertile, outer variously sterile. Carpels 5, free; carpels with 2 ovules attached basally side by side, only 1 ovule developing; style basal on inner angle of carpel, slender. Fruit a hard 1-seeded nut.

A monotypic family of tropical shores.

SURIANA L.

Maritime shrubs. Leaves alternate, simple, generally crowded towards the ends of the branches. Inflorescences of solitary and axillary flowers or few-flowered panicles, near the ends of the branches; flowers bisexual. Calyx deeply 5-lobed, persistent. Petals 5. Stamens 10; filaments densely hairy towards the base, alternate filaments shorter. Ovary of 5 free carpels, densely hairy, ovules 2 per carpel; styles 5, free, lateral. Fruiting carpels dry, 1-seeded.

1 species, extending throughout the tropics.

Suriana maritima L., Sp. Pl.: 284 (1753); F.M.S.: 42 (1877); Baker in B.M.I.K. 1894: 147 (1894); Schinz in A.S.N.G. 21: 85 (1897); Voeltzkow in A.S.N.G. 26: 550 (1902); Dupont, Report: 35 (1907); Hemsley in B.M.I.K. 1919: 117 (1919); F.M.C. 105: 7 (1950); F.Z. 2: 211, fig. 37 (1963).

Much-branched shrub up to 2 m tall; branches greyish-tomentose. Leaves tending to be in terminal rosettes, spreading by day, appressed at night, linear-spathulate to oblanceolate, 1–3 × 0.3–0.5 cm, apex obtuse to sub-acute, base wedge-shaped and continuous with the rather short petiole, margins entire, very shortly pubescent on both sides, slightly glandular towards the base. Inflorescence as long as or shorter than the leaves; branches of inflorescence densely pubescent; bracts 3–8 mm long, similar in shape to the leaves and densely pubescent. Calyx green, lobes lanceolate, 5–8 mm long, apex acuminate, glandular-pubescent. Petals yellow, broadly obovate, 5 × 3 mm. Stamens with 5 long filaments 4.5 mm long and 5 short filaments 3.5 mm long. Ripe fruiting carpels obovoid, 3–4 mm long, indehiscent, free but closely appressed, becoming dark with age. Fig. 7/8–9.

Can be confused with coastal forms of *Pemphis acidula* (Family 29), which has white flowers, opposite leaves and brittle twigs.

---

* For discussion of its morphology, family status and presumed relationships see Gutzwiller in Engl., Bot. Jahrb. 81: 1–49 (1961).

Distr. ALDABRA: W, P, M, S, Esprit, Moustique; ASSUMPTION; COSMOLEDO: W, M; ASTOVE; pantropical strand plant.

Notes. A frequent species of sandy shores, often extending further down the beach into the spray zone than other terrestrial plants. Now being destroyed by tortoises on the south and east coasts of Aldabra. Also recorded from the tops of windward cliffs and the landward side of mangrove communities. Flowers throughout the year if moisture sufficient, but primarily during the wet season.

Vernac. 'matelot' or 'suriana'.

16. OCHNACEAE

Trees, shrubs and herbs. Leaves alternate, usually simple; stipules present. Flowers in cymes, racemes or panicles, regular, bisexual. Sepals usually 4–5, usually overlapping, free or almost so. Petals usually 4–5, distinct, contorted or overlapping. Stamens 5, 10 or many, free, hypogynous, sometimes some modified to staminodes; anthers 2-celled, opening by terminal pores or longitudinal slits. Ovary superior and borne on a short stalk or disk, 1–15-celled, often lobed or apocarpous; ovules 1-many per cell; style 1, stigmas 1–5. Fruit fleshy or capsular, usually separating into distinct, fleshy, fruitlets in a ring on an enlarged receptacle.

A rather small tropical family, some species with ornamental potentialities, but seldom planted.

OCHNA L.

Small shrubs to medium-sized trees. Leaves serrate to entire; petiole present; stipules entire or fringed to deeply 2-fid, soon falling. Flowers terminal, solitary or arranged in panicles, racemes or umbels; bracts scale like, soon falling. Sepals usually 5, rarely 4, persistent, becoming red and leathery in fruit. Petals 5, rarely more, falling. Stamens 20 or more, free; anthers dehiscing by longitudinal slits or terminal pores. Carpels 5–15, free, ovules 1 per carpel; styles elongated, stigma slightly expanded. Fruit of 1 or more 1-seeded druplets, usually dark coloured and inserted on a red, fleshy receptacle.

A genus of c 85 species in the Old World Tropics from Africa to southeast Asia.

Ochna ciliata Lam., Encycl. Méth. Bot. 4: 511 (1798); Baker in B.M.I.K. 1894: 147 (1894); Schinz in A.S.N.G. 21: 88 (1897); Voeltzkow in A.S.N.G. 26: 551 (1902); Dupont, Report: 35 (1907).

O. fryeri Hemsley in J. Bot. 54, Suppl. 2: 7 (1916) & B.M.I.K. 1919: 118 (1919). Types: Aldabra, *Abbott* s.n. (K, syn., US, isosyn.) *Dupont* s.n., *Fryer* 32 (both K, syn.).

O. mauritiana sensu Hemsley in B.M.I.K. 1919: 118 (1919), not Lam. (1798).

Diporidium ciliatum (Lam.) H. Perr. in Not. Syst. 10: 33 (1941); F.M.C. 133: 33 (1951).

Spreading trees, 3–5 m high; bark smooth, greyish. Leaves elliptic-lanceolate, (4–)6–12 × (1.5–)2–5 cm, apex acute to obtuse, margins spiny-toothed, leathery, shiny dark green above, paler beneath with a conspicuous pinkish midrib; petiole 3–8 mm long; stipules minute, up to 1 mm long, acuminate. Flowers fragrant, in dense showy panicles 2–3 cm long; pedicels 1–1.5 cm long, articulated up to about a third of the length from the base. Sepals 5, greenish-yellow when young, becoming red and reflexed with age, ovate, 5–8 mm long. Petals 5, golden yellow, broadly ovate, 1 cm long. Anthers dehiscing by a terminal pore. Drupelets up to 10, black, oblong-ovoid, 8–10 × 5–7 mm, attached to the receptacle at one end; seeds 1, ovoid, 8–9 × 5–6 mm. Fig. 11/1–2.

Distr. ALDABRA: W, P, M, S; Esprit, Michel; Madagascar.

Notes. An important constituent of the inland mixed scrub. Flowers mainly in the early wet season just before or at the same time as the new leaves are produced. Seeds locally dispersed by turtle doves and tortoises.

Vernac. 'bois mangue' or 'bois bouquet'.

17. MELIACEAE

Trees and shrubs. Leaves usually alternate, pinnately compound or rarely simple; stipules absent. Flowers usually in cymes or panicles, regular, usually bisexual, perianth biseriate. Sepals 4–5, small, usually united at base. Petals 4–5, distinct or united, or united with staminal tube. Stamens 8–10, rarely 5; filaments usually variously united into a short tube, rarely nearly or quite free, upper corners of flat filaments often projecting as awl-shaped appendages. Ovary superior, subtended by a disk, usually 2–5-celled, ovules 2 or rarely more per cell, placentation axile. Fruit a capsule, drupe, or berry; seeds often winged.

A moderate-sized tropical family with several valuable timber trees, such as mahogany.

1. Leaves 3-foliolate, leaflets thin; calyx subtruncate or minutely toothed; staminal tube united part-way, cup-like, free part of filament 2-toothed; fruit a small drupe   **1. Malleastrum**

1. Leaves with predominantly more than 3 leaflets, leaflets leathery or more or less leathery; calyx with rounded lobes, separated almost to base; staminal tube urn-shaped, 8-lobed at top; fruit a very large, globose capsule
2. Xylocarpus

1. MALLEASTRUM (Baill.) Leroy

Shrubs or small trees. Leaves 1-foliolate to pinnately-compound with an odd, terminal leaflet. Flowers in irregular cymes. Calyx shallowly cup-shaped, 4—5 toothed. Petals 4—5, margins touching. Stamens or staminodes flat, 4—10, united at base or as much as half or two thirds, usually alternately short and long; disk usually present, free or united to base of ovary. Ovary conical, 1—5-celled, ovules 2 per cell; style with capitate stigma surrounded by a membranous collar. Fruit indehiscent; seeds 1 per cell, aril absent, endosperm absent, cotyledons thick, plano-convex.

Confined to Madagascar except for 1 atypical Aldabra species.

Malleastrum leroyi Fosberg in K.B. 29: 255, fig. 1 (1974). Type: Aldabra, Gionnet Channel area, *Fosberg* 49664 (US, holo., K, iso.).

Shrub or small tree, 3—4 m high; bark pale grey-brown, youngest parts lightly silky-haired, glabrescent. Leaves uniformly 3-foliolate; leaflets narrowly divergent on short petiolules, elliptic to oblong, up to 11 × 4.5 cm, usually smaller, apex obtuse to rounded, base attenuated, base of lateral leaflets somewhat asymmetric, margins entire, venation not strong; petioles 2—4.5 cm long. Flowers fragrant apparently dioecious; cymes loose, irregularly 4—6 times branched, branching tending to be somewhat congested distally, lightly silky-haired; peduncle up to 8 cm long; buds broadly oblong, blunt. Female flowers: calyx cup-like, subtruncate, minutely 4—5-toothed, sparsely and minutely pubescent, especially the teeth; petals 4—5, greenish-yellow, oblong, 2.5 × 1 mm, apex subacute, margins minutely but densely mealy-ciliolate; disk not evident; staminodes 4—9, subequal, weakly coherent, firmly united only at base or in basal third but length of united portion varies in different flowers, filaments strap-shaped, with 2, long, awl-shaped appendages at summit, anthers sterile, oblong-lanceolate, c 0.6 mm long, very shortly stalked between the appendages; ovary ovoid-conical, c 1 mm long, 2-celled, densely silky-haired; style very short, falling early, stigma capitate, surrounded by a membranous collar. Male flowers: similar; calyx somewhat more prominently toothed; petals apparently 4; stamens 6, filaments strap-shaped, united in lower third, anthers linear-oblong, c 0.8—0.9 mm long, dehiscing by longitudinal slits, stalk c 0.2 mm long; vestigial ovary oblong, c 1 mm long, stiffly pubescent, vestigial stigma irregularly capitate, collar not evident. Immature fruit somewhat obliquely oval or mango-shaped, 9 × 7 very sparsely, stiffly and minutely pubescent; seeds 1 per cell. Fig. 9/1.

This plant was tentatively referred, at one time, to *Vepris* Comm. ex Juss. (Rutaceae). However, if this disposition is correct the distinction between Meliaceae and Rutaceae seems to disappear.

Distr. ALDABRA: P, M (east), S (east); endemic.
Notes. A rather infrequent constituent of mixed scrub. Flowers in early to mid wet season.
Vernac. 'bois trois feuilles'.

2. XYLOCARPUS Koen.

Large shrubs or small trees; loosely branched. Leaves pinnately compound, leaflets 3–5(–7), more or less leathery, terminal leaflet frequently shed so that leaflets appear evenly paired. Flowers in loose, axillary or subterminal, panicles or racemes; bracts absent; flowers 4-merous. Sepals scale-like, obtuse, united at base, much shorter than the petals. Petals 4, united, concave. Stamens 8, united into a tube closely surrounding the ovary, with 8 terminal appendages alternating with the anthers. Ovary 4-celled; style very short, stigma discoid, thick, barely exserted. Fruit large, globose, woody, seeds numerous large, irregularly angular, corky, crowded together.

A small Indo-Pacific genus of 2 or 3 species, growing in edges of mangrove swamps.

1. Leaflets rounded or obtuse at apex; inflorescence much shorter than leaves
1. X. granatum
1. Leaflets definitely pointed; inflorescence as long as or longer than leaves.
2. X. moluccensis

1. Xylocarpus granatum Koen. in Naturf. 20: 2 (1784); Fryer in Trans. Linn. Soc. Lond. II, Zool. 14: (1914); F.Z. 2: 295, fig. 57/A (1963); Fosberg in P.T.R.S. B, 260: 220, 225 (1971); Renvoize in P.T.R.S. B, 260: 230 (1971); Macnae in P.T.R.S. B, 260: 241 (1971).

Carapa obovata Bl., Bijdr. 1: 179 (1825); Hemsley in B.M.I.K. 1919: 118 (1919).
C. moluccensis sensu F.M.S.: 48 (1877), not Lam. (1783).

Spreading shrub or small tree, evergreen, glabrous. Leaflets, mostly 3 or 5, terminal leaflet usually falling early or abortive, oblong to somewhat obovate, up to 9 × 4 cm, apex rounded to obtuse, base broadly wedge-shaped, leathery, veins 5–6 pairs, not very conspicuous; petiolules short, thick. Racemes or panicles in upper axils or subterminal, short, few, to 5 cm long; pedicels 5–6 mm long, at right angles to rhachis; flowers fragrant. Sepals orbicular, united at base. Petals creamy-yellow, oval, c 4 mm long, apex obtuse. Staminal tube sub-urn-shaped, terminal lobes elliptic, 1 mm or less

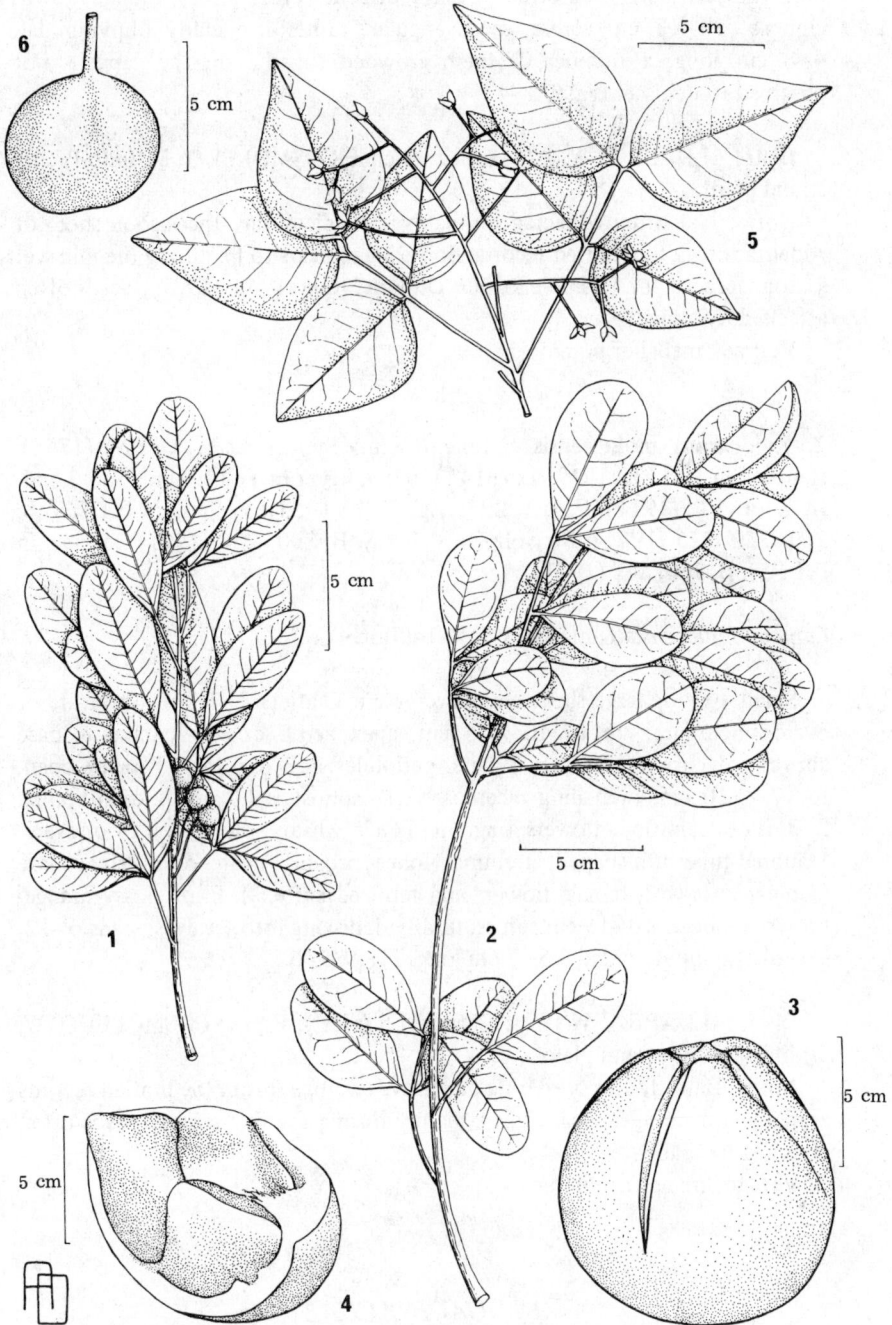

Fig. 9   1, *Malleastrum leroyi*, habit. 2, *Xylocarpus granatum*, habit;
3, fruit; 4, seeds. 5, *Xylocarpus moluccensis*, habit; 6, fruit.

long. Stigma barely exserted, fleshy, discoid. Fruit glossy orange-brown, globose, 10–18 cm across; seeds angular, rather irregularly obpyramidal, 4–8 cm long; a number of them crowded together in the capusle, not observed to dehisce. Fig. 9/2–4.

Distr. ALDABRA: W, P, S, Esprit, Michel; COSMOLEDO: M; Indo-Pacific strand plant.

Notes. Not often collected, probably sporadic along the lagoon shore of Aldabra. From the limited records flowering appears to be during the mid wet season. Record for Cosmoledo fide Dupont, Report: 11 (1907). Seeds often attacked by crabs.

Vernac. 'manglier pomme'.

2. **Xylocarpus moluccensis** (Lam.) Roem., Syn. Hesper. 1: 124 (1846); Hemsley in B.M.I.K. 1919: 117, 142 (1919); Fryer in Trans. Linn. Soc. Lond. II, Zool. 14: (1911); F.Z. 2: 297, fig. 57/B (1963); Fosberg in P.T.R.S. B, 260: 220, 225 (1971); Renvoize in P.T.R.S. B, 260: 230 (1971); Macnae in P.T.R.S. B, 260, 241 (1971).

Carapa moluccensis Lam., Encycl. Méth. Bot. 1: 621 (1783).

Small tree or large shrub, semi-evergreen. Leaflets 3–7, broadly ovate or ovate-subcordate, up to 10 × 5 cm, apex acute or sub-acuminate, base abruptly decurrent, veins 4–6 pairs; petiolules 1–3 mm long. Panicles open, to 17 × 10 cm, branching racemosely, at almost right angles, each branch 2–3 times ramified; flowers fragrant. Petals white, oblong-oval or obovate. Staminal-tube urn-shaped, terminal lobes broadly ovate or nearly square (some plants with female flowers and reduced anthers). Fruit glossy orange-brown, globose, 10–15 cm across, tardily dehiscing into 4 valves; seeds 6–12, irregularly obpyramidal, 3.5–7 cm long. Fig. 9/5–6.

Distr. ALDABRA: W (West Channel), S (west), Esprit; COSMOLEDO: W, M; Indo-Pacific strand plant.

Notes. Rare, found bordering mangrove swamps. From the limtied records available flowering appears to be mainly during the dry season. Seeds often attacked by crabs.

Vernac. 'manglier pomme'.

18. ICACINACEAE

Trees, shrubs or climbers. Leaves usually alternate, simple; stipules absent. Flowers usually bisexual, rather small, regular, 4–5 merous, Calyx lobed. Petals distinct or rarely united, margins touching. Stamens alternate with petals, free, anthers 2-celled. Ovary superior, 1-celled, ovules usually 2, apically attached; style with 2–5 stigmas or stigma lobes. Fruit a drupe or rarely a samara, endocarp often irregular; seeds 1–2.

A small, tropical family .

APODYTES E. Mey. ex Arn.

Trees or shrubs. Leaves alternate, margins entire. Flowers in terminal, rarely axillary, panicles. Calyx 5-lobed. Petals 5, free, linear, glabrous. Stamens 5, alternating with the petals. Ovary 1-celled, ovules 2. Fruit 1-seeded, hard with a fleshy lateral appendage.

15 tropical and subtropical species from Africa to the Far East and Australia.

Apodytes dimidiata E. Mey. ex Arn. in Hook., Lond. J. Bot. 3: 155 (1840); F.M.C. 119: 16 (1952); F.Z. 2: 343 (1963); F.T.E.A. Icac.: 4, fig. 2 (1968).

A. mauritiana (Miers) Benth. & Hook.f., Gen. Pl. 1: 351 (1862); F.M.S.: 48 (1877); Schinz in A.S.N.G. 21: 83 (1897); Voeltzkow in A.S.N.G. 26: 549 (1902); Dupont, Report: 40 (1907); Hemsley in J. Bot. 54, Suppl. 2: 8 (1916) & in B.M.I.K. 1919: 118 (1919).

Shrubs or small trees, up to 5 m high. Leaves elliptic, 5–9 × 2–4 cm, apex blunt, leathery, dark green and shiny above, dull below; petioles red, 5–10 mm long. Panicles branched, pyramidal, up to 10 cm long; pedicels short; flowers numerous, small, sweet-scented. Calyx very small, 0.5 mm long, pubescent. Petals white, becoming black on drying, 4 mm long. Fruit black with a red lateral appendage at maturity; asymmetrical, oblong-obovoid, laterally compressed, 1 cm long, glabrous or sparsely pubescent; seed oblong, compressed, 4 × 2.5 mm. Fig. 10/1–3.

Distr. ALDABRA; W, P, M, S, Esprit, Michel; tropical Africa to India.
Notes. A constituent of the inland scrub. Flowers mainly during the mid wet season; flower buds can be formed and remain dormant through the dry season. Fruit eaten and seed locally dispersed by turtle doves, bulbul and white-eyes.
Vernac. 'bois Marie' or 'la fouche petite feuille' or 'bois none'.

19. CELASTRACEAE

Shrubs, small trees, or woody climbers. Leaves alternate or opposite, simple, often with crenulate margins; stipules absent or small, soon falling. Flowers in cymose inflorescences or fasciculate, small, bisexual or rarely functionally unisexual, regular. Sepals 4–5, free or basally united, overlapping. Petals 4–5 or rarely absent, free, alternate with sepals. Stamens 4–5, alternate with petals, rarely 10, inserted beneath a disk. Ovary superior, 2−5-celled, often 3-celled, placentation axile, ovules usually 2 per cell; style 1, short, stigma capitate or shortly lobed. Fruit various, often a loculicidal capsule; seeds usually arillate.

A widely distributed medium-sized family, with few economic species except ornamentals.

1. Stamens erect; style divided into 2 branches, each with a 2-lobed stigma; fruit a small, dehiscent capsule   **1. Maytenus**
1. Stamens becoming deflexed; style short, undivided, with a capitate stigma; fruit an indehiscent, fleshy drupe   **2. Mystroxylon**

1. MAYTENUS Molina

Shrubs or small trees. Leaves spirally arranged, simple. Inflorescence a small, many-flowered, pedunculate or sessile, axillary fascicle; flowers small, bisexual. Sepals 5, united at the base. Petals 5, spreading, slightly larger than the sepals. Stamens 5. Ovary 2−3-celled, ovules 2 per cell; style divided into 2 branches, each branch bearing a 2-lobed stigma. Fruit a small dehiscent capsule.

A genus of c 200 species throughout the tropics and subtropics.

Maytenus senegalensis (Lam.) Exell in Bol. Soc. Brot. II, 26: 223 (1952); F.Z. 2: 367. Fig. 76/A (1966).

Celastrus senegalensis Lam.,   Encycl. Méth. Bot. 1: 661 (1785); Baker in B.M.I.K. 1894: 147 (1894); Dupont, Report: 35 (1907); Hemsley in J. Bot. 54, Suppl. 2: 8 (1916).
Gymnosporia senegalensis (Lam.) Loes. in Engl., Bot. Jahrb. 17: 541 (1893); Voeltzkow in A.S.N.G. 26: 550 (1902).
Celastrus senegalensis sensu Hemsley in B.M.I.K. 1919: 118 (1919), partly.

Shrubs or small trees, 0.5−8 m high, branches glabrous with greyish bark, short spines occasionally present. Leaves obovate to narrowly obovate, 2−4.5 (−6) × (0.5−)1−1.5(−4) cm, apex obtuse, rounded or shallowly notched, margins serrate-crenate to almost entire, glabrous, pale green, glaucous, leathery; petiole 2−7 mm long, glabrous. Inforescence of many flowers, each 3−4 mm in diameter; peduncle and pedicels 2−4 mm long. Sepals green, ovate, 0.75 mm long, with sparse short hairs on the margins, otherwise glabrous. Petals cream or white, oblong, 1.5−2 mm long, sometimes with sparse short hairs on the margins, otherwise glabrous. Stamens short, not exceeding the petals. Ovary ovoid, glabrous. Fruit a brown or reddish, dry, dehiscent capsule, 3−4 mm in diameter; seeds 4, dark glossy reddish brown, ovoid, 3 × 2 mm, with a fleshy rose-pink aril obliquely covering the lower third to two-thirds. Fig. 10/4−5.

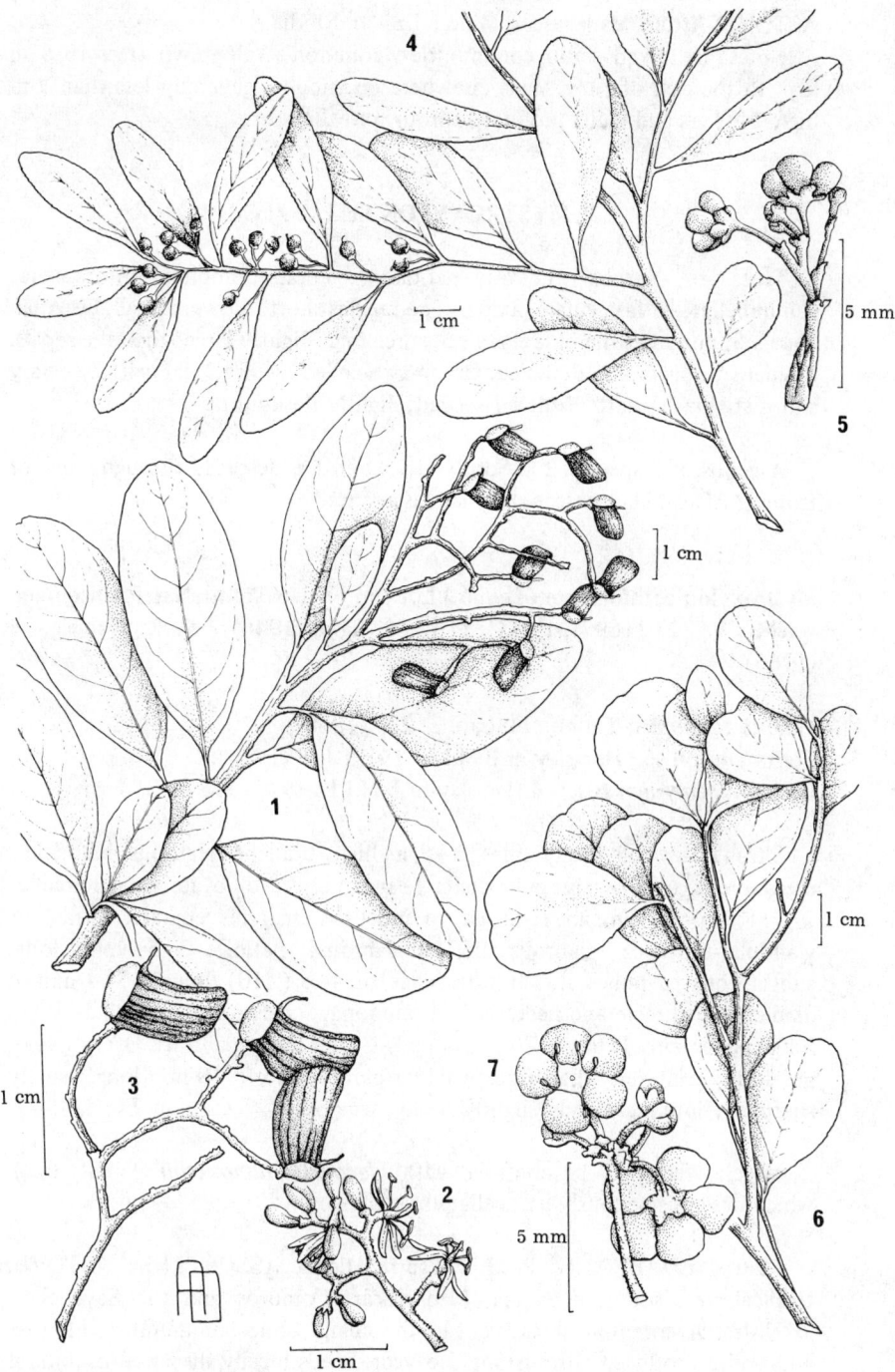

Fig. 10 1, *Apodytes dimidiata*, habit; 2, flowers; 3, fruit. 4, *Maytenus senegalensis*, habit and fruit; 5, flowers. 6, *Mystroxylon aethiopicum*, habit; 7, flowers.

Distr. ALDABRA: W, P, M, S, Esprit; ASSUMPTION; COSMOLEDO: M; ASTOVE; Africa, Madagascar, Middle East and India.

Notes. In mixed scrub communities, common. Well-grown trees to 8 m high to the east of Anse Mais, elsewhere on Aldabra generally less than 3 m high. Flowers and fruits produced throughout the year.

2. MYSTROXYLON Eckl. & Zeyh.

Trees or shrubs. Leaves petiolate, spiral, simple. Inflorescence subumbellate or fasciculate, axillary; peduncles short. Flowers small, bisexual. Sepals 5, united at the base. Petals 5, spreading, slightly larger than the sepals. Stamens 5, becoming deflexed. Ovary 2—3-celled, ovules 2 per cell; style very short, stigma captiate. Fruit a 1-seeded, slightly fleshy drupe.

A genus of 3 species, 2 in South Africa and 1 widespread through much of tropical Africa, Madagascar and the Mascarenes.

Mystroxylon aethiopicum (Thunb.) Loes. in Engl. & Prantl, Nat. Pflanzenfam. Nachtr. 1: 223 (1897); F.M.C. 116: 36, fig. 6 (1946); F.Z. 2: 376, fig. 78 (1966).

Cassine aethiopica Thunb., Fl. Cap. 2: 227 (1818).
Elaeodendron sp.; Hemsley in B.M.I.K. 1919: 118 (1919).
Celastrus senegalensis sensu Hemsley in B.M.I.K. 1919: 118 (1919), partly.

Shrubs or small bushy trees, 1—5 m high, branches glabrous, branchlets subglabrous to sparsely pubescent. Leaves elliptic or ovate to subcircular, 2—7 × 1.5—5 cm, apex obtuse to indented, margins crenate, serrate or glandular-toothed, glabrous or subglabrous; petiole 3—5 mm long, subglabrous to pubescent. Inflorescence of 4—6 (—10) flowers, 3—4 mm in diameter, peduncle and pedicels 2—4 mm long. Sepals green, ovate, 0.5—1 mm long, pubescent. Petals yellow, ovate, 1—2 mm long, glabrous. Ovary ovoid, glabrous. Fruit red when ripe, ovoid or globose, up to 10 mm long, usually sharply pointed, glabrous, slightly fleshy; seed ovoid, 7 × 4 mm. Fig. 10/6—7.

Vegetatively may be confused with *Flacourtia ramontchii* (Family 6/1), which has conspicuously lenticellate branches.

Distr. ALDABRA: W, P, M, S, Esprit, Michel; ASSUMPTION; ASTOVE; tropical and southern Africa, Madagascar, Comoros and the Seychelles.

Notes. A common constituent of the inland scrub communities. Flowers and fruits produced throughout the year. Seeds locally dispersed by bulbul and tortoises.

Vernac. 'ti bané' or 'bois moset'.

20. RHAMNACEAE

Trees, shrubs or lianes, twining or tendril climbers, sometimes spiny. Leaves alternate or rarely opposite, simple, often 3-nerved; stipules usually present, small, sometimes spinose. Inflorescences usually axillary, cymose or corymbose; flowers usually bisexual, small, 4- or usually 5-merous, somewhat perigynous. Petals free, often clawed, blade often concave and enveloping each stamen. Stamens 5, opposite the petals; intrastaminal disk usually present. Ovary superior or semi-inferior, 2—4-celled, with 1, rarely 2, basally attached ovules per cell; styles 1—2. Fruit a capsule, drupe, or samara; seeds often hard.

A medium-sized family, well represented in tropical and subtropical areas but with many temperate species in some regions.

1. Twining vine with coiled tendrils   **2. Gouania**
1. Trees or shrubs without tendrils
  2. Dense, spiny shrubs or trees   **3. Scutia**
  2. Loose, spreading, sprawling, unarmed shrubs   **1. Colubrina**

1. COLUBRINA Brongn.

Unarmed trees or shrubs. Leaves alternate. Flowers tiny, bisexual; in small axillary thyrses or fascicles. Sepals and petals 5. Ovary 3-celled. Fruit orbicular, breaking septicidally into 3 cocci at maturity; cocci dehiscent, 1-seeded.

A genus of 31 tropical and subtropical species, mostly in the Americas.

Colubrina asiatica (L.) Brogn., Mém. Fam. Rham.: 62 (1826); F.M.S.: 52 (1877); Baker in B.M.I.K. 1894: 147 (1894); Schinz in A.S.N.G. 21: 86 (1897); Voeltzkow in A.S.N.G. 26: 550 (1902); Dupont, Report: 35 (1907); Hemsley in B.M.I.K. 1919: 118 (1919); F.M.C. 123: 18, fig. 5 (1950); F.Z. 2: 430, fig. 90 (1966); F.T.E.A. Rham.: 3, fig. 1 (1972).

Ceanothus asiaticus L., Sp. Pl.: 196 (1753).

Shrub, up to 3 m, stems long, sparsely branched, leafy, scrambling or spreading. Leaves ovate, 3—8 × 1.5—5.5 cm, apex acuminate, base rounded, margins toothed, glabrous; petioles 1—2 cm long. Flowers on pedicels 3—4 mm long, elongating in fruit to 7 mm long. Sepals yellowish green, triangular, 1.5—2 mm long. Petals yellowish green, 1.5 mm long, enveloping the stamens. Fruit green when young, becoming brown at maturity, orbicular, shallowly 3-lobed, 7 mm in diameter; seeds 3, greyish or reddish brown, more or less circular, 3-sided, 1 side domed, 2 sides flat, 4—6 × 4—6 mm. Fig. 12/1.

Distr. ALDABRA: W, P, M, S; COSMOLEDO; W, M; ASTOVE; an Indo-Pacific strand plant; introduced into the West Indies.

Notes. Occurs frequently in inland mixed scrub. Flowering mainly during the wet season but unseasonal rains can stimulate flowering. Mature fruits may remain on the plants for several months; seed locally dispersed by tortoises, also water dispersed.

Vernac. 'bois savon' or 'savonier'.

2. GOUANIA Jacq.

Climbing shrubs or lianes with coiled tendrils. Leaves alternate, entire or toothed. Flowers small, bisexual, in terminal clusters or spikes. Sepals and petals 5. Fruit a 3-winged or 3-angled capsule separating into 3 indehiscent 1-seeded cocci.

A genus of 26 tropical and subtropical species, mostly in the Americas.

Gouania scandens (Gaertn.) R.B. Drummond in F.Z. 2: 435, fig. 88/D (1966); F.T.E.A. Rham.: 11 (1972).

Retinaria scandens Gaertn., Fruct. 2: 187, fig. 120/4 (1791).
Gouania retinaria DC., Prodr. 2: 40 (1825); Hemsley in J. Bot. Suppl. 2: 10
    (1916) & in B.M.I.K. 1919: 118 (1919), name superfluous.
G. tiliaefolia sensu F.M.S.: 52 (1877); Schinz in A.S.N.G. 21: 86 (1897);
    Voeltzkow in A.S.N.G. 26: 550 (1902); Dupont, Report: 35 (1907),
    Hemsley in J. Bot. 54, Suppl. 2: 10 (1916) & in B.M.I.K. 1919: 119 (1919),
    not Lam. (1789).

Liane or scrambling, twining shrub; stems with dense reddish hairs when young, becoming glabrescent with age. Leaves broadly ovate, 4–7 × 3–6 cm, apex acute to acuminate, base rounded or slightly lobed, margins slightly toothed, sparsely hairy above, hairy below on the veins, becoming glabrescent with age; petioles 1.5–3 cm long, sparsely to densely hairy; stipules linear, 5 mm long, soon falling. Flowers in sparse spikes, 4–8 × 1 cm, elongating in fruit to 12 cm; each spike subtended by a tendril, axes of spikes, peduncles and pedicels densely hairy. Sepals and petals greenish white, more or less triangular, 1–1.5 mm long. Capsules varying considerably in shape and size according to age; young capsules barely 3-ribbed, obovate, up to 0.8 cm long, glabrous; becoming 3-angled, more or less orbicular, up to 1 cm long; at full maturity 3-winged, compressed orbicular, wings up to 1.5 cm long; seeds 3, shiny grey-brown, broadly ellipsoid, slightly laterally compressed, 3–4 mm long. Fig. 12/2-3.

The identification of this species from Aldabra is most difficult, due mainly to the variation in the capsules (see description). Even immature fruits, when dried, fracture in a similar manner to mature fruits and are likely to mislead one into recognizing more species than are justified. *G. tiliaefolia*

Lam. from Madagascar and Mauritius differs in having tougher leaves, denser flowering spikes and 3-angled, not winged, capsules.

Fruiting specimens may be confused with those of the non-tendril climber *Dioscorea bemarivensis* (Family 67), whose winged fruits lack the ellipsoid central body.

Distr. ALDABRA: W, P, S; COSMOLEDO: W; Tanzania, Mozambique, Mauritius and Rodrigues.

Notes. Occurs in the inland mixed scrub. Locally abundant in the north-west of Aldabra but only small populations or isolated plants elsewhere. Flowers mainly during the wet season. Mature fruits may remain on the plant for several months. Seed locally dispersed by tortoises.

Vernac. 'liane charretiers'.

3. SCUTIA Commers. ex Brongn.

Trees or large shrubs; often armed with straight or hooked spines. Leaves usually opposite, oval to oblong, margins entire or toothed, glabrous. Flowers small, bisexual, in axillary fascicles or umbels. Sepals and petals 5, arising from a cuplike or discoid base. Fruit an ovoid berry; seeds 2–4.

A genus of 9 species in tropical America, Africa and India.

Scutia myrtina (Burm. f.) Kurz in J. As. Soc. Beng. 44(2): 168 (1875); F.M.C. 123: 4, fig. 1 (1950); F.Z. 2: 428, fig. 89 (1966); F.T.E.A. Rham.: 21, fig. 7 (1972).

Rhamnus myrtina Burm.f., Fl. Ind.: 60 (1768).
Scutia commersonii Brongn., Mém. Fam. Rham.: 56, fig. 4 (1826); F.M.S.: 51 (1877); Baker in B.M.I.K. 1894: 14 (1894); Schinz in A.S.N.G. 21: 86 (1897); Voeltzkow in A.S.N.G. 26: 550 (1902); Dupont, Report: 35 (1907); Hemsley in B.M.I.K. 1919: 118 (1919).

Dense spreading shrubs or small trees, up to 5 m high; branchlets angular or ribbed, glabrous, armed with paired or single, recurved, axillary stipular spines up to 8 mm long. Leaves opposite, elliptic, ovate, obovate or round, 2–6 × 1–4 cm, apex acute, rounded or indented and often with a short, sharp, flexible point, margins entire or wavy, leathery, dark green, shiny above; petiole 1–5 mm long. Flowers on short pedicels up to 2 mm long, in small, condensed, axillary cymes. Sepals triangular, 1.5–2 mm long. Petals 1 mm long, 2-lobed at the apex. Fruit black when ripe, up to 8 mm in diameter, glabrous; seeds 2, grey, plano-convex, 6 × 6 × 1.5 mm. Fig. 12/4.

Sterile specimens can be confused with *Pisonia aculeata* (Family 55.3), which has thicker stems; longer internodes, larger leaves acute at the apex and petioles up to 2 cm long.

Distr. ALDABRA: W, M, P, S; Esprit, Michel;ASSUMPTION; ASTOVE; East Africa to the Cape, Madagascar and the Mascarene Islands, tropical Asia east to Burma.

Notes. A frequent constituent of the inland mixed scrub. Flowers mainly during the wet season, throughout the year if sufficient moisture is available. Fruit eaten and seed locally dispersed by blue pigeon, turtle dove and bulbul.

Vernac. 'tiré bonnet', 'bambara' or 'bois senti'.

21. SAPINDACEAE

Trees, shrubs or tendrilliferous climbers, rarely herbaceous. Leaves alternate, rarely opposite, compound or rarely simple, stipules present or absent. Inflorescence usually cymose paniculate or racemose; flowers 4- or 5-merous, unisexual or bisexual, regular or irregular. Sepals free. Petals free or absent, alternate with sepals, overlapping in bud; disk present. Stamens 4–10, usually 10, rarely many, inserted within disk. Ovary superior, 1–4-celled, usually 3-celled; ovules 1–2 per cell; style usually 1. Fruit usually a capsule, often fleshy, or a drupe, berry or samara; seed often arillate.

A large pantropical family.

1. Leaves simple, fruit winged   **2. Dodonaea**
1. Leaves compound, fruit fleshy
   2. Leaves 2-pinnate; fruits ovoid, up to 1 cm diameter, black or dark purple
      when mature   **3. Macphersonia**
   2. Leaves 3-foliolate; fruits obovate, up to 6 mm long, red or orange when
      mature   **1. Allophylus**

1. ALLOPHYLUS L.

Trees or shrubs. Leaves (1-) 3-foliolate. Inflorescence paniculate or racemose. Flowers small, unisexual or bisexual, male and hermaphrodite occurring on the same inflorescence. Sepals 4, overlapping, the outer 2 larger than the inner 2. Petals 4, spatula-shaped, unequal-sided. Stamens 8. Ovary 2–3-lobed, 1 ovule per cell; style 2–3-fid. Fruit of 1, or more rarely 2, drupes; seeds obovoid.

This is a large, complex genus of c 255 species occurring throughout the tropics of the world.

Allophylus aldabricus Radlk. in Sitz. Akad. München 38: 218. 236 (1909). Type: Aldabra, *Abbott* s.n. (K, holo.).

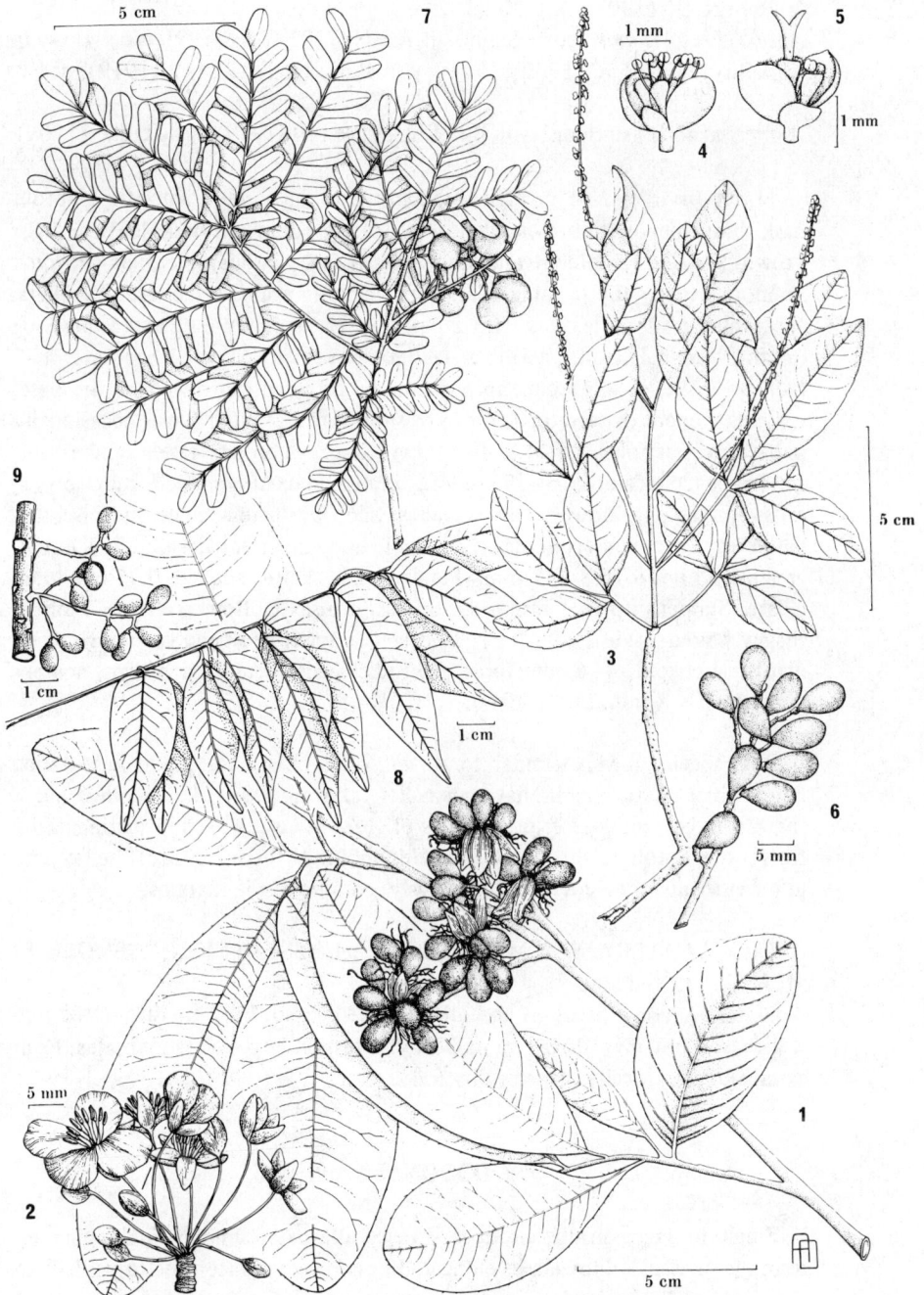

Fig. 11 1, *Ochna ciliata*, habit and fruit; 2, flowers. 3, *Allophylus aldabricus*, habit; 4, male flower; 5, hermaphrodite flower; 6, fruit. 7, *Macphersonia hildebrandtii*, habit and fruit. 8, *Operculicarya gummifera*, leaf; 9, fruit.

Schmidella africana sensu Baker in B.M.I.K. 1894: 147 (1894), Dupont,
Report: 35 (1907), not DC. (1824).
Allophylus africanus sensu Schinz in A.S.N.G. 21: 86 (1897); Voeltzkow in
A.S.N.G. 26: 550 (1902); Hemsley in B.M.I.K. 1919: 119 (1919), not P.
Beauv. (1819).
A. racemosus sensu Hemsley in B.M.I.K. 1919: 119 (1919), not Radlk. (1909).

Shrub or tree, 2–4 m high, branches with light grey, smooth, glabrous
bark, branchlets with brown, glabrous or rarely pubescent bark, lenticels light
brown, usually prominent and numerous. Leaves 3-foliate, leaflets subequal
or more usually the terminal leaflet half as long again as the lateral leaflets,
subsessile or lateral leaflets with a petiolule up to 2 mm long and 5 mm long
on the terminal leaflet; terminal leaflet obovate to elliptic-lanceolate, 4–7
(–9) × 2–3.5 (–4.5) cm, apex acute or rounded, narrowed at the base,
margins crenate or serrate, glabrous or subglabrous; young leaves occasionally
pubescent; petiole 1.5–4.5 (–5) cm long, glabrous, rarely pubescent.
Inflorescence racemose, 5–17 cm long, branched or unbranched, subglabrous,
rarely pubescent; flowers very small, sessile, in clusters, male and bisexual.
Outer sepals cream, subcircular, 1 × 1 mm, inner sepals broadly elliptic,
greenish-white, 0.75 × 0.5 mm. Petals cream, spatula-shaped, 0.75 mm long,
ciliate. Stamens up to 1 mm long, or less in bisexual flowers. Ovary 2-lobed,
shaggy haired; style 1 mm long, 2-fid at the apex. Fruit red or orange when
mature, obovoid, 4–6 mm long, glabrous; seeds 1, reddish brown, acutely
obovoid, 6 × 4 mm. Fig. 11/3–6.

This species is very similar to *A. alnifolius* (Bak.) Radlk. in East Africa
from which it differs in having smaller, obovate fruit. *A. alnifolius* has a
slightly larger, globose fruit. Species of *Allophylus* are readily distinguished
from other 3-foliate shrubs by the veins ending at the margins of the leaflets;
in other shrubs they curve and fade away, parallel to the margins.

Distr. ALDABRA: W, P, M, S, Esprit, Michel; ASSUMPTION; COSMOLEDO:
M; ASTOVE; endemic.
Notes. A constituent of the inland mixed scrub. Flowers during the mid
wet season but may flower sporadically in response to unseasonal rains. Fruit
eaten and seed locally dispersed by fody.

2. DODONAEA Mill.

Small to large shrubs or trees. Leaves alternate, simple, rarely pinnate,
resinous or viscid. Flowers in elongated or contracted racemose or paniculate
clusters, regular, unisexual or more rarely bisexual. Sepals 3–7, free or united
at the base. Petals absent. Stamens 5–8, filaments very short, in female
flowers stamens absent or sterile. Ovary 2–4(–6)-celled, ovules 2 per cell;
style often twisted, 2–6-divided. Fruit a 2–6-celled, winged capsule; seeds 1
or 2 per cell.

A genus of c 50 species, mostly Australian, 1 pantropical, 1 or 2 in Madagascar.

Dodonaea viscosa Jacq., Enum. Syst. Pl. Ins. Carib.: 19 (1760); F.M.S.: 61 (1877); F.Z. 2: 542, Fig. 117 (1966).

Shrubs or small trees, up to 10 m tall; branchlets angular and resinous. Leaves narrowly elliptic, up to 10 × 3 cm, apex acute, tapering at the base, margins entire, glabrous, viscid; petiole 2–3 mm long. Flowers in terminal or subterminal clusters, 2–4 cm long; pedicels up to 5 mm long, slender, glabrous. Sepals greenish-yellow, ovate, 2 mm long. Stamens 6, anthers 3 mm long. Ovary (2–) 3 (–4)-celled; style (2–) 3 (–4)-lobed. Fruit subcircular in outline, 2 cm in diameter, apex and base notched, with 2–3 membranous wings, 4–6 mm broad; eventually breaking up into indehiscent, winged sements; seeds black, ovoid, 3 mm long. Fig. 7/10–12.

Distr. ALDABRA: Michel; pantropical.
Notes. Only known from 1 collection on Isle Michel, *S. Hnatiuk* 731876 (US).
Vernac. 'bois de reinette'.

3. MACPHERSONIA Bl.

Trees or shrubs. Leaves pinnate; leaflets sessile or shortly petiolate, opposite or alternate. Inflorescence paniculate or racemose. Flowers unisexual, male and female on different plants. Sepals 5, petaloid. Petals 5, very small. Stamens 7–8. Ovary (2–) 3-celled, ovules 1 per cell, ovary small and undeveoped in male flowers. Fruit baccate, 1-seeded.

8 species centred on Madagascar and extending to East Africa.

Macphersonia hildebrandtii O.Hoffm., Sert. Pl. Madagasc.: 14 (1881); F.Z. 2: 534, fig. 113 (1966).

Albizia fastigiata sensu Baker in B.M.I.K. 1894: 148 (1894); Voeltzkow in A.S.N.G. 26: 550 (1902); Dupont, Report: 36 (1907), not (E. Mey.) Oliv. (1871).
Macphersonia madagascariensis sensu Hemsley in J. Bot. 54, Suppl. 2: 10 (1916) & B.M.I.K. 1919: 119 (1919), not Blume (1847).

Tree up to 4 m high; bark smooth, light brown to grey; branchlets pubescent, with yellowish hairs, to almost glabrous. Leaves up to 25 cm long, rhachis with yellowish hairs to almost glabrous; pinnae (3–) 5–7 (–8) pairs, leaflets (4–) 8–16 (–20) pairs opposite or alternate, subsessile, obliquely oblong (6–) 10–15 (–18) × 5–8 mm, apex obtuse or indented, margins entire, shiny dark green above, subglabrous or glabrous. Inflorescence an axillary raceme 4–10 (–15) cm long, becoming pendulous in fruit. Sepals pink, ovate, 2 mm long, ciliate. Petals obovate, 1 mm long, ciliate. Stamens 8. Ovary glabrous; style up to 0.8 mm long. Fruit dark purple or black at maturity, ovoid, 8–10 mm in diameter, glabrous; seeds 1, depressed ovoid, 6 × 5 mm, with thin arillode. Fig. 11/7.

May occur and be confused with *Calliandra alternans* (Family 24: 2/1), which it superficially resembles; the latter has only 1 pair of pinnae.

Dist. ALDABRA: S (east); ASSUMPTION; COSMOLEDO: W. M; ASTOVE; coast and islands of East Africa, Madagascar and the Comoros.
Notes. Not common, restricted in Aldabra to the inland scrub at the eastern end of South Island.
Vernac. ? 'tamarind marron' or 'tamarin bâtard'.

22. ANACARDIACEAE

Resinous trees and shrubs. Leaves usually alternate, pinnate, 3-foliolate, simple or more rarely simple by reduction; stipules absent. Inflorescence usually paniculate; flowers small, bisexual but often functionally unisexual, regular or rarely irregular. Sepals 3–5, basally united. Petals 3–5, usually free. Stamens usually 10, in 2 whorls, or partly reduced to staminodes, free or filaments basally united, on or under the rim of an intrastaminal disk. Ovary 1–(rarely 5–)-celled, 1 functional ovule; style usually 1. Fruit commonly a drupe, mesocarp usually resinous, often rather dry, seeds 1.

A medium-sized tropical family; many cultivated or otherwise, economically important species; some may cause severe dermatitis in susceptible people.

OPERCULICARYA Perr.

Small to medium sized dioecious or polygamous trees. Leaves pinnate with an odd terminal leaflet, restricted to the apex of the branches and falling early. Flowers very small, bisexual or unisexual, arranged in clusters on a spicate, branched inflorescence. Sepals 4–5, free. Petals 4–5, free, larger than the sepals. Stamens 4–5, alternating with the petals. Ovary 1-celled. Fruit a 1-seeded drupe.

A genus of 4 species, restricted to Madagascar, with 1 species extending to Aldabra.

Operculicarya gummifera (Sprague) Capuron in Adansonia II, 14: 571 (1975).

Poupartia gummifera Sprague in Bull. Herb. Boiss. II, 5: 408 (1904); F.M.C. 114: 10, fig. 3/1–5 (1946); Fosberg in K.B. 29: 258 (1974).
Odina wodier sensu Hemsley in B.M.I.K. 1919: 119 (1919) [as *O. woodier*], not Roxb. (1832).

Small deciduous trees, up to 5 m high (15 m in Madagascar), bark smooth, light grey on the trunk to red on the branches. Leaves appearing after the flowers, leaflets 6–7 pairs, with an odd terminal leaflet, rhachis 10–25 cm long, glabrous to pubescent; leaflets obovate, oblong or elliptic, 5–10 × 1.5–3 cm, apex acuminate, base asymmetrical, membranous to papery, glabrous or sparsely pubescent. Inflorescences terminal, sparsely branched. Male flowers: slender inflorescence branches up to 14 cm long; flowers arranged in small clusters 5 mm apart, 1 mm in diameter, sessile, 5-partite; stamens 10. Female flowers: inflorescences up to 4 cm long, the branches stouter than in the male inflorescence. Fruit greenish-brown to purplish, ovoid, 7–8 × 5–7 mm; seeds reddish-brown, kidney-shaped – obovoid, 8 × 6 mm, somewhat angular. Fig. 11/8–9.

When leafless may be easily confused with *Euphorbia pyrifolia* (Family 62/2, 5) from which it may be readily distinguished by the semi-circular leaf scars; these are narrow in *Euphorbia*.

Distr. ALDABRA: M (east), S; Madagascar.
Notes. A locally frequent constituent of the mixed scrub and in clumps of scrub on platin, especially at the eastern end of South Island.

23. MORINGACEAE

Trees or shrubs. Leaves alternate, clustered at the ends of branches, 2–3-pinnate; stipules minute or absent. Flowers in cymose panicles, irregular, bisexual. Sepals 5, unequal, fused at the base. Petals 5, fused at the base, 2 posterior petals smaller, 1 anterior larger than 2 lateral. Stamens 5, staminodes 3–5, on edge of shallow hypanthium; anthers 1-celled. Ovary superior, borne on a short stalk, 1-celled, ovules many, on 3 parietal placentas; style slender, stigma minute. Fruit an elongated capsule, triangular in cross-section, dehiscing by 3 valves; seeds many, large, usually with 3 wings.

A very small family of the Old World tropics with 1 genus.

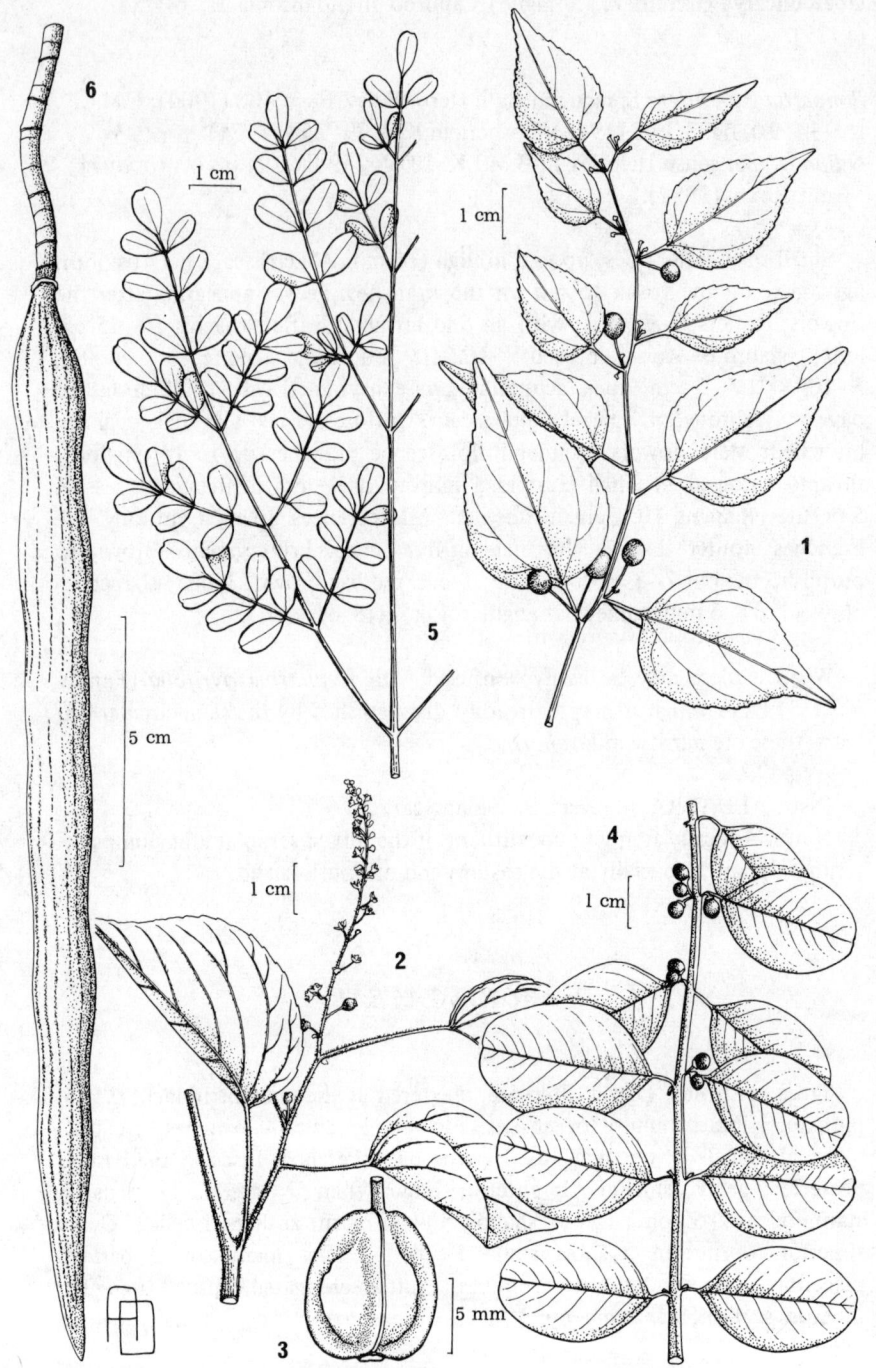

Fig. 12   1, *Colubrina asiatica*, habit and fruit. 2, *Gouania scandens*, habit;
3, fruit. 4, *Scutia myrtina*, habit. 5, *Moringa oleifera*, habit;
6, fruit.

MORINGA Adans.

Generic description as for the family.

A genus of c 12 species, mainly in Africa, Madagascar and India, with 1 species introduced throughout the tropics.

Moringa oleifera Lam., Encycl. Méth. Bot. 1: 398 (1785).

M. pterygosperma Gaertn.,Fruct. 2: 314 (1791); F.M.S.: 11 (1877); Schinz in A.S.N.G. 21: 84 (1897); Voeltzkow in A.S.N.G. 26: 550 (1902); Hemsley in B.M.I.K. 1919: 119 (1919), name illegit.

Small trees 3–6 m high; trunk with grey, smooth bark; branches brown; branchlets with conspicuous horseshoe-shaped leaf scars. Leaves 3-pinnate, 20–40 cm long, rhachis glabrous or pubescent; leaflets elliptic to obovate, 1–2(–3) × 0.6–1.2(–1.5) cm, glabrous, sessile or petiolule up to 2 mm long. Inflorescence 10–30 cm long, branched. Sepals petaloid, white tinged with pink, narrowly elliptic, 1–1.5 cm long, spreading or reflexed at maturity. Petals white or cream tinged with red, narrowly obovate, 1–2 cm long, upper pair smaller, lateral pair ascending, lower longer. Stamens up to 0.5–1 cm long, fertile stamens stouter than the sterile. Ovary narrowly oblong, up to 5 mm long; style as long as the ovary, shortly hairy. Fruit brown, up to 45 cm long, smooth, with valves up to 2 cm wide, longitudinally ribbed; seeds orbicular, up to 1 cm diameter with pale papery wings. Fig. 12/5–6.

Distr. ALDABRA: W, M, S, Esprit; ASSUMPTION; COSMOLEDO: M; ASTOVE; a native of India, widely cultivated throughout the tropics.
Notes. On Aldabra the few trees around the settlement on West Island have been pollarded at about 1–2 m. Flowers mainly during the wet season but may flower throughout the year if sufficient moisture available.
This species has been introduced throughout the tropics as a vegetable, the roots being thick and soft and used as a relish resembling horseradish; the seeds yield a fine oil – ben oil; leaves and flowers used as potherb, fruit sliced, cooked and eaten when young.
Vernac. 'brède morongue', 'brède mome', 'horseradish tree' or 'drumstick tree'.

24. LEGUMINOSAE

Trees, shrubs, lianes or herbs. Leaves alternate, compound, rarely simple, 1-foliolate or absent; stipules usually present. Flowers regular or irregular. Sepals 5, usually united. Petals 5, rarely less or absent, usually free, equal or unequal. Stamens 10 or various reduced, rarely more, free or fused. Ovary superior, of 1 carpel with the ovules arranged along 1 side; style 1. Fruit very diverse, typically a pod (legume) opening by 1 or 2 sutures or indehiscent.

A cosmopolitan family which includes many crop plants and is important ecologically because of the association with nitrogen-fixing bacteria in the root nodules.

This family is usually divided into 3 subfamilies as follows:

1. Flowers regular or nearly so; sepals and petals usually united at the base
   24:2. Mimosoideae
1. Flowers irregular, sepals free or united, petals usually free
   2. Upper petal (standard) enclosed by the others in bud; sepals often free, stamens free, 10 or variously reduced   **24.1. Caesalpinoideae**
   2. Upper petal (standard) enclosing the others in bud; sepals united at the base; stamens 10, usually 9 united with 1 free   **24:3. Papilionoideae**

24:1 CAESALPINOIDEAE

Trees, shrubs rarely lianes or herbs. Leaves 1- or 2-pinnate. Inflorescence usually racemose, flowers irregular, rarely small. Sepals usually free, margins overlapping. Petals usually 5, mostly free, overlapping in bud, the upper overlapped by the laterals. Stamens usually 10 or fewer, free or joined below; anthers without apical gland. Fruit various; seeds usually without areoles (elliptic or oblong-shaped areas bounded by fine lines).

1. Densely spiny, climbing or scrambling shrubs or herbs with erect spikes of yellow or greenish-yellow flowers   **1. Caesalpinia**
1. Trees, shrubs or herbs, not armed with spines or prickles
   2. Erect annual or perennial shrubs or prostrate perennial herbs with yellow flowers   **2. Cassia**
   2. Trees
      3. Inflorescence large and showy with red and orange flowers
         3. Delonix
      3. Inflorescence modest, of drooping, elongate clusters of red and cream flowers   **4. Tamarindus**

1. CAESALPINIA L.

Shrubs usually scrambling or climbing, or trees, usually armed with spines or prickles. Leaves 2-pinnate, without an odd terminal leaflet, leaflets

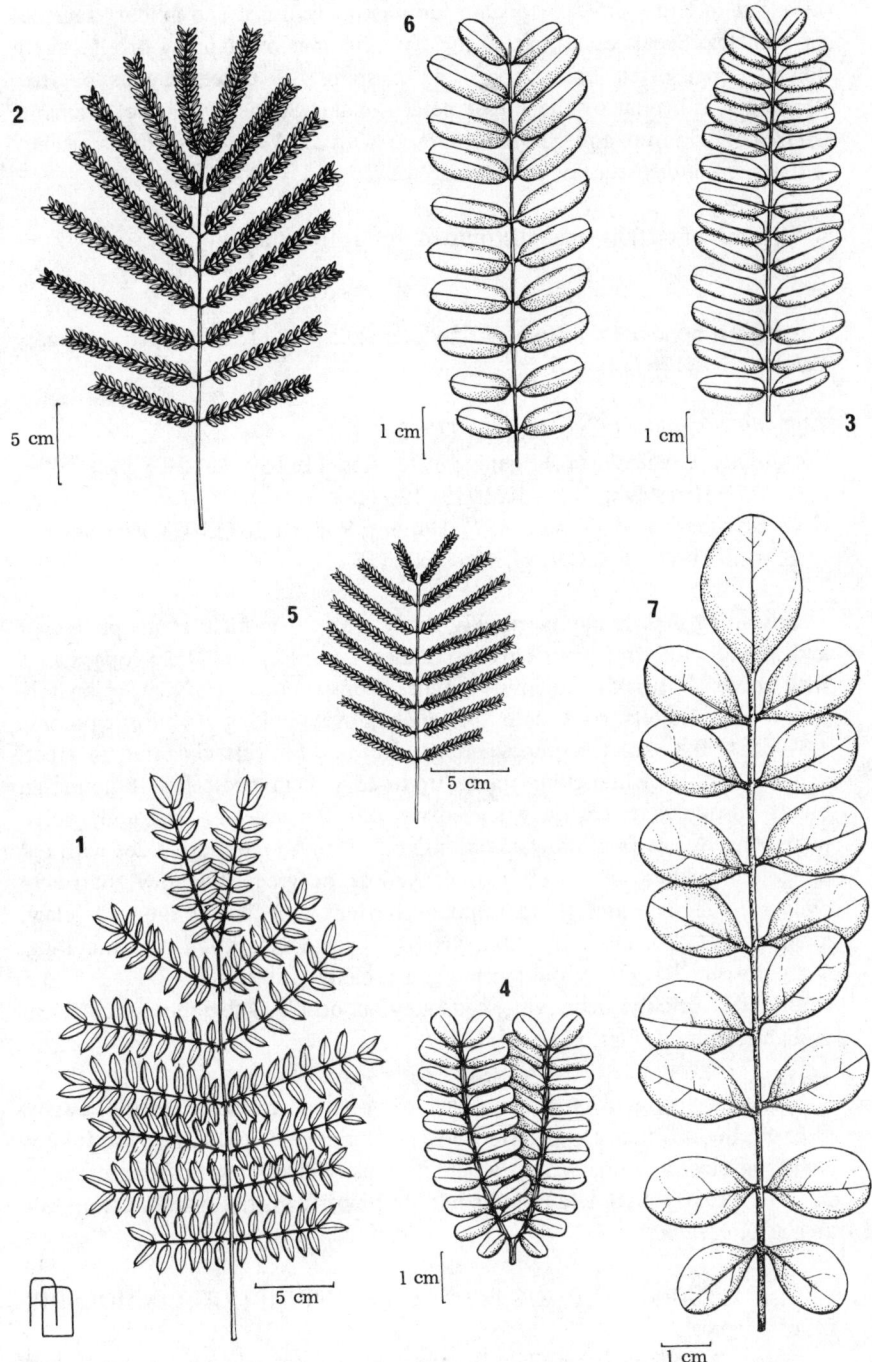

Fig. 13 Leaves of *Leguminosae*. 1, *Caesalpinia bonduc*. 2, *Delonix regia*. 3, *Tamarindus indica*. 4, *Calliandra alternans*. 5, *Dichrostachys microcephala*. 6, *Abrus precatorius*. 7, *Sophora tomentosa*.

opposite, few to many; stipules minute to conspicuous. Flowers usually arranged in large, showy, elongate terminal or terminal and axillary racemes or panicles. Sepals 5, united at the base, margins overlapping or almost so, lowest sepal often boat-shaped and clasping the others. Petals 5, free, spreading, subequal or the upper petal smaller. Stamens 10, free, filaments alternately longer and shorter. Ovary with 2–10 ovules. Fruit usually a flattened, indehiscent or dehiscent, 2-valved pod.

A genus of c 200 species distributed throughout the tropics.

Caesalpinia bonduc (L.) Roxb., Fl. Ind. ed. 2, 2: 362 (1832); F.T.E.A. Leg.-Caesalp.: 37 (1967).

Guilandina bonduc L., Sp. Pl.: 381 (1753).
Caesalpinia bonducella (L.) Fleming in As. Res. 11: 159 (1810); F.M.S.: 88 (1877); Hemsley in B.M.I.K. 1919: 120 (1919).
C. sepiaria sensu F.M.C.: 88 (1877); Dupont, Report: 36 (1907); Hemsley in B.M.I.K. 1919: 120 (1919), not Roxb. (1832).

Spreading or scrambling, prickly shrub, up to 2 m high; stems pubescent and armed with straight or curved prickles. Leaves up to 50 cm long, rachis with hooked prickles below; stipules conspicuous, leaf-like, pinnately compound; leaflets 6–9 pairs per pinna, ovate-oblong or elliptic-oblong, 1.3–4.5 × 0.8–2.2 cm, pubescent or glabrous. Flowers clustered in erect, axillary, simple or branching spikes, up to 25 × 3 cm; pedicels 4–6 mm long; bracts conspicuous, narrowly lanceolate, 0.8–1.6 cm long, red and golden pubescent. Sepals free, obovate-oblong, 6.5–9 mm long, both sides with red or golden pubescence. Petals free, yellow or greenish-yellow, narrowly obovate, 12–13 mm long, upper shorter with a pronounced claw, membranous, sparsely hairy. Stamens 10, 5 well-developed, 1–1.2 mm long, 5 slightly smaller. Ovary pubescent. Pod broadly elliptic, 4.5–7.5 × 3.5–4.5 cm, densely prickly, dehiscent; seeds grey, globose or subglobose, 1.5–2 cm in diameter, hard. Figs. 13/1, 15/1, 16/1.

Some specimens from Aldabra have been named *C. major* (Medic) Dandy & Exell; the differences however between this species and *C. bonduc* are not always as clear as some writers would have us believe – see Fosberg in Taxon 22: 162 (1973). I (S.A.R.) prefer here to identify all the material from our area as *C. bonduc*.

Dist. ALDABRA: W, P, M, S, Esprit, Michel; ASSUMPTION; COSMOLEDO: M; pantropical.
Notes. A widely distributed strand plant, found only on the north and west facing coasts of the main islands of Aldabra despite viable seed being washed up elsewhere on the island. Flowers during the mid to late wet season. Seeds sea-dispersed.
Vernac. 'cadoque'.

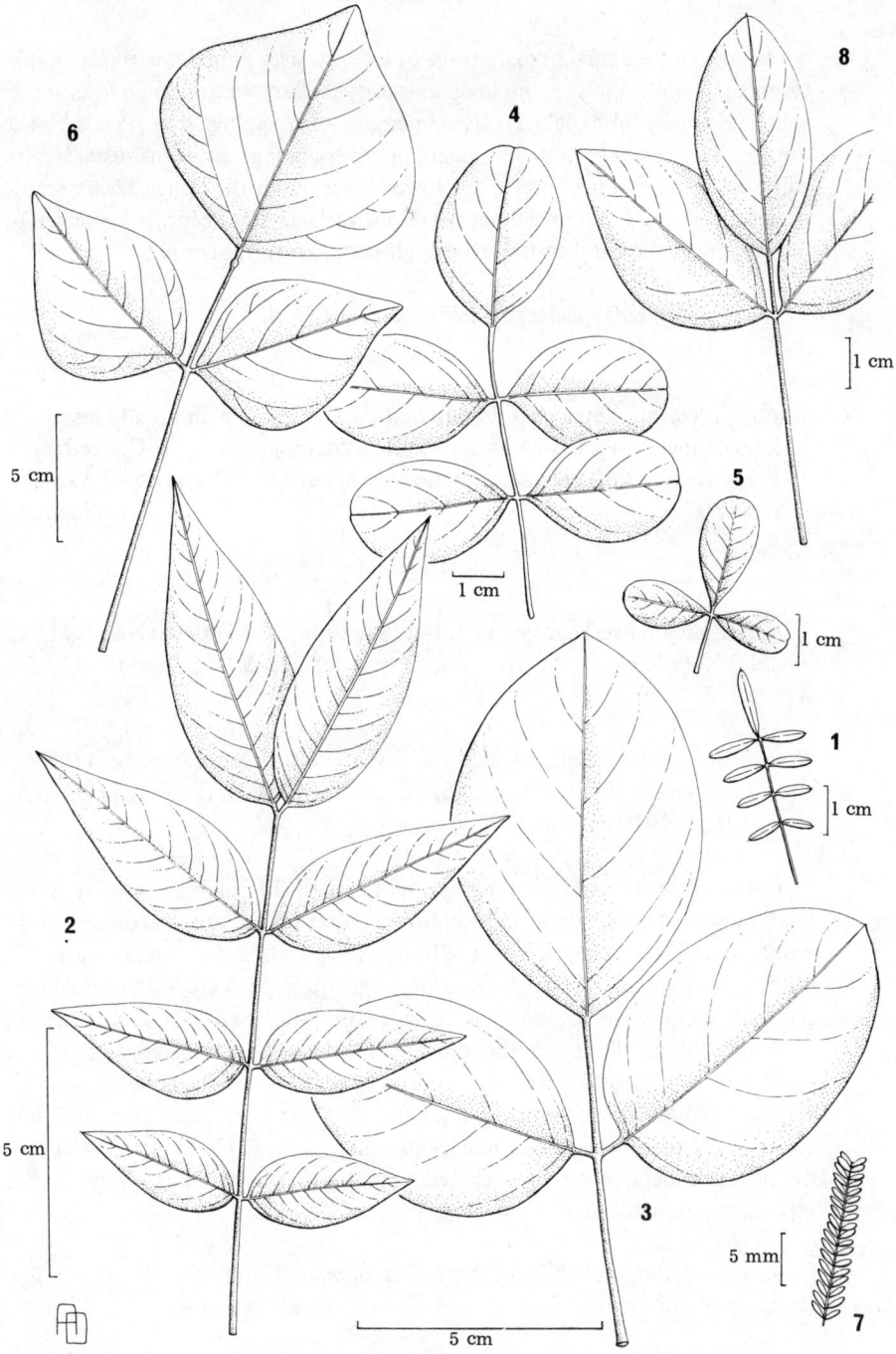

Fig. 14   Leaves of *Leguminosae*. 1, *Cassia aldabrensis*. 2, *Cassia occidentalis*.
3, *Canavalia rosea*. 4, *Clitorea ternatea*. 5, *Crotalaria laburnoides*.
6, *Erythrina variegata*. 7, *Tephrosia pumila* var. *aldabrensis*.

2. CASSIA L.

Annual or perennial herbs, shrubs or trees. Leaves pinnate, without an odd terminal leaflet. Flowers in long or short, axillary or terminal racemes or solitary, usually bisexual. Sepals 5, margins overlapping, free. Petals 5, the upper often slightly smaller, margins overlapping. Stamens usually 10, subequal or of varying length, the lower longer than the upper. Ovary sessile or shortly stalked, ovules numerous. Fruit variable, dehiscent and 2-valved or indehiscent, cylindrical or flat, woody, leathery or membranous.

A genus of c 600 species, mainly in the New World.

1. Low, prostrate, somewhat woody herb, up to 30 cm high usually less; leaflets oblong, 2—4 x 0.5—1 mm; pods 2—3 cm long   **1. C. aldabrensis**
1. Erect herb up to 2 m high, leaflets ovate-acuminate, 3—9 x 1.5—3.5 cm; pods 10—11 cm long   **2. C. occidentalis**

1. **Cassia aldabrensis** Hemsley in J. Bot. 54, Suppl. 2: 12 (1916) & in B.M.I.K. 1919: 120 (1919). Types: Aldabra, *Thomasset* 231 & 262, *Dupont* 123 (all K, syn.).

C. mimosoides sensu Baker in B.M.I.K. 1894: 147 (1894); Schinz in A.S.N.G. 21: 85 (1897); Voeltzkow in A.S.N.G. 26: 550 (1902); Hemsley in B.M.I.K. 119: 120 (1919), not L. (1753).

Prostrate or ascending, perennial herb, up to 30 cm high, usually much less; branches wiry, dark reddish-brown, sparsely ciliate, becoming quite woody at the base. Leaves 5—15(—20) mm long; leaflets 4—10 pairs, narrowly oblong, 2—4 × 0.5—1 mm, glabrous or subglabrous. Flowers solitary, axillary; pedicels slender, pubescent, up to 1.5 cm long. Sepals narrowly ovate, 3—4 mm long, ciliate on the outside. Petals yellow or orange, obovate, 4—7 mm long. Stamens 10, 4 almost as long as the petals, 6 much shorter. Stigma slightly shorter than the petals, hairy on the lower part, glabrous above, curved inwards. Pods linear-oblong, beaked, 2—3 cm × 0.3—0.4 cm, dehiscent, chocolate brown, covered with short spinous hairs; seeds 6—12, brown, quadrangular 2 × 1 mm. Figs 14/1, 15/2, 16/2.

Readily distinguished from *Tephrosia pumila* (Family 24:3/8) which has white, pink or mauve flowers and an odd terminal leaflet.

Distr. ALDABRA: W, P, M (west), S (east); ASSUMPTION; endemic.
Notes. Found on the main islands of Aldabra except western portion of South Island. Grows in crevices in platin and champignon. Flowers mainly during the mid-late season but may flower throughout the year if there is sufficient mositure.
Vernac. 'cassie'.

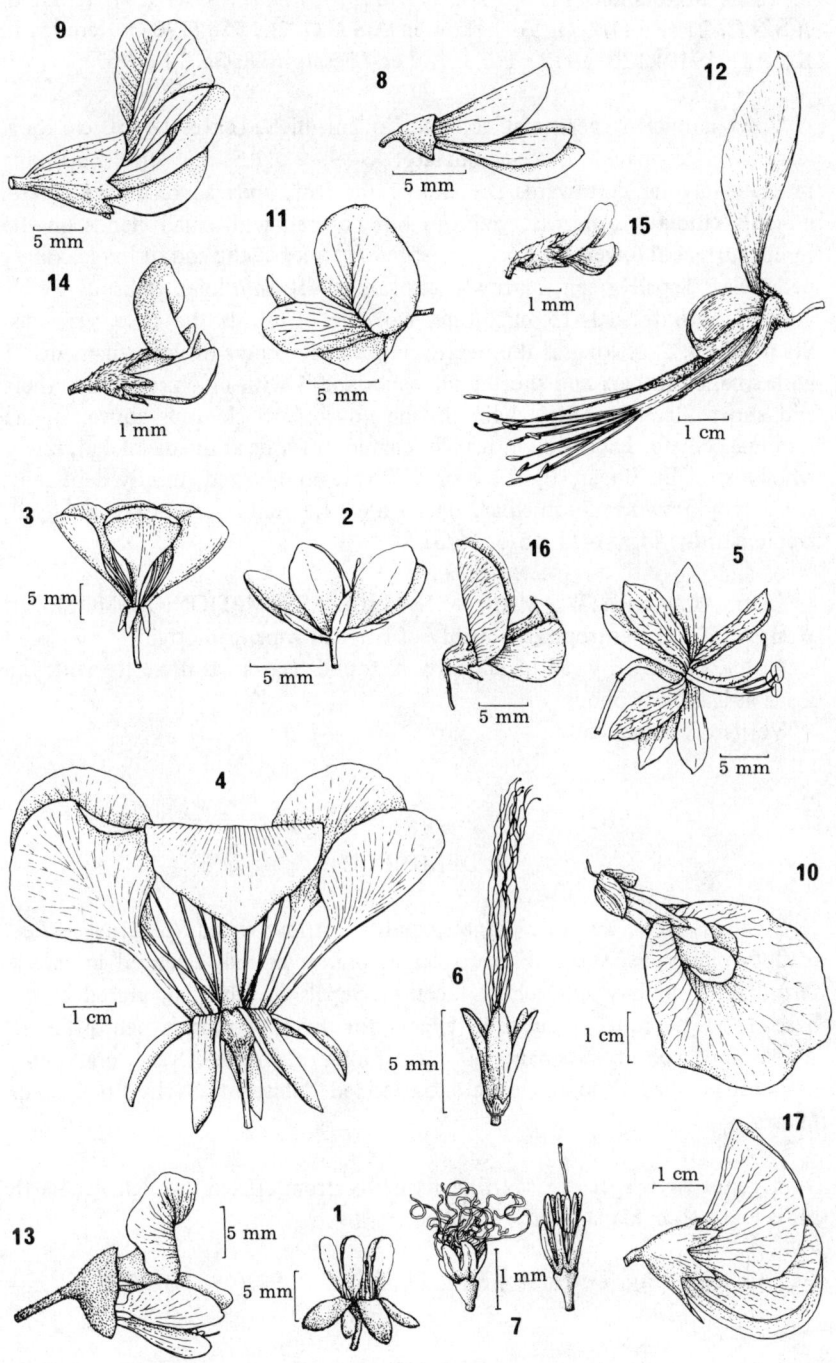

Fig. 15   Flowers of *Leguminosae*. *1, Caesalpinia bonduc*. *2, Cassia aldabrensis*. *3, Cassia occidentalis*. *4, Delonix regia*. *5, Tamarindus indica*. *6, Calliandra alternans*. *7, Dichrostachys microcephala*. *8, Abrus precatorius*. *9, Canavalia rosea*. *10, Clitorea ternatea*. *11, Crotalaria laburnoides*. *12, Erythrina variegata*. *13, Sophora tomentosa*. *14, Tephrosia pumila* var. *aldabrensis*. *15, Teramnus labialis*. *16, Vigna marina*. *17, Vigna unguiculata*.

2. **Cassia occidentalis** L., Sp. Pl.: 377 (1753); F.M.S.: 89 (1877); Schinz in A.S.N.G. 21: 85 (1897); Voeltzkow in A.S.N.G. 26: 550 (1902); Hemsley in B.M.I.K. 1919: 120 (1919); F.T.E.A. Leg-Caesalp.: 78, fig. 14 (1967).

Erect annual or perennial herb, up to 2 m high. Leaves 10–20 cm long; leaflets 4–5 pairs, ovate-acuminate, 3–9 × 1.5–3.5 cm, becoming progressively larger towards the end of the leaf, apex acute or acuminate, margins ciliolate, otherwise glabrous but covered with small glands on the under surface. Flowers in short, congested racemes at the end of long, axillary peduncles. Sepals green, narrowly obovate, 6–10 mm long, glabrous. Petals yellow, obovate, 10–15 mm long, narrowed towards the base, glabrous. Stamens 10, 2 as long as the petals, with large anthers on long filaments, 4 with smaller anthers and shorter filaments and 4 with much reduced anthers and short filaments. Style hairy in the lower part, glabrous above, stigma terminal, ciliate. Pods brown, usually curved upwards at the distal end, rarely wholly straight, linear, 10–11 × 0.8–1 cm, compressed, tardily dehiscent; seeds grey-brown, suborbicular, up to 5 × 4.5 mm, transversely arranged, areole elliptic. Figs. 14/2, 15/3, 16/3.

Distr. ALDABRA: W,P,M,Michel, Sylvestre; ASSUMPTION; COSMOLEDO: W,M; ASTOVE; pantropical, possibly of tropical American origin.

Notes. Usually a weed of cultivation, found near areas of settlement; the seeds at least, poisonous.

Vernac. 'cassie puante'.

3. DELONIX Raf.

Spineless trees. Leaves 2-pinnate, pinnae without an odd terminal leaflet; leaflets numerous, small. Flowers large, orange or red, arranged in showy terminal or axillary corymbose racemes. Sepals 5, subequal, united at the base. Petals 5, equal or subequal except for the uppermost which differs in shape and colour. Stamens 10, free, long exserted. Ovary containing numerous ovules. Fruit an elongate, flattended, dehiscent, 2-valved pod; seeds many.

A genus of 8 species occurring naturally from Africa through Arabia to India and also in Madagascar.

Delonix regia (Boj. ex Hook.) Raf., Fl. Tellur. 2: 92 (1837); F.T.E.A. Leg.-Caesalp.: 23 (1967).

Poinciana regia Boj. ex Hook. in Bot. Mag. 56: t. 2884 (1829).

Spreading trees, 6–12(–18) m high. Leaves 20–30 × 10–15 cm, rachis sparsely to densely pubescent; leaflets numerous, oblong, obtuse to acute, 8–10 × 2–4 mm, sparsely to densely pubescent on both surfaces or subglabrous; stipules conspicuous, leaf-like, compound, soon falling. Flowers

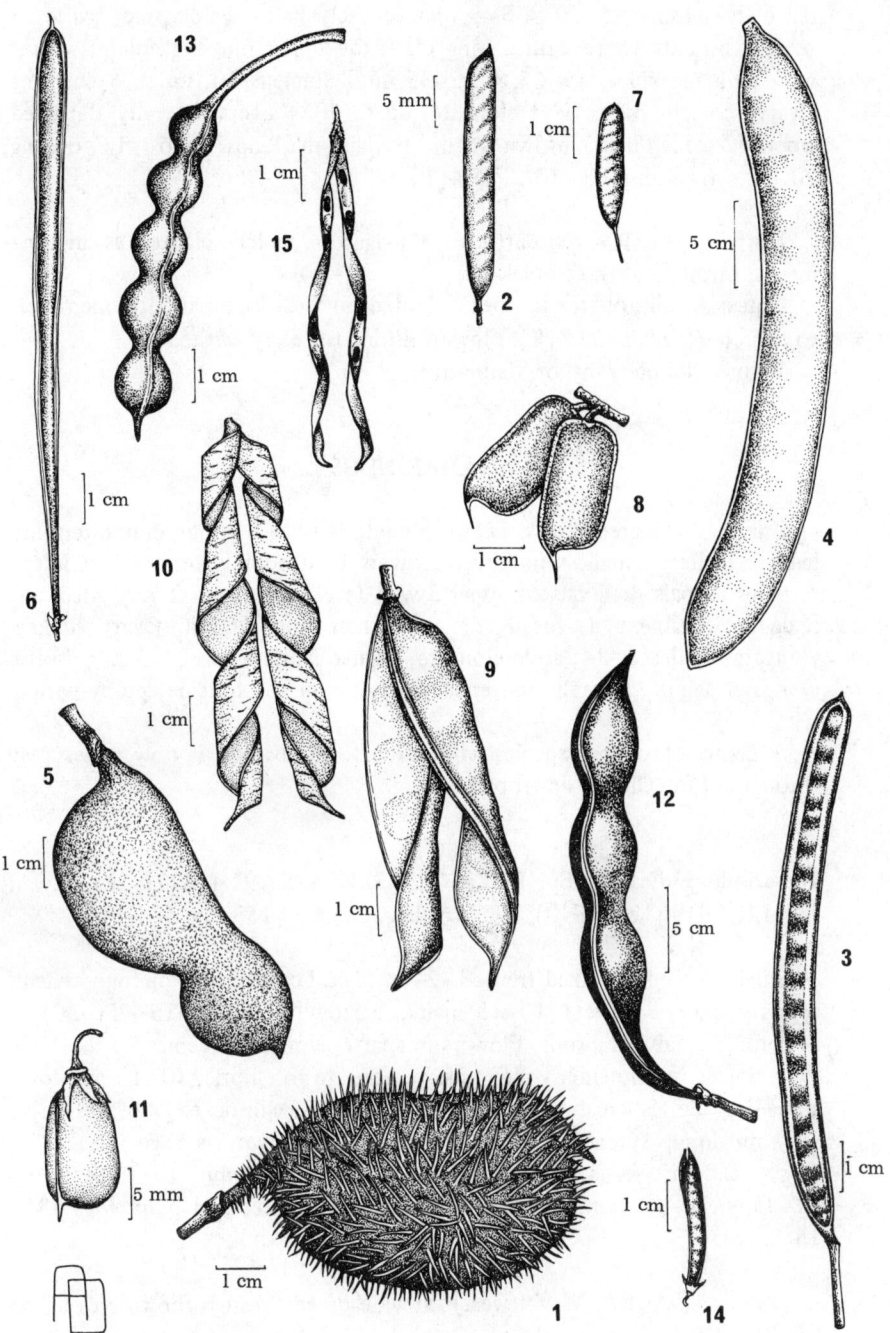

Fig. 16 Fruits of *Leguminosae*. 1, *Caesalpinia bonduc*. 2, *Cassia aldabrensis*.
3, *Cassia occidentalis*. 4, *Delonix regia*. 5, *Tamarindus indica*.
6, *Calliandra alternans*. 7, *Dichrostachys microcephala*. 8, *Abrus
precatorius*. 9, *Canavalia rosea*. 10, *Clitorea ternatea*. 11, *Crotalaria
laburnoides*. 12, *Erythrina variegata*. 13, *Sophora tomentosa*.
14, *Tephrosia pumila* var. *aldabrensis*. 15, *Teramnus labialis*.

in large, showy, axillary racemes. Sepals free, orange outside, scarlet inside, narrowly-oblong, 25–30 × 5–8 mm, sparsely hairy. Petals free, scarlet or orange, broadly ovate with a long claw, the upper one spathulate, whitish with pink markings, 45–65 × 25–35 mm. Stamens 10, up to 5 cm long. Ovary sparsely hairy. Pod elongate, up to 70 × 6 cm, laterally flattened, woody; seeds light brown with purple markings, narrowly oblong, 20–22 × 6–8 mm. Figs 13/2, 15/4, 16/4.

Distr. ALDABRA: W; native of Madagascar, widely planted as an ornamental throughout the tropics.

Notes. A solitary tree has been planted on West Island at Settlement next to church, *Renvoize* 719 (K). Flowers during the early wet season.

Vernac. 'flamboyant' or 'flame-tree'.

4. TAMARINDUS L.

Unarmed evergreen trees. Leaves pinnate, pinnae without an odd terminal leaflet; leaflets, small, numerous. Flowers in drooping, terminal or lateral racemes. Sepals 4. Petals 5; upper 3 well-developed, lower 2 very small and scale-like. Stamens 3, fused for about half their length. Ovary stalked, elongate, ovules 8–14; style elongate, stigma capitate. Fruit an indehiscent, elongate, slightly laterally-flattened pod, dry on the outside, pulpy within.

A genus of only 1 species which due to cultivation is now widespread throughout the Old World tropics.

Tamarindus indica L., Sp. Pl.: 34 (1753); F.M.S.: 91 (1877); Hemsley in B.M.I.K. 1919: 120 (1919); F.T.E.A. Leg.-Caesalp.: 153, fig. 32 (1967).

Small to medium sized trees, 3–24 m high. Leaves 5–14 cm long, rhachis sparsely hairy; leaflets 10–18 pairs, narrowly oblong, 15–25(–30) × 5–8 mm, usually glabrous. Flowers in sparse, elongate racemes up to 15 cm long. Sepals red outside, yellow inside, obovate to elliptic, 10–12 mm long, spreading. Petals creamy yellow with red veins, elliptic to obovate, up to 12 mm long, spreading. Pod irregular, dull brown, sausage-like, up to 14 × 3 cm; seeds chestnut brown 1–10, rhombic to trapeziform, 11–17 × 10–12 mm, embedded in a gummy, sweet-acid pulp. Figs 13/3, 15/5, 16/5.

Distr. ALDABRA: W, W (west), S; widespread through the tropics of the Old World.

Notes. A few trees occur near settlement areas; 1 large remote specimen was found in mixed scrub of the eastern end of South Island. Flowers mainly during the wet season. Fruits edible, the acid pulp used to make a cool drink.

Vernac. 'tamarinier' or 'tamare'.

24:2 MIMOSOIDEAE

Trees, shrubs, lianes, rarely herbs. Leaves 1- or 2-pinnate, or modified to phyllodes or absent. Inflorescence usually spikes, racemes or heads of usually small, regular (3−)5(−6)-merous flowers. Sepals usually united, margins touching but not overlapping in bud. Petals usually united below, margins touching but not overlapping in bud. Stamens 4−10 or numerous, free or joined below; anthers sometimes with an apical gland. Fruit various; seeds usually with areoles (elliptic- or oblong-shaped areas bounded by fine lines).

1. Inflorescence globose; flowers all bisexual and all the same colour, pinkish-
orange **1. Calliandra**
1. Inflorescence oblong, flowers at the apex pale yellow, bisexual and
different in colour to flowers at the base, which are neuter and pale pink
2. Dichrostachys

1. CALLIANDRA Benth.

Small trees, shrubs, or perennial herbs. Leaves 2-pinnate, pinnae without an odd terminal leaflet; stipules usually persistent. Flowers in globose heads, axillary or aggregated into terminal compound inflorescences. Calyx and corolla 4, 5 or 6-toothed. Stamens numerous and very showy, projecting beyond the corolla, fused at the base. Ovary sessile, ovules numerous. Fruit a compressed, elongate pod, thickened at the margins, splitting elastically from the apex into 2 valves.

100 species, mainly in tropical America.

1. **Calliandra alternans** Vahl ex Benth. in Trans. Linn. Soc. 30: 548 (1875).

Pithecellobium ambiguum Hemsley in J. Bot. 54, Suppl. 2: 13 (1916) & in
B.M.I.K. 1919: 120 (1919); Renvoize in P.T.R.S. B, 260: 231 (1971).
Type: Aldabra, *Fryer* 39 (K, holo.).

Shrubby trees, up to 5 m high; bark smooth, light grey. Leaves 2-pinnate, pinnae 1 pair, 3.5−5.5 cm long, sparsely hairy to subglabrous, petiole 2−15 mm long, grooved above and with a conspicuous gland at the junction of the pinnae; leaflets opposite or subopposite, 8−10(−11) pairs, obliquely oblong, 10−12 × 3−6 mm, dark green and shiny above, rhachis grooved above, narrowly winged between the leaflets. Flowers sessile, clustered in globose heads on shortly hairy peduncles 1−3 cm long, often with 1 or 2 small scale-like bracts scattered along the peduncle. Calyx 5-toothed, up to 1 mm long. Corolla pinkish-orange, 4-toothed, 5 mm long. Stamens numerous, up to 15 mm long, yellow or orange. Style up to 17 mm long. Pod linear, 16 × 0.8 cm, tapering at the base, tardily dehiscent; seed, not seen. Figs 13/4, 15/6, 16/6.

Superficially resembles *Macphersonia hildebrandtii* (Family 21/3) which has 3–5 pairs of pinnae, not 1 pair.

Distr. ALDABRA: M, S (east); Madagascar.

Notes. A locally abundant constituent of the inland mixed scrub. From the few observations made, flowers during the early dry season.

2. DICHROSTACHYS (DC.) Wight & Arn.

Small trees or shrubs with or without spines. Leaves 2-pinnate, pinnae with several to many pairs of small leaflets. Flowers small, densely clustered in oblong or elongate, axillary spikes, upper part of the spike with bisexual flowers, lower part broader, with differently coloured, neuter flowers. Sepals 5, fused to form a shortly 5-toothed calyx. Petals 5, longer than the sepals, fused below. Stamens 10, all fertile in bisexual flowers, reduced to elongate staminodes without anthers in neuter flowers. Pods clustered to solitary, narrowly oblong, flattened, contorted or straight, indehiscent or dehiscent.

A genus of c 20 species in the Old World tropics.

Dichrostachys microcephala Renvoize in K.B. 26: 437 (1972). Type: Aldabra, South Island, Takamaka, *Renvoize* 1059 (K, holo.).*

Gagnebina tamariscina sensu Hemsley in B.M.I.K. 1919: 121 (1919), not (Lam.) DC. (1825).

Desmanthus commersonianus sensu Baker in B.M.I.K. 1894: 148 (1894); Schinz in A.S.N.G. 21: 84 (1897); Voeltzkow in A.S.N.G. 26: 550 (1902); Hemsley in B.M.I.K. 1919: 121 (1919), not Baill. (1883).

Small non-spiny trees or shrubs, up to 4 m high; branches sparsely pubescent. Leaves pubescent, pinnae 8–18 pairs, 1–2 cm long; rachis 6–9 cm long, 1 large gland usually present between the 2 basal pinnae, small glands present or absent between the higher pinnae; leaflets numerous, oblong, 1–1.5 mm long. Flowers clustered in erect, small, oblong spikes, 5–10 mm long; peduncles 5–10 mm long. Upper, bisexual flowers pale yellow; lower, neuter flowers pale pink. Calyx 0.5 mm long, shortly toothed. Petals narrowly oblong or ovate, 1–2 mm long. Anthers linear, 1.2 mm long; staminodes contorted, c 5 mm long, projecting beyond the petals. Ovary shortly hairy. Pods solitary, flattened, narrowly oblong, straight or slightly curved, 25–30 × 5 mm, apiculate, dehiscent; seeds 6–8, flattened, quadrangular, 3 × 2.5 mm. Figs. 13/5, 15/7, 16/7.

---

* Fosberg maintains that this species, in spite of its andromonoecious character, should remain with the Madagascar genus *Gagnebina* Baill., a group with a distinctive habit and fruit morphology, which seems closest to the very similar American genus *Desmanthus* Willd.

Distr. ALDABRA: W, P, M, S, Esprit; ASTOVE; Madagascar.

Notes. Frequent constituent of the inland mixed scrub. Common only in the north-west and west of Aldabra, where the tortoise density is low; seedlings and young plants easily damaged by tortoises. Flowers mainly during the wet season but may respond slowly to unseasonal rain. Leafless during periods of drought.

Vernac. 'trèfle'.

24:3. PAPILIONOIDEAE

Usually herbs, sometimes trees, shrubs or lianes, usually unarmed. Leaves usually pinnate or 3-foliolate. Flowers irregular, usually solitary or in racemose or paniculate inflorescences. Sepals united, tubular. Petals free or rarely united, overlapping in bud. Stamens usually 10, often united into groups of 9 and 1 or 5 and 5. Seeds without areoles.

1. Trees or shrubs, branchlets with prickles   **5. Erythrina**
1. Trees, shrubs or herbs without prickles
  2. Plants climbing, scrambling or prostrate:
    3. Plant twining; leaves pinnate; seeds conspicuous, red and black
                                                 1. Abrus
    3. Plants scrambling or trailing; leaves pinnate with an odd terminal leaflet or 3-foliolate; seeds not red and black
      4. Leaves 3-foliolate
        5. Calyx 2-lobed; petals pinkish-mauve; stamens 10, all fertile
                                         2. Canavalia
        5. Calyx 4- or 5-lobed
          6. Slender perennial; standard petals white, pink or purple;
             5 sterile stamens alternating with 5 fertile   **9. Teramnus**
          6. Robust perennials; standard petals white, yellow or pale
             purple; all stamens fertile   **10. Vigna**
      4. Leaves pinnate, flowers white, blue or blue and white   **3. Clitoria**
  2. Plants erect
    7. Trees or shrubs; leaves pinnate, silvery-grey; flowers yellow
                                        7. Sophora
    7. Herbs, sometimes shrubby
      8. Leaves pinnate; small perennial herbs
        9. Leaflets 3–4, alternate; erect herb; flowers orange-brown
                                    6. Indigofera
        9. Leaflets 7–11, pairs opposite; decumbent herb; flowers mauve, pink and white   **8. Tephrosia**
      8. Leaves 3-foliolate; erect annual herb; flowers yellow **4. Crotalaria**

1. ABRUS Adans.

Shrubby climbers. Leaves pinnate, without an odd terminal leaflet, apex of the rhachis terminating in a small bristle. Flowers arranged in terminal or

axillary elongate clusters. Calyx shortly 5-toothed. Petals 5, longer than the
sepals, standard ovate, wings oblong, curved, keel longer and broader than the
wings. Stamens 9, united into a tube and split from above. Ovary subsessile,
ovules numerous; style short and incurved. Fruit an oblong or linear,
compressed, dehiscent, 2-valved pod; seeds few to many, subglobose or
ellipsoid, shiny.

A genus of 6 species distributed throughout the tropics and subtropics.

Abrus precatorius L., Syst. Nat. ed. 12, 2: 472 (1767), F.T.E.A. Leg.-Papilion.:
114 (1971).

Twining perenniel plants; stems tough, wiry. Leaves 5–12 cm long, rachis
sparsely hairy; leaflets 9–16 pairs, oblong, 1–2.5 × 0.4–0.7 cm, sparsely
hairy. Flowers in dense clusters, (2–)4–7(–10) cm long; peduncle 4–10 cm
long, stout. Calyx green, 2–3 mm long, obscurely toothed, hairy. Petals pale
pink or mauve, 8–10 mm long, glabrous or very sparsely hairy. Pod oblong,
(2–)2.5–3.5(–4.3) × 0.8–1.4 cm, smooth or rough, pubescent, beaked;
seeds 3–5, scarlet with black basal area, ovoid, 5–8 × 4–5 mm, shiny,
remaining attached to the opened valves for some time.

subsp. **africanus** Verdc. in Mitt. Bot. München 7: 328 (Mar. 1970) & in K.B.
24: 241 (Apr. 1970); F.T.E.A. Leg.-Papilion.: 114 (1971).

A. precatorius sensu F.M.S.: 78 (1877); Baker in B.M.I.K. 1894: 147 (1894);
Schinz in A.S.N.G. 21: 85 (1897); Voeltzkow in A.S.N.G. 26: 550 (1902);
Dupont, Report: 36 (1907); Hemsley in B.M.I.K. 1919: 120 (1919).

Pod (2–)2.5–3(–3.5) cm long, rough, with low ridges and protuberances.
Differs from the Indian subsp. *precatorius* which has smooth and slightly
longer pods. Figs. 13/6, 15/8, 16/8.

Dist. ALDABRA: W, P, M, S, Esprit; ASSUMPTION; tropical Africa,
Seychelles, Madagascar, Mauritius, introduced into Australia and America.
   Notes. Occurs occasionally in mixed scrub and then usually close to areas
of settlement. Flowers during mid to late wet season. The record for
Assumption is based on Dupont, loc. cit., who also suggests that it is a recent
introduction. Seeds contain toxins.
   Vernac. 'liane réglisse' or 'réglisse'.

2. CANAVALIA DC.

Climbing or prostrate herbs. Leaves pinnately 3-foliolate. Flowers clustered
at the ends of long axillary peduncles. Calyx 5-lobed, 2-lipped, the upper
larger than the lower. Petals 5, longer than the sepals; standard orbicular,
reflexed; wings narrow; keel broader than the wings, incurved. Stamens 10,

united into a tube. Ovary shortly stalked, ovules numerous; style incurved. Fruit an oblong or broadly linear, 2-valved dehiscent pod; seed usually several or numerous.

A genus of 40 species distributed throughout the warmer parts of the world.

Canavalia rosea (Sw.) DC., Prod. 2: 404 (1825); F.T.E.A. Leg.-Papilion.: 576 (1971).

Dolichos roseus Sw., Prodr. Veg. Ind. Occ.: 105 (1788) & in Fl. Ind. Occ. 3: 1243 (1806).
Canavalia obtusifolia (Lam.) DC., Prodr. 2: 404 (1825); F.M.S.: 80 (1877).

Climbing or trailing perennial. Leaves pinnately 3-foliolate, terminal leaflet generally orbicular, lateral leaflets elliptic, 7–11 × 4–11 cm, sub-glabrous to sparsely hairy. Flowers clustered in robust, erect spikes 10(–20) cm long; peduncles stout, up to 20 cm long. Sepals green, 10–15 mm long, fused for most of their length, sparsely pubescent on the outer surface. Petals pinkish-mauve, standard elliptic, 2–3 cm long, apex 2-lobed. Stamens 3 cm long. Pod yellowish-green, broadly linear, laterally flattened, 10–17 × 2.5–3 cm, usually somewhat curved, with a double ridge on the concave edge, dehiscent; seeds brown with darker mottles, elliptic-orbicular, 15–20 × 7–14 mm. Figs. 14/3, 15/9, 16/9.

Distr. ALDABRA: W, S, (north-west); ASSUMPTION; pantropical strand plant.
Notes. Herbarium records only for Assumption, *Frazier* 23 & 33, *Stoddart* 1046 (all K). Only known population on Aldabra at Anse Anglais much affected by tortoise activity. Plants generally found growing amongst debris just above high tide level. Viable seed found amongst beach drift; seeds can float for at least 8 weeks and subsequently germinate. Flowers during the early dry season.

3. CLITORIA L.

Climbing or erect herbs or shrubs, rarely trees. Leaves pinnate, 3- or 1-foliolate. Flowers solitary or paired or in sparse racemes; upper bracts small, fused at the base; bracteoles large, conspicuous. Calyx 5-lobed, upper 2 lobes joined only at the base. Petals 5, standard large, orbicular, much exceeding the other perals, wings spathulate, keel blunt. Stamens 10, united into a tube with 1 free. Ovary shortly stalked, ovules 2–many; style curved. Fruit a linear-oblong, compressed pod dehiscing by 2 valves; seeds several, sub-globose or ellipsoid, compressed.

A tropical and subtropical genus of 30–40 species.

Clitoria ternatea L., Sp. Pl.: 753 (1753); F.M.S.: 81 (1877); F.T.E.A. Leg.-
Papilion.: 515, fig. 75 (1971).

Perennial, climbing or trailing herb; rootstock woody; stems slender.
Leaves pinnate, with an odd terminal leaflet; leaf axis up to 9 cm long,
stipules lanceolate, 4–10 mm long, persistent; leaflets 5–7, elliptic, oblong or
circular, 1–6.7 × 0.3–4 cm, apex acute, rounded or shallowly notched,
sparsely to densely pubescent. Flowers axillary, solitary or paired; pedicels up
to 9 mm long. Calyx up to 2 cm long, pubescent. Petals white, margined with
blue or entirely blue, standard held lowermost, oblong-ovate, up to 2 ×
3.5 cm. Pods linear-oblong, up to 12 × 1 cm, dehiscent; seeds 8–10, olive,
pale brown to deep reddish-brown with darker mottling, ellipsoid, oblong or
oblong-kidney-shaped, 4.5–7 × 3–4 × 2–2.5 mm. Figs 14/4, 15/10, 16/10.

Distr. ALDABRA: W; pantropical, its true distribution obscured by
cultivation.
Notes. Only known around areas of settlement, probably introduced as an
ornamental. Flowers in the late wet mid-dry season if moisture sufficient.
Vernac. 'liane madame', 'liane ternate' or 'butterfly pea'.

4. CROTALARIA L.

Shrubs or herbs. Leaves simple or digitately 3, 5 or 7-foliolate; stipules
present or absent. Flowers usually in terminal or leaf-opposed racemes,
occasionally solitary. Calyx 5-lobed or sometimes 3-lobed. Petals 5, standard
orbicular or ovate, wings obovate or oblong, keel curved inwards and
produced into a beak. Stamens 10, united into a tube and split from above.
Ovary sessile or stalked, ovules 2-many; style incurved. Fruit a globose or
oblong, often inflated, non-septate pod, dehiscing by 2 valves; seeds few to
many.

A tropical and subtropical genus of c 350 species.

Crotalaria laburnoides Klotzsch in Peters, Reise Mossamb. Bot. 1: 57 (1862);
F.T.E.A. Leg.-Papilion.: 966 (1971).

Erect annual herb, up to 60 cm high. Leaves 3-foliolate, leaflets narrowly
obovate to elliptic, central leaflet slightly larger than the lateral leaflets,
1.5–4 × 0.5–1.5 cm, glabrous above, hairy beneath; petiole 15–32 cm long;
stipules linear, 2–4 mm long. Flowers clustered in terminal spikes up to
8(-12) cm long. Calyx pale green, 5–8 mm long, lobes acute to acuminate,
pubescent outside. Petals yellow, keel up to 1 cm long, narrowly beaked,
standard obovate-oblong, up to 1 cm long. Stamens up to 1 cm long. Ovary
shortly stalked, hairy. Pod oblong, inflated, 1–2(–2.5) cm long, pubescent;
seeds orange-brown, obliquely heart-shaped, 2.5 mm long. Figs. 14/5, 15/11,
16/11.

Distr. ALDABRA: S (south); COSMOLEDO: M; East Africa from Somalia to Mozambique and Comoro Is.

Notes. On Aldabra found growing growing in sand near Dune de Messe, *Stoddart* 1023, *Merton* s.n. and *R. & S. Hnatiuk* 730206 (all K), or in short turf on pavé near Anse Quive. Flowers mid wet season.

5. ERYTHRINA L.

Trees or shrubs, rarely herbs; armed with strong prickles. Leaves pinnately 3-foliolate; stipules persistent or soon falling. Flowers in showy, pyramidal, raceme-like inflorescences. Calyx 1–5-toothed, sometimes split to the base. Petals 5, longer than the sepals, standard elongate, longer than the wings and keel, wings short, keel longer or shorter than the wings. Stamens 10, united in a tube for half their length, 1 stamen free from the base upwards. Ovary stalked, ovules numerous; style curved inwards. Fruit an elongate, 2-valved, dehiscent or subdehiscent pod with constrictions between the ovoid seeds.

A genus of c 200 species distributed throughout the tropics and subtropics.

Erythrina variegata L., Herb. Amb.: 10 (1754) & in Amoen. Acad. 4: 122 (1759); F.T.E.A. Leg.-Papilion.: 549 (1971).

E. indica Lam., Encycl. Méth. Bot. 2: 391 (1786); F.M.S.: 82 (1877).

Trees 6–12 m high; branchlets densely covered with short, straight prickles. Leaves pinnately 3-foliolate, terminal leaflet longer than those of the lateral pair; petiole 2–20 cm long, prickles absent; leaflets very broadly ovate, 9–25 × 6.5–22 cm, apex acute to acuminate, rather thin, subglabrous; rachis between the terminal and lateral leaflets 1–7 cm long. Flowers in dense conspicuous spikes, 10–20 cm long, covered with deciduous brown hairs, usually appearing before the leaves. Calyx spindle-shaped in bud, splitting dorsally as the flowers open, brownish-red, covered with deciduous hairs. Petals scarlet or dark crimson, standard elliptic, 5–6 cm long, curved upwards, shortly clawed. Pod up to 20 × 2–3 cm, constricted between the seeds, tardily dehiscent; seeds 3–14, pinkish, oblong, 1.5–2.5 × 0.8–1.2 × 0.8–1.2 cm. Figs 14/6, 15/12, 16/12.

Distr. ALDABRA: W, M, S (north-west); Tanzania, Zanzibar, Seychelles, Mascarene Is., China, Taiwan, Fiji, Samoa and Society Is.; widely cultivated.

Notes. A well-known strand species, often planted as an ornamental. On Aldabra known on West Island at Settlement, where it may have been planted but could be naturally introduced at Anse Anglais and near Anse Malabar. Massive flowering during the mid-dry season. The Aldabra plants, are all the non-variegated var. *orientalis* (L) Merrill.

Vernac. 'mourouc' or 'nourouc', sometimes wrongly called 'flamboyant'.

6. INDIGOFERA L.

Annual or perennial herbs and shrubs. Leaves simple, 3-foliolate or usually pinnate with an odd terminal leaflet; stipules present. Flowers usually in axillary racemes. Corolla usually falling early. Stamens 10, 9 filaments united, 1 free. Fruit a pod, usually dehiscent; seeds 1-many.

A tropical and sutropical genus of c 700 species.

Indigofera sp.

Small herb, possibly perennial, 4–10 cm tall; stem slender, densely covered with appressed, silvery-white, stiff hairs. Leaves alternate, 5–15 mm long, leaflets alternate (3-)4, elliptic-obovate, 2–7 × 1.5–2 mm, terminal leaflet slightly larger and more elongate, densely covered with silvery-white, stiff hairs above, more densely beneath; stipules awl-shaped, 2 mm long. Racemes axillary, 1–2 cm long, longer than the leaves; flowers 6–12, crowded above, lax below; peduncle 5–10 mm long; pedicels 1 mm long. Calyx 3.5 mm long, deeply dissected with awl-shaped teeth, densely hairy. Corolla orange-brown, 4.5–5 mm long, soon falling; standard obovate, hairy outside. Pods reflexed, linear-oblong, 10 × 2 mm, densely covered with appressed, silvery-grey, stiff hairs; seeds c 8, immature.

The fragmentary gatherings clearly show that this species belongs to Sect. *Alternifolae* but cannot be matched exactly. In Gillett's revision of the African *Indigofera* (Kew Bull. Add. 1 (1958)) it comes nearest to *I. alternans* DC. from southern Africa, but differs principally in the fewer leaflets and the calyx shorter than the corolla. There are a number of species in this group in eastern Africa and Madagascar and it is possible that this species will prove to be endemic when better known.

·Distinguished by the organge-brown flowers from the mauve, pink or white flowered *Tephrosia pumila* (Family 24: 3/8).

Distr. ALDABRA: S; not matched elsewhere.

Notes. Only known from 2 fregmentary collections, *Hnatiuk* 73065 & 731758 (both K) from tortoise turf and crevices in the coastal champignon near Cinq Cases; more material required. Flowers open during the late afternoon.

7. SOPHORA L.

Trees or shrubs, rarely perennial herbs. Leaves pinnate with an odd terminal leaflet; leaflets large and few to small and many. Flowers in terminal or axillary racemes. Calyx tubular, upper 2 lobes often fused. Petals 5, standard orbicular, usually narrowed into a short claw, wings oblong, keel oblong and almost straight. Stamens 10, free or united at the base only. Ovary shortly stalked, ovules numerous; style usually curved. Fruit an elongate, indehiscent pod, constricted between the seeds.

A genus of c 80 species distributed throughout the warm temperate and tropical parts of the world.

Sophora tomentosa L., Sp. Pl.: 373 (1753); Hemsley in B.M.I.K. 1911: 120 (1919); F.T.E.A. Leg.-Papilion.: 44 (1971).

subsp. **tomentosa**; F.T.E.A. Leg.-Papilion.: 44 (1971).

Large shrubs, (0.7–)1–5(–10) m tall; branches greyish pubescent. Leaves pinnate with an odd terminal leaflet, 10–30 cm long, leaflets 5–9 pairs, apical leaflet often slightly larger than the lateral ones, leaflets broadly elliptic, 2–4 × 1.5–2.5 cm, grey pubescent, becoming subglabrous above when mature. Flowers in dense, spike-like, terminal racemes, up to 14 cm long. Calyx bell-shaped, 5–7 mm long, shortly 5-toothed, grey pubescent. Petals yellow, standard 1.5–2 cm long. Pod cylindrical with bulges at intervals, 8–20 cm long, grey tomentose, tardily dehiscent; seeds 3–9, glossy dark brown, hard-shelled, subglobose, 5–7 mm in diameter. Figs. 13/7, 15/13, 16/13.

Distr. ALDABRA: W, P, M, S, Esprit, Michel; ASSUMPTION; an Indo-Pacific strand plant (subsp. *occidentalis* (L.) Brummitt is a south Atlantic strand plant).
Notes. Occurs mainly along the northern coasts of Middle and South Islands just inland of the coastal cliff; a few records from dense *Pemphis* scrub on Middle Island. The seeds can float for at least 8 weeks and subsequently germinate. Few observations but flowers during the dry season.

8. TEPHROSIA Pers.

Annual or perennial herbs or softly woody shrubs. Leaves pinnate with an odd terminal leaflet, rarely palmate, leaflets few to many. Flowers in terminal or lateral, raceme-like inflorescences. Calyx 5-lobed. Petals 5, free, longer than the sepals. Stamens 10, more or less equal, 9 united for most of their length, 1 united for up to half of its length. Ovary sessile, ovules few to many; style often curved and flattened. Fruit an ovate, oblong to linear, dehiscent, 2-valved pod.

A widespread tropical and subtropical genus of 400 species.

Tephrosia pumila (Lam.) Pers., Syn. Pl. 2: 330 (1807); F.M.S.: 71 (1877); F.T.E.A. Leg.-Papilion.: 184 (1971).

Galega pumila Lam., Encycl. Méth. Bot. 2: 599 (1786).

Small, spreading, much-branched, erect or prostrate annual or short-lived perennial herbs; branches with spreading or appressed hairs. Leaves pinnate, with an odd terminal leaflet, 2–6 cm long, rachis sparsely to densely hairy; stipules narrowly triangular, up to 4 mm long; leaflets 7–13, narrowly obovate, 5–20 mm long, apical leaflet longer, grey-green, hairy on both sides. Flowers white, pink, mauve or purple, arranged in sparse, elongate racemes, up to 4 cm long. Pod linear, 2.5–4.2 cm long, dehiscent, valves twisting when dry; seeds 9–15, square or rhomboid, 2 × 1.2 mm.

This species is widely distributed in the Old World tropics from southern and eastern tropical Africa eastwards to Indonesia.

var. **aldabrensis** (J.R. Drummond & Hemsley) Brummitt in Bol. Soc. Brot. II, 41: 260 (1968); F.T.E.A. Leg.-Papilion.: 185 (1971). Type: Aldabra, *Dupont* 11 (K, holo.).

T. aldabrensis J.R. Drummond & Hemsley in J. Bot. 54, Suppl. 2: 11 (1916); Hemsley in B.M.I.K. 1919: 119 (1919).
T. purpurea sensu Baker in B.M.I.K. 1894: 147 (1894); Schinz in A.S.N.G. 21: 85 (1897); Voeltzkow in A.S.N.G. 26: 550 (1902); Dupont, Report: 36 (1907), not (L.) Pers. (1807).

Low, prostrate, scrambling or erect perennial herb, stems up to 30 cm long. Leaves 1–2.5(-4) cm long; leaflets 7–11, narrowly obovate, 5–8 (-12) mm long, apex sharply tipped, densely appressed hairy below, subglabrous above. Flowers solitary or in sparse terminal clusters of 2 or 3 flowers. Calyx 3–5 mm long, deeply toothed, appressed-hairy. Petals mauve, pink or white, standard 7–9 mm long, shortly hairy on the back. Pod linear, 25–30 × 3–4 mm, appressed-hairy, style and calyx persistent; seeds small, 10–12, brown, flattened, rhomboid, 1.5 mm in diameter. Figs 14/7, 15/14, 16/14.

Readily distinguished from *Cassia aldabrensis* (Family 24:1/2) which has orange flowers and no odd terminal leaflet to the pinnae and *Indigofera* sp. (Family 24:3/6) which has orange-brown flowers and pinnae with an odd terminal leaflet.

Distr. ALDABRA: W, P, M, S, Esprit; ASSUMPTION; COSMOLEDO: M; ASTOVE; Kenya, Tanzania and Zanzibar.
Notes. Found in both inland mixed scrub and coastal communities. Flowers during the morning and early afternoon. Flowering during the wet season but can occur throughout the year if sufficient moisture available.
Vernac. 'indigo sauvage'.

9. TERAMNUS P.Br.

Climbing or trailing perennial herbs, rarely small erect shrubs. Leaves subdigitate or pinnate, 3-foliolate; stipules present. Flowers in lateral, elongate, raceme-like inflorescences. Calyx 4—5-lobed according to whether the upper 2 lobes are free or united. Petals small, standard obovate or orbicular. Stamens 10, all united or 1 free, 5 fertile alternating with 5 sterile. Ovary linear, ovules many. Fruit a linear, beaked pod; seeds 8, oblong, ovoid or subglobose.

A genus of 8 species widespread throughout the tropics.

Teramnus labialis (L.f.) Spreng., Syst. Veg. 3: 235 (1826); F.T.E.A. Leg.-Papilion.: 535 (1971).

Glycine labialis L.f., Suppl. Pl.: 325 (1781).

Climbing trailing or prostrate perennial herbs; rootstock sometimes woody; stems up to 3 m long, slender, with appressed to spreading white to reddish-brown hairs or glabrescent, occasionally rooting at the nodes. Leaves 3-foliolate; petiole 1—5 cm long; stipules narrowly lanceolate, 2—3 mm long; leaflets very variable, round to narrowly lanceolate, 1—8 × 0.5—4 cm, glabrous or hairy; petiolules 2 mm long; rachis 1—12 mm long. Flowers in slender, few-flowered, elongate inflorescences, up to 10 cm long; pedicels slender, 1.3—4 mm long. Calyx 2—6 mm long, with 5, elongate, acute teeth, glabrescent or hairy. Standard white, pink or purple, obovate, 5 × 3.5 mm; wings pale mauve; keel white. Pods linear, 2.5—6 × 0.2—0.4 cm, glabrescent or with sparse appressed or spreading hairs; beak 2—3 mm long; seeds yellow-brown to dark purplish-brown, oblong or almost cylindrical, 2—3 × 1.2—2 × 1.2—1.5 mm, smooth or granular.

This species is widespread from tropical and southern Africa to the Philippines, New Guinea and Guam. It also occurs in the West Indies and Guyana.

subsp. **arabicus** Verdc. in K.B. 24: 272 (1971); F.T.E.A. Leg.-Papilion.: 537, fig. 80/1—12 (1971).

T. labialis sensu F.M.S.: 79 (1877).

Leaflet oblong, elliptic or rhomboid, 1.5—7 × 1—3.3 cm, apex acute or subobtuse, appressed hairy above. Pods linear, 2.5 × 0.3 cm, dehiscent; seeds dark brown, oblong or almost cylindrical, 2—3 × 1.2—2 × 1.2—1.5 mm, granular. Fig 15/15, 16/15.

Distr. ALDABRA: W; E (east) Africa from Sudan to Rhodesia, Arabia, Principe, Mauritius, Réunion, Seychelles, Comoro Is., Madagascar, also introduced into the West Indies and Guyana.

Notes. Occurs in disturbed ground around Settlement, *Fosberg* 48835 (K, US), probably introduced by man. From the few observations made, flowers during the early to mid dry season.

10. VIGNA Savi

Climbing, prostrate or erect herbs or shrubs. Leaves pinnate, 3- or 1-foliolate; stipules various. Flowers in terminal or axillary raceme-like inflorescences, subumbellate clusters of fascicles; bracts and bracteoles soon falling. Calyx 5-lobed, 2-lipped, lower lip 3-lobed, upper lip 2-lobed or almost entire. Petals 5, longer than the calyx, standard orbicular or ovate, wings obovate or oblong, keel truncate, obtuse or beaked, the beak sometimes sharply curved upwards. Stamens 10, fused into a tube with 1 free. Ovary sessile, ovules 3-many; style straight or curved. Fruit a linear or linear-oblong, cylindrical or flattened, straight or curved pod dehiscing by 2 valves; seeds several, kidney-shaped or quadrate.

A large tropical genus of c 150 species.

1. Leaflets obovate, apex and base rounded   **1. V. marina**
1. Leaflets ovate to lanceolate, apex acute or acuminate, base entire or
   3-lobed   **2. V. unguiculata**

1. Vigna marina (Burm.) Merrill, Interpret. Rumph. Herb. Amboin.: 285 (1917); F.T.E.A. Leg.-Papilion.: 626 (1971).

Phaseolus marinus Burm., Ind. Alter Univ. Herb. Amboin. [18] (1769).
Vigna lutea (Sw.) A. Gray, Bot. Wilkes U.S. Expl. Exped.: 452 (1854); F.M.S.: 85 (1877).

Scrambling or climbing perennial herbs, several metres long. Leaflets 3, rounded-obovate, 3.5–9.5 × 2.5–7.5 cm, sparsely hairy; petioles up to 11.5 cm long; stipules ovate, 2.5 mm long, 2-lobed at the base. Calyx up to 4 mm long, upper lip entire, ciliate. Standard yellow, obovate, 1.2–1.3 × 1.4 cm. Pods linear-cylindrical, slightly curved, inflated, up to 6 × 0.9 cm, slightly constricted between the seeds; seeds 2–6, yellow-brown or red-brown, oblong, slightly narrowed at one end, 6–7 × 5–6 × 4.5–5 mm. Fig. 15/16.

Distr. ALDABRA: W; COSMOLEDO; a pantropical strand plant.
Notes. A rare constituent of the strand flora. Few collections from the beach near settlement on Aldabra, *Hnatiuk* 731961 & 732059 (both US) and two collections from Cosmoledo, *Frazier* 573 and *Fosberg & McKenzie* 49855 (both US).

2. Vigna unguiculata (L.) Walp., Rep. 1: 779 (1842); F.T.E.A. Leg.-Papilion.: 642 (1971).

Dolichos unguiculatus L., Sp. Pl.: 725 (1753).
Vigna sinensis (L) Hassk., Cat. Pl. Hort. Bogor.: 279 (1844); F.M.S.: 85 (1877).

Annual or perennial, trailing or climbing herbs. Leaflets 3, ovate, rhomboid or lanceolate, lateral leaflets oblique, up to 16.5 × 12.5 cm, although usually much less, apex acute or acuminate, all entire or terminal leaflet 3-lobed at the base and the laterals lobed on the outer margin, glabrous or sparsely pubescent. Inflorescences axillary, few-flowered; bracts soon falling; bracteoles persistent, spathulate, 3–5 mm long. Calyx up to 2 cm long, upper lip 2-lobed, ciliate. Standard white, yellow or pale purple, circular, up to 3.3 cm long, glabrous; wings blue; keel white or pale blue. Pods linear-cylindrical, up to 10 × 1.1 cm; seeds white to dark red or black, often mottled with black or brown, oblong or kidney-shaped, 3.5–5(–12) × 2–3.5 (–6.5) × 2.2 (–4.5) mm. Fig. 15/17.

Distr. ALDABRA: W; ASTOVE; pantropical, several varieties cultivated.
Notes. Found only around Settlement in Aldabra, *Fosberg* 49528 (US) and Astove, *Ridgway* 46 (US).
Vernac. 'cow pea'.

25. BREXIACEAE

Trees or shrubs. Leaves usually alternate, simple, leathery; stipules present or absent. Flowers axillary, solitary or in cymes or false umbels. Sepals 4–6, margins touching or overlapping, persistent or falling. Petals 4–6, margins overlapping and clawed or margins touching, persistent or falling. Stamens 4–6, hypogynous to perigynous, free, anthers large, opening lengthwise. Ovary superior, 4–7-celled, ovules 2-many per cell. Fruit a capsule, drupe or berry.

A small family of 3 genera.

BREXIA Thouars

Small, glabrous trees or shrubs. Leaves alternate, entire or spinous toothed; stipules tiny, falling. Flowers in umbel-like cymes, sometimes on the old wood. Sepals with margins overlapping, falling. Petals leathery, spreading, margins overlapping and twisted in the bud. Stamens 4–5, slightly perigynous, inserted between the lobes of the disk; filaments slightly dilated near the base. Ovary elongate-ovoid, 5–10-angled, 5–7-celled, ovules numerous per cell; style stout, simple, stigma 5–7-lobed. Fruit tough, 5–7-celled or becoming 1-celled, indehiscent; seeds many.

1 species in Madagascar, Seychelles and East African coast.

Brexia madagascariensis (Lam.) Ker-Gawl., Bot. Reg. 9, t. 730 (1823); F.M.S: 97 (1877); Hemsley in J. Bot. 54, Suppl 2: 13 (1916); F.T.E.A. Brex.: 1. Fig. 1 (1968).

Venana madagascariensis Lam., Illustr. Gen. 2: 99 (1797), t. 131 (1792).

Evergreen shrub or small tree, 2–6 (–10) m high. Leaves variable, even on the same plant, narrowly oblong or linear-oblong to broadly obovate, 3.5–35 × 2–7.5 cm, apex rounded, base rounded or tapering, margins entire, obscurely or spinously toothed, (the sole Aldabra specimen has entire margins); petiole 1–2 cm long. Flowers in clusters of 3–12; peduncles flattened 1–9 cm long; pedicels up to 2 cm long. Sepals ovate, 2.5 × 3.5–4 mm, rounded. Petals thick, greenish or yellowish-white, elliptic-oblong, 12–17 × 9–12 mm, obtuse. Fruit ovoid, oblong-spindle-shaped or cylindrical, 4–10 × 1.9–3 cm, prominently 5-ribbed, woody; seeds brown or blackish, irregularly compressed-ellipsoid, 4.5–7.5 × 3–3.5 mm, keeled. Fig. 17/1–4.

The Madagascan material requires revision in order to determine the validity of the species recognised by Perrier in Bull. Soc. Bot. Fr. 80: 198 (1933). The Seychelles specimens may be subspecifically distinct.

Distr. ALDABRA: S; Tanzania, Zanzibar, Madagascar, Comoro Is., Seychelles.

Hab. A single, stunted bush found growing in a pit in very rough coastal champignon between Cinq Cases and Takamaka, *Renvoize* 957 (K, US). The fruits are capable of floating in the sea for several months without the seeds losing their viability.

Vernac. 'bois cateau' or 'bois cato'.

CRASSULACEAE

The widespread succulent herb *Bryophyllum pinnatum* (Lam.) Oken is known from a number of low coral islands in the Western Indian Ocean, Farquhar, Coetivy, Amirantes etc, but surprisingly has not yet been discovered in the Aldabra group although it may be expected to occur there. It is readily recognised by the lax inflorescence of pendulous, 4-merous flowers with a red, cylindrical corolla tube up to 3 cm long.

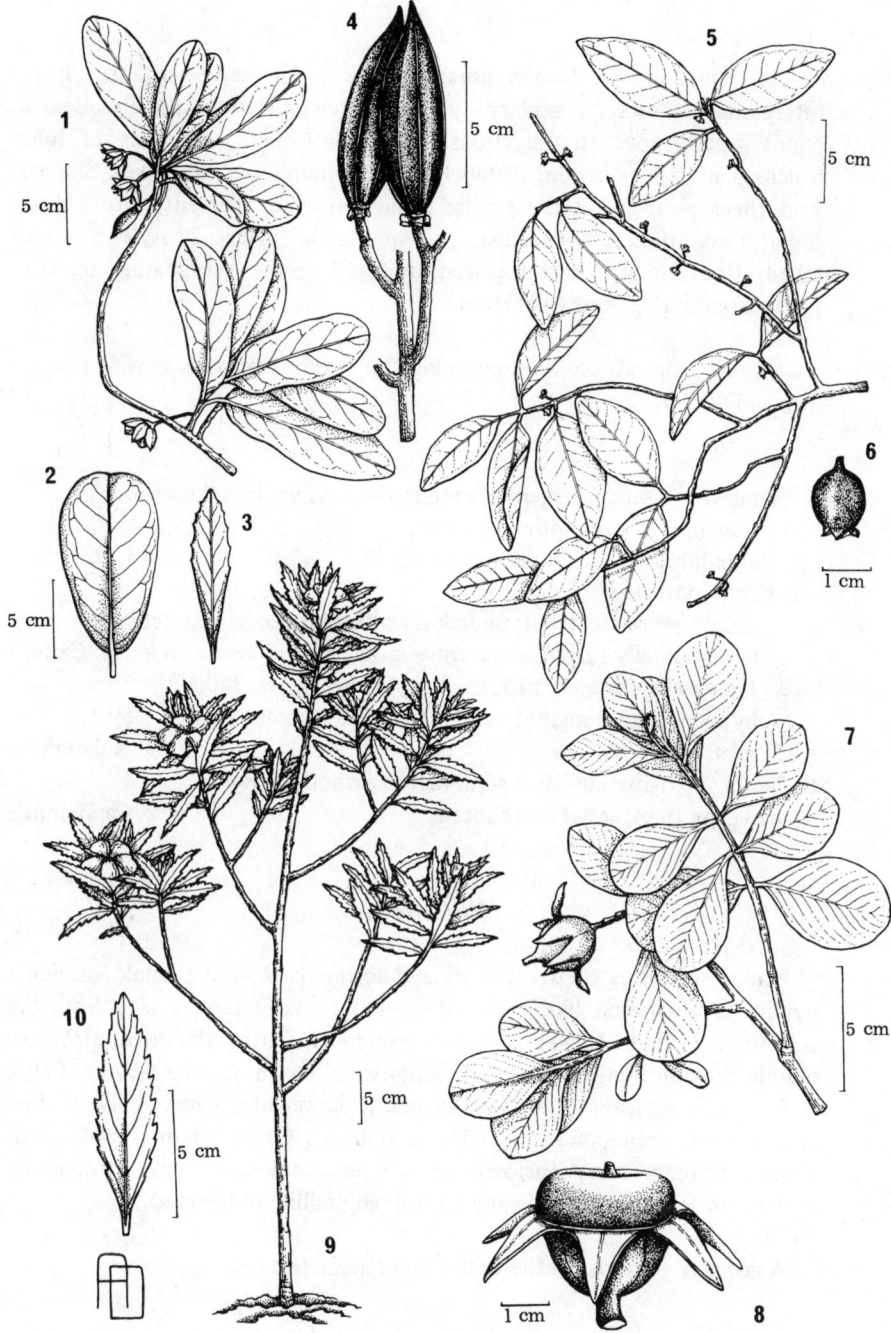

Fig. 17 1, *Brexia madagascariensis*, habit; 2 & 3, leaves; 4, fruit. 5, *Eugenia elliptica* var. *levinervis*, habit; 6, fruit. 7, *Sonneratia alba*, habit; 8, fruit. 9, *Turnera ulmifolia*, habit; 10, leaf.

26. RHIZOPHORACEAE

Trees and shrubs. Leaves usually opposite, simple, persistent; stipules interpetiolar. Flowers in axillary, cymose clusters or solitary, usually bisexual, regular, perigynous to epigynous. Calyx 3—14-lobed, margins of lobes touching in bud, persistent. Petals free, same number as the sepals. Stamens 2—4 times as many as the petals; anthers 4-celled, filaments short. Ovary superior to inferior, 1—5-celled, placentas axile, ovules 2; style 1, stigma lobed. Fruit usually baccate; seed in some genera germinating on tree, producing floating, fleshy embryos.

A small tropical family, containing the true mangroves, as well as some dry-land genera.

1. Plants viviparous, seed germinating in the fruit while still on the
     tree; mangrove trees with aerial roots
   2. Calyx lobes 10—15; aerial roots knee-like   **1. Bruguiera**
   2. Calyx lobes 4—6
      3. Leaves obovate, apex rounded; root-like hypocotyl distinctly
           longitudinally ridged; aerial roots stout, finger-like   **3. Ceriops**
      3. Leaves elliptic; apex with an abrupt spiny tip; root-like
           hypocotyl not longitudinally ridged; aerial roots as prop
           roots only   **4. Rhizophora**
1. Plants not viviparous, seed germination orthodox; not
     mangrove trees, aerial roots absent   **2. Cassipourea**

1. BRUGUIERA Lam.

Mangrove shrubs or trees; roots and lower parts of the trunk forming a pyramidal buttressed base; knee-like aerial roots present, arising from the surrounding mud. Leaves opposite, evergreen, leathery, entire, glabrous, petiole present. Flowers bisexual, solitary or few; bracteoles absent. Calyx 8—15-lobed, persistent, lobes awl-shaped to lanceolate, acute. Petals 2-lobed at the apex, each petal embracing 2 stamens. Ovary inferior, 2—4-celled. Fruit a leathery berry enclosed by the persistent calyx; seed germinating in the fruit and both seedling and fruit finally falling to the mud.

A genus of 6 species in the Indian and Pacific Oceans.

Bruguiera gymnorhiza (L.) Lam., Encycl. Méth. Bot. 4: 696 (1796); F.M.S.: 110 (1877); Hemsley in B.M.I.K. 19: 121 (1919); F.M.C. 150: 34, fig. 10 (1954); F.T.E.A. Rhizoph.: 6, fig. 3. (1956).

Rhizophora gymnorhiza L., Sp. Pl.: 443 (1753).

Shrubs or trees, up to 10 m tall. Leaves borne at the apex of the branches, dark green, elliptic, 5–15 × 3–6 cm, apex acute; petiole 2–3 cm long. Flowers solitary, sub-terminal; peduncles 1 cm long. Calyx pale green when young, becoming bright red at maturity, lobes 10–15, linear, 1.5–2 cm long, leathery, glabrous, fused at the base. Petals white, soon turning brown, oblong, 1.5 cm long, membranous, margins hairy, apex 2-lobed, each lobe with 3 or 4 short bristles and a single bristle from the sinus. Fruit obconical, the root-like hypoctyl cigar-shaped, blunt, up to 25 cm long, slightly longitudinally ridged or angular.

Distr. ALDABRA: W, P, M, S, Esprit, Moustique, Michel; COSMOLEDO: M; ASTOVE; an Indo-western Pacific mangrove. Fig. 18/1–3.

Notes. A constituent of the lagoon mangrove community; flowers throughout the year.

Vernac. 'grand manglier' or 'manglier latte'.

2. CASSIPOUREA Aubl.

Trees or shrubs; aerial roots absent. Leaves opposite, leathery or membranous; stipules interpetiolar, falling early. Flowers solitary or in clusters. Calyx deeply or slightly 4–7-lobed, bell-shaped or the lobes spreading, hairy or glabrous. Petals free, narrowly to broadly spathulate, apex divided or lacerated, inflexed, glabrous or hairy. Stamens 8–40. Ovary superior to sub-inferior, 2–4-celled. Fruit a dehiscent capsule; seeds 2–4, aril present.

A genus of c 80 species in tropical America, the West Indies, tropical Africa, Mascarenes and India.

Cassipourea thomassetii (Hemsley) Alston in B.M.I.K. 1925: 255 (1925). Type: Aldabra, *Thomasset* 224 (K, holo.).

Weihea thomassetii Hemsley in J. Bot. 54, Suppl. 2: 14 (1916) & in B.M.I.K. 1919: 121 (1919).

Shrubs or small trees up to 4 m tall; branchlets grey-brown, leafscars semi-circular, raised. Leaves elliptic-lanceolate to slightly obovate, 5–11 × 2–6 cm, apex blunt to acuminate, margins entire and wavy or toothed towards the apex, somewhat greyish-green and glabrous above, brown and very sparsely hairy below, midrib yellow above, yellow-green, prominent below; petiole 5 mm long. Flowers small, borne in small, axillary clusters; pedicels 3 mm long, hairy. Calyx deeply 4–5-lobed, lobes ovate-lanceolate, 4 mm long, spreading or reflexed at maturity, shortly hairy. Petals narrowly spathulate, 5 mm long, divided at the apex. Stamens 15, as long as the petals. Ovary 3-celled, hairy. Fruit not known. Fig. 18/4–5.

Distr. ALDABRA: W, P, M (west); endemic.

Notes. A rather rare constituent of the inland mixed scrub. Observations few but flowering observed during early-mid wet season.

3. CERIOPS Arn.

Mangrove trees or shrubs; roots and lower part of the trunk forming a pyramidal base, aerial roots present as stout, finger-like projections from the mud. Leaves opposite, evergreen, entire, leathery, glabrous. Flowers bisexual, in small cymose clusters; bracteoles paired, fused and cup- like at the base of the flower, persistent. Calyx 5–6-lobed, lobes lanceolate, spreading in fruit. Petals united below, lobes oblong, apex truncate and appendaged. Stamens 10–12. Ovary inferior, 3-celled. Fruit a persistent leathery berry; seed germinating in the fruit, the seedling finally falling to the mud, leaving the fruit on the tree.

A genus of 2 species, in estuaries and on muddy seashores of the Indian and western Pacific oceans.

Ceriops tagal (Perr.) C.B. Robinson in Philipp. J. Sci. 3: 306 (1908); F.T.E.A. Rhizoph.: 5, fig. 2 (1956).

Rhizophora tagal Perr. in Mém. Soc. Linn. Paris 3: 138 (1825).
Ceriops candolleana Arn. in Ann. Nat. Hist. 1: 368 (1838); F.M.S.: 109 (1877); Dupont, Report: 36 (1907); Hemsley in B.M.I.K. 1919: 121 (1919).

Shrubs or trees, up to 15 m tall. Leaves borne at the ends of the branches, elliptic to obovate, 3–8 × 2–5 cm, apex obtuse, base wedge-shaped, glossy dark green above, yellowish-green below; petiole 1.5–2.5 cm long. Flowers 4–8, borne in small terminal cymose clusters; peduncle stout, 15 mm long, pedicels 4 mm long, bracteoles 2 mm long. Calyx reddish-brown, fused only at the base, lobes 5, lanceolate, 4 mm long, apex acute, leathery. Petals white, oblong, 3 mm long, apex truncate with 3 club-shaped appendages. Stamens 10. Fruit obconical, the root-like hypocotyl lanceolate, slender, 25 cm long when fully extended, distinctly longitudinally ridged. Fig. 18/6–8.

Distr. ALDABRA: W, P, M, S, Esprit, Michel, Moustique; ASSUMPTION; COSMOLEDO: M; Indo-western Pacific mangrove.

Notes. An abundant constituent of the lagoon mangrove communities.

Vernac. 'manglier jaune' or 'manglier girofflier'.

4. RHIZOPHORA L.

Mangrove trees or shrubs; aerial roots developed as prop-roots and adventitious roots from upper nodes. Leaves opposite, evergreen, entire, leathery,

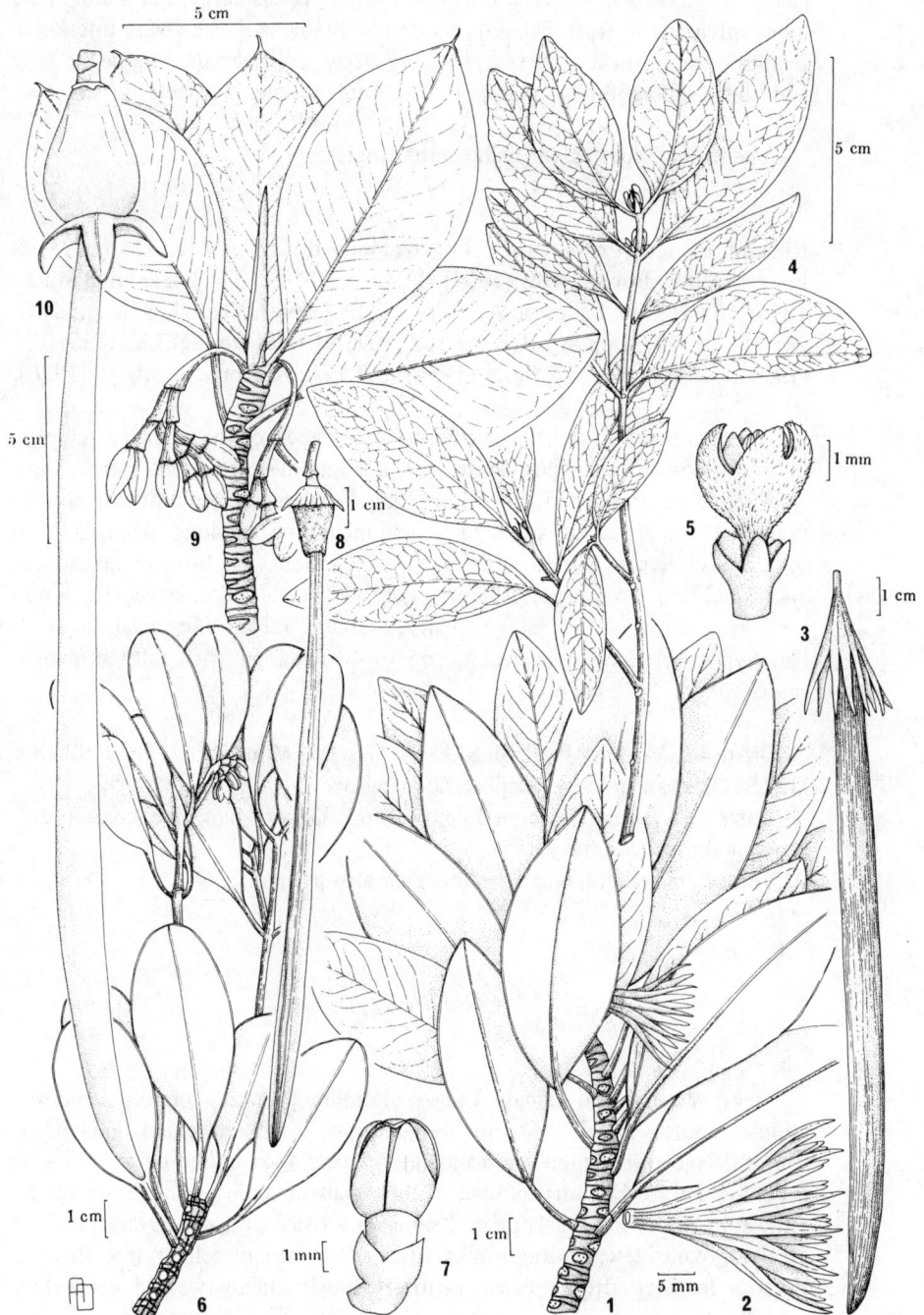

Fig. 18   1, *Bruguiera gymnorhiza*, habit; 2, flower; 3, fruit. 4, *Cassipourea thomassetii*, habit; 5, flower. 6, *Ceriops tagal*, habit; 7, flower; 8, fruit. 9, *Rhizophora mucronata*, habit; 10, fruit.

glabrous. Flowers in clusters; bracteoles paired, persistent. Calyx 4-lobed, the lobes spreading in fruit. Petals 4, lanceolate. Stamens 8–12. Ovary inferior to semi-inferior. Fruit a persistent, leathery berry, seed germinating in the fruit and the seedling finally falling to the mud, leaving the fruit on the tree.

A genus of 7 species, throughout the tropics.

Rhizophora mucronata Lam., Encycl. Méth. Bot. Illustr. t. 396 (1797) & Encycl. Meth. Bot. 6: 189 (1804); F.M.S.: 109 (1877); Baker in B.M.I.K. 1894: 148 (1894); Schinz in A.S.N.G. 21: 88 (1897); Voeltzkow in A.S.N.G. 26: 551 (1902); Dupont, Report: 36 (1907); Hemsley in B.M.I.K. 1919: 121 (1919); F.M.C. 150: 32, fig. 9 (1954); F.T.E.A. Rhizoph.: 2, fig. 1 (1956).

Trees up to 25 m tall. Leaves borne at the apex of the branches, elliptic, 7–17 × 4.5–8.5 cm, apex acute with a sometimes deciduous apical point, glabrous, dark green; petiole 2–3 cm long. Flowers fragrant, borne in stoutly branched, sub-apical axillary cymes; peduncle 3–5 cm long, pedicels 1 cm long. Calyx cream, fused only at the base, lobes 4, broadly lanceolate, 1–1.3 cm long, apex acute, leathery, glabrous. Petals white, lanceolate, 8 mm long, apex acute, fleshy, margins hairy. Stamens 8. Fruit obconical, the root-like hypocotyl slender, lanceolate, up to 40 cm long when fully extended, smooth. Fig. 18/9–10.

Distr. ALDABRA: W, P, M, S, Esprit, Michel, Moustique; COSMOLEDO: M; ASTOVE; an Indo-western Pacific mangrove.
Notes. An abundant constituent of the lagoon mangrove community. Flowers throughout the year.
Vernac. 'manglier hauban' or 'manglier gros poumon'.

27. COMBRETACEAE

Trees, shrubs, and lianas. Leaves alternate or rarely opposite, simple; stipules absent. Flowers usually 4–5-merous, usually bisexual, epigynous, spikes or racemes, simple or paniculate. Sepals 4–8, united below, margins touching. Petals the same number as the sepals or absent, small. Stamens the same number as the sepals or in 2 series and twice as many. Ovary inferior, 1-celled, ovules few, on long stalks attached to apex of cell; style 1, slender. Fruit a leathery drupe, often variously longitudinally winged or keeled; seeds 1.

A small tropical family.

1. Petals present; calyx tubular-bell-shaped, with 2 bracteoles fused to calyx
   1. Lumnitzera
1. Petals absent; calyx cup-like, without bracteoles **2. Terminalia**

1. LUMNITZERA Willd.

Small trees or shrubs. Leaves sessile or subsessile, fleshy or leathery. Flowers bisexual, 5-merous, regular, in short terminal spikes. Calyx produced beyond the inferior ovary to form a tube bearing 2 persistent bracteoles and terminating in persistent lobes. Petals 5, falling early. Stamens 5 or 10. Style slender, persistent. Ovary containing 2–5 ovules. Fruit indehiscent, compressed-ellipsoid, tough, crowned by the persistent calyx.

A genus of 2 mangrove species, from East Africa to Polynesia.

Lumnitzera racemosa Willd. in Neue Schrift. Ges. Naturf. Fr. Berlin 4: 187 (1803); Dupont, Report: 36 (1907); Hemsley in B.M.I.K. 1919: 121 (1919); F.M.C. 151: 3, fig. 1 (1954); F.T. E.A. Combret.: 93, fig. 13 (1973).

Small trees or shrubs to 9 m; bark rough. Leaves spirally arranged, narrowly obovate to elliptic, 2–8 × 1–3 cm, apex rounded, narrowed to the base, yellow-green; sessile. Spikes axillary, 2–7 cm long; flowers sessile or subsessile. Calyx tubular, compressed, 6–8 mm long, contracted just above the bracteoles. Sepals ovate, 1 mm long. Petals white or cream (? sometimes pink) or yellow, elliptic or ovate-elliptic, 4 × 1 mm, glabrous. Stamens 10, equalling or slightly exceeding the petals. Fruit compressed-ellipsoid, 10–12 × 3–5 mm, green.

var. **racemosa**: F.T.E.A.: 95 (1973).

Petals white or cream. Fig. 19/1.

The yellow-petaled var. *lutea* (Gaud.) Excell is known only from Timor.

Distr. ALDABRA: S (east); ASTOVE; an Indo-western Pacific mangrove species.
Notes. Found as low bushes surrounding shallow saline pools or at Takamaka occurs as robust, gnarled trees at the junction of the mangrove swamps and the *Pemphis* scrub. Flowers mainly during the wet season, but unseasonal rain can stimulate flowering.
Vernac. 'manglier à petites feuilles'.

2. TERMINALIA L.

Trees or shrubs. Leaves spirally or alternately arranged, often crowded at the ends of the branches. Flowers small, green or white, in long or short spikes, all bisexual or some male towards the apex. Male flowers apparently stalked through abortion of the ovary, bisexual flowers sessile or subsessile. Calyx cup-like, shallowly 5-lobed. Petals absent. Stamens 10. Ovary inferior. Fruit variable in shape and size, usually woody, sometimes winged or ridged.

A genus of c 200 species, throughout the tropics and subtropics.

1. Small trees up to 5 m; leaves up to 6 cm long; flowers in short spikes up to 4 cm long   **1. T. boivinii**
1. Large, spreading trees up to 10 m; leaves 15–30 cm long; flowers in long spikes up to 15 cm long   **2. T. catappa**

1. Terminalia boivinii Tul. in Ann. Sci. Nat. II, 6: 95 (1856); F.M.C. 151: 59, fig. 15 (1954); F.T.E.A. Combret.: 78 (1973).

T. fatraea sensu Baker in B.M.I.K. 1894: 148 (1894); Schinz in A.S.N.G. 21: 88 (1897); Voeltzkow in A.S.N.G. 26: 551 (1902); Dupont, Report: 37 (1907); Hemsley in B.M.I.K. 1919: 121 (1919), not (Poir.) DC (1828).

Shrubs or small trees up to 3 m high; main shoots and lateral shoots straight, lateral shoots terminating in a spur shoot and bearing 4–6 lateral spurs 2–10 mm long; leaves and inflorescences clustered on spur shoots. Leaves alternate, narrowly obovate to obovate, 2.5–6 × 1–3.5 cm, apex obtuse, somewhat papery, dark green and shiny above, brownish and dull beneath, glabrous; petiole up to 1 cm long. Spikes 2–4 cm long, including the 5–15 mm long peduncles which elongate slightly in fruit. Flowers yellowish, subsessile, all bisexual. Calyx 1 mm long, 3 mm across, glabrous, or nearly so. Fruit pale yellowish-green or-brown, ellipsoid, 8–12 × 4.5–6 mm, glabrous not ridged. Fig. 19/2–3.

Although very similar to *T. fatraea* (Poir.) DC. from Madagascar, with which it is often confused, it is readily distinguished by its smaller fruits, without ridges.

Distr. ALDABRA: W, P, M, S, Michel; ASSUMPTION, ASTOVE; Kenya, Tanzania, Pemba, Zanzibar, Mozambique and Madagascar.
Notes. An important constituent of the mixed scrub communities. Flowers mainly during the wet season but can flower throughout the year if sufficient moisture available; may become leafless during times of drought. Fruit eaten and locally dispersed by pigeon, bulbul and tortoise.
Vernac. 'bois faune'.

2. Terminalia catappa L., Mant. 2: 519 (1771); F.M.S.: 111 (1877); Dupont, Report: 37 (1907); F.M.C. 151: 63 (1954); F.T.E.A. Combret.: 74 (1973).

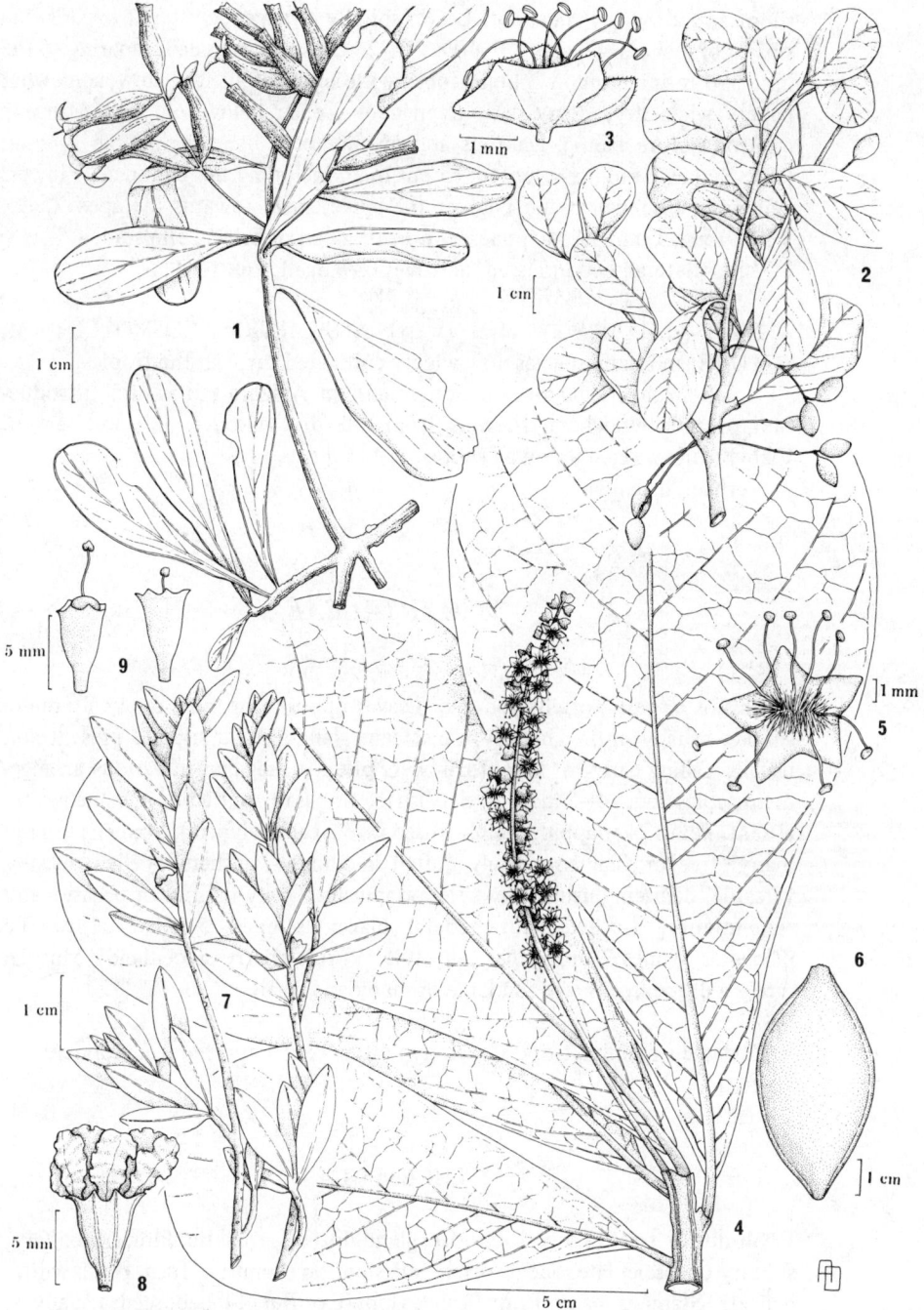

Fig. 19 1, *Lumnitzera racemosa*, habit. 2, *Terminalia boivinii*, habit;
3, flower. 4, *Terminalia catappa*, habit; 5, flower; 6, fruit.
7, *Pemphis acidula*, habit; 8, flower; 9, long and short style forms.

Spreading tree, up to 10 m high; branches forking to give an upper short shoot and a lower long shoot. Leaves spirally arranged, clustered towards the tips of branches, obovate, 15–40 × 10–20 cm, apex rounded, tapering to the base and terminating in 2 blunt lobes with a pair of glands below, somewhat papery when dry, glossy dark green above, paler below, old leaves turning crimson before falling, glabrous above, pubescent beneath; petiole 1–2 cm long. Spikes pendulous, up to 15 cm long, including the peduncle. Flowers yellowish, bisexual at the base of the spike, male towards the apex. Calyx 3 mm long, 1 mm across, pubescent. Fruit green or brown, ellipsoid, 4.5–6 × 3.5 cm, bilaterally compressed, narrowly 2-winged. Fig. 19/4–6.

Distr. ALDABRA: W, M, S, (west); ASSUMPTION; COSMOLEDO: M; ASTOVE; an asiatic species now widely cultivated through the tropics.

Notes. Probably planted at Settlement on Aldabra but natural introduction possible elsewhere. Flowers from late dry season to late wet season. Viable fruits washed up on all shores.

Vernac. 'badamier'.

28. MYRTACEAE

Shrubs or trees; often aromatic. Leaves opposite or more rarely alternate, simple, usually entire, usually translucent gland-dotted; stipules present and usually falling early, or absent. Flowers bisexual, regular, in various arrangement, epigynous, 4–5-merous. Sepals free or basally united or in some genera absent. Petals overlapping, rarely fused into a cap (calyptra). Stamens usually many, free or filaments rarely united in bundles; anthers 2-celled, usually versatile, dehiscing usually by slits, or apically. Ovary inferior or occasionally semi-inferior, 1-celled with parietal placentas or 4–5-celled with axile placentas, ovules 2–many per cell; style 1. Fruit a berry or loculicidal capsule, rarely a drupe; seeds few and large or many and small.

A large, mainly tropical family, often an important component of vegetation.

EUGENIA L.

Shrubs or trees. Leaves opposite, gland-dotted, aromatic. Flowers axillary, solitary or fasciculate. Sepals often persistent on summit of fruit. Petals white, delicate. Stamens many. Fruit fleshy, globose or fluted-lobed; seeds usually 1, large.

Pantropical genus, even in a narrow sense including numerous illdistinguished species.

Eugenia elliptica Lam., Encycl. Méth. Bot. 3: 206 (1789); Fosberg in K.B. 31: 133 (1978).

E. cotinifolia sensu F.M.S.: 114 (1877) partly, not Jacq. (1768).

Known from Mauritius but is probably more widespread. It is characterized by oval-elliptic, nearly sessile leaves, prominent venation and 4–7 pedicels per node.

var. **levinervis** Fosberg in K.B. 31: 134 (1978). Type: Aldabra, Takamaka Well, *Fosberg* 49342 (US, holo., A, EA, K, L, MO, NY, P, all iso.).

Eugenia sp.; Grubb in P.T.R.S. B, 260: 359 (1971).

Glabrous shrub; young internodes faintly quadrangular. Leaves broadly elliptic, up to 10 × 5.7 cm, apex more or less acute to obtuse, base somewhat rounded, margins entire, thick-papery, sometimes folded, veins very faint, intra-marginal vein undulate; petiole 3–5 mm long. Flowers 1–2 per axile; pedicels 3–4 (–8) mm long. Calyx lobes 4, orbicular-ovate, strongly glandular-punctate, falling from fruit. Petals not seen. Stamens many (as estimated by scars on disk), apparently short (only very few available); anthers twisted when dry. Fruit purplish black, globose, c 1 cm diameter; seed 1, globose (not seen). Fig. 17/5–6.

Distr. ALDABRA: S; endemic.

Notes. A rare constituent of scrub communities. Known from the type locality at Takamaka, *Fosberg* 49342 & 49349 (both K & US), *Fosberg & Grubb* 49601 (US), *Renvoize* 1079 (K, US) and *Hnatiuk* 731926 (US); also south of Anse Cédres, *Gibson* 1 (K). Very susceptible to woolly coccid infestation.

29. LYTHRACEAE

Herbs, shrubs and trees. Leaves opposite, simple, usually entire; stipules usually absent. Flowers variously arranged, often axillary, usually regular, bisexual, perigynous. Sepals 4–8, on edge of a tubular receptacle. Petals alternate with the sepals (or absent), often clawed, blade crumpled in bud, usually crumpled, like tissue-paper, when expanded. Stamens usually twice the number of petals, 1 whorl alternate with them, (rarely reduced or many). Ovary superior, 2–6-celled, ovules several to many per cell; style and stigma 1. Fruit a capsule; seeds without endosperm.

Small family, mostly tropical, containing a few commonly planted ornamentals, several widespread weeds.

PEMPHIS J.R. & G. Forst

Small trees or shrubs, much branched. Leaves small, entire. Flowers axillary. Sepals 6, fused below, with a spur between each 2 calyx teeth. Petals 6, free, obovate-oblong. Stamens 12, in 2 ranks. Ovary globose; style slender, stigma capitate. Capsule globose, enclosed by the persistent calyx, splitting transversely near the apex when ripe.

A genus of only 2 species, 1 of which is widespread.

Pemphis acidula Forst., Char. Gen.: 68, fig. 34 (1776); F.M.S.: 101 (1877); Baker in B.M.I.K. 1894: 148 (1894); Schinz in A.S.N.G. 21: 88 (1897); Voeltzkow in A.S.N.G. 26: 551 (1902); Dupont, Report: 36 (1907); Hemsley in B.M.I.K. 1919: 122 (1919); F.M.C. 147: 20 (1954).

Much-branched, dense, small trees or shrubs, up to 5 m tall, usually much less; trunk and branches erect, decumbent or scrambling, often gnarled; bark rough, black or greyish, branchlets pubescent, growth rings often present. Leaves tending to be crowded towards the ends of the branchlets, ovate-lanceolate to obovate, 10–35 × 3–10 mm, apex acute, grey-green, leathery or slightly fleshy, pubescent; petiole 1 mm long. Flowers solitary, pedicels up to 5(–12) mm long, pubescent. Calyx teeth short, broad, 1 mm long, apex acute. Petals white or cream, 5 mm long, falling early. Capsule, including hypanthium, 5–8 mm long, cap dark brown; seeds reddish-brown, wedge-shaped, 3 × 2 mm, winged. Fig. 19/7–9.

Professor D. Lewis of University College, London, has recently studied heterostyly in this species; he has recorded 3 different style lengths for Aldabra.

Coastal forms of this species can be confused with *Suriana maritima* (Family 15), which has yellow flowers and alternate, spathulate leaves. The two species may sometimes be found growing together.

Distr. ALDABRA: W, P, M, S, Esprit, Michel, Moustique; ASSUMPTION; COSMOLEDO: W, M; ASTOVE; an Indo-Pacific strand plant.

Notes. A major constituent of coastal and inland scrub, often forming pure stands on champignon. There are notable differences in habit between populations on Aldabra. Near the coast the plants are generally low, prostrate bushes, sometimes with fleshy leaves. Inland they are straggling, erect trees, always with leathery leaves. Can flower throughout the year provided sufficient moisture available.

Vernac. 'bois d'amande'.

30. SONNERATIACEAE

Trees or shrubs. Leaves opposite, simple, entire, leathery; stipules absent. Calyx leathery, 4–8-lobed, lobes with margins touching, not overlaping, persistent in fruit. Petals free, alternate with the calyx-lobes, or absent. Stamens 12 to many, inflexed in bud; anthers 2-celled. Ovary superior or partly inferior, enclosed by the calyx-tube, 4–many-celled, ovules numerous, placentation axile; style 1, elongate, stigma capitate. Fruit a berry or capsule; seeds many, small.

A small Old World family of 2 genera.

SONNERATIA L.f.

Trees or shrubs. Leaves glabrous, leathery. Calyx 6–8-lobed, lobes as long as the tube, margins touching, not overlapping in bud. Petals 6–8, not exceeding the calyx-lobes, broad and wrinkled or narrow and smooth. Stamens numerous. Ovary depressed-globose, fused at the base with the calyx tube, multicelled, ovules numerous per cell; style straight with a subcapitate stigma. Fruit a multicelled berry, free at maturity from the enveloping calyx, and usually borne on a short stipe; seeds many.

A small genus of 5 species, occurring in mangrove swamps from eastern Africa to the western Pacific.

Sonneratia alba Sm. in Rees, Cycl. 33, No. 2 (1816); F.M.C. 148: 2, fig. 1 (1954); F.T.E.A. Sonnerat.: 1, fig. 1 (1968).

S. acida sensu F.M.S.: 102 (1877); Dupont, Report: 36 (1907); Hemsley in B.M.I.K. 1919: 122 (1919), not L.f. (1781).

Evergreen shrub or tree, up to 20 m; pneumatophores stout, finger-like. Leaves obovate, oval or almost circular, 4–12 × 2–10 cm, apex rounded or notched, tapering at the base, midrib prominent; petiole stout, 5–10 mm long. Flowers solitary or in 3's at the shoot apex, scented. Calyx green, 2.5–3.5 cm long, tube somewhat expanded, almost bell-like, lobes 6–8, magenta-pink inside, green outside, shorter or longer than the tube, becoming reflexed in fruit. Petals white or tinged pink, soon falling, strap-shaped, 13–20 mm long, inconspicuous and resembling the showy, white filaments. Ovary 14–18-celled. Fruit green, depressed-globose, 2–3 × 3–4 cm, bearing the persistent style base; seeds wedge-shaped, 2 × 2 × 0.7 mm. Fig. 17/7–8.

Distr. ALDABRA: W, Esprit; COSMOLEDO: M; an Indo-western Pacific mangrove.
Notes. A rate constituent of the mangrove swamps. On Aldabra recorded only from Passe Femme and Ile Esprit. From the few observations available

appears to flower throughout the year. The flowers open at night, on the following morning the petals and stamens fall.

Vernac. 'manglier fleurs'.

31. TURNERACEAE

Herbs or shrubs, rarely trees. Leaves alternate, simple, often glandular; stipules small or absent. Flowers bisexual, regular, perigynous. Sepals, petals and stamens 5, inserted on hypanthium. Ovary superior, 1-celled, placentation parietal, ovules many; styles 3, stigmas fringed. Fruit a loculicidal capsule; seeds many, arillate.

A small, tropical, mostly American family.

TURNERA L.

Shrubs or herbs. Leaves entire, or margins serrate or deeply divided, often bearing 2 glands at the junction of the blade and petiole; stipules small or absent. Flowers axillary, solitary, rarely clustered, yellow. Calyx tubular to bell-shaped, 5-lobed. Petals 5, free, inserted at the base of the calyx lobes.

A genus of 60 species restricted to tropical and sub-tropical America, with the exception of 1 widespread species.

Turnera ulmifolia L., Sp. Pl.: 271 (1753); F.M.S.: 104 (1877); F.M.C. 142: 3 (1950).

Shrubby, perennial herb, up to 1.5 m high. Leaves broadly lanceolate to ovate-lanceolate, 5–10 × 1–2(–2.5) cm, apex acute, attenuate at the base, margins toothed, pubescent above and beneath; petiole 5–12 mm long, 1 pair of glands at the junction with the blade, 1 gland midway. Flowers solitary, their pedicels fused to the adjacent petioles. Calyx tubular, 15–25 mm long, divided for half its length into 5 pointed lobes, pubescent. Petals yellow, obovate, 1.5–2.5 cm long, apex sharply pointed, indented or rounded, spreading. Capsule orbicular, up to 1 cm long, dehiscent; seeds numerous, yellow-brown, irregularly oblong, 2–3 mm long, longitudinally ribbed Fig. 17/9–10.

Distr. ALDABRA, W, M (west), S (west); COSMOLEDO; ASTOVE; an American weed, now widespread through the tropics.

Notes. A weed of disturbed ground around habitation. Known on Aldabra from Settlement, Anse Mais, and western Middle Island. Flowers throughout the year if moisture available. Attempts are being made to eradicate this weed.

Vernac. 'coquette' or 'la coquette'.

32. PASSIFLORACEAE

Usually climbers, sometimes erect trees or shrubs, usually with tendrils axillary or opposite the leaves. Leaves alternate, usually simple; stipules small, sometimes falling. Flowers regular, axillary, usually bisexual, 4- or usualy 5-merous. Perianth differentiated into calyx and corolla, corolla rarely lacking; perianth with a usually multifid corona present on a hypanthium. Stamens 5 or more. Ovary borne on a short stalk (gynophore) which sometimes also bears the stamens (androgynophore), 1-celled, with 3 parietal placentae, ovules many; styles 3–5, spreading or united, stigmas capitate or discoid. Fruit a berry or capsule; seeds many, usually arillate.

A medium-sized tropical family, a number of species with edible fruit, some ornamentals, noted for their complicated flowers.

PASSIFLORA L.

Herbaceous or woody climbers with axillary tendrils, rarely herbs, shrubs or small trees. Leaves usually alternate, simple, often lobed; stipules minute to large. Flowers with a complex structure, solitary, paired or rarely clustered; bracts and bracteoles present. Calyx fused at the base or forming a long or short tube, 5-lobed. Petals 5, borne at the base of the calyx lobes and alternating with them. Corona present, consisting of a complex series of filaments of various shapes; nectary flap sometimes present within the corona. Stamens 5, spreading above, usually fused below and enclosing within a tube the stalk (gynophore) bearing the ovary. Ovary globose or ovoid; styles 3, spreading. Fruit an indehiscent berry.

A genus of c 500 species, mostly confined to the Americas.

1. Bracts and stipules finely divided, glandular; flowers large and showy, up to 5 cm across; fruit globose, 3 cm in diameter, orange when ripe
   1. P. foetida
1. Bracts and stipules simple; flowers small, up to 1.5 cm across; fruit globose, 1–1.3 cm in diameter, black   **2. P. suberosa**

1. Passiflora foetida L., Sp. Pl.: 959 (1753); F.M.S.: 105 (1877); F.M.C. 143: 47 (1945); F.T.E.A. Passiflor.: 13 (1975).

Herbaceous climber; stems glabrous; stipules clasping, deeply divided, glandular. Leaves suborbicular to ovate, up to 12 × 12 cm, shallowly to deeply 3-lobed, glabrous except for margins often with glandular hairs; petioles up to 6 cm long. Flowers large, 2–5 cm across, solitary; bracts conspicuous, finely divided, glandular. Calyx fused at the base, calyx lobes and petals similar, white, obovate-oblong, membranous. Corona of 3 or more whorls of white and purple filaments, 2 outer whorls long, the innermost short; nectary flap membranous. Fruit a large orbicular berry, 3 cm in diameter, orange, green or dark red when ripe.

var. **hispida** (DC. ex Triana & Planch.) Gleason in Bull. Torrey Bot. Club 58: 408 (1931): F.T.E.A. Passiflor.: 14 (1975).

Plant with long, rather rigid hairs; leaves usually 3-lobed; divisions of bracts very numerous, appearing interwoven; ovary glabrous; fruit orange. Fig. 20/1–2.

This is an extremely variable species and the above description is especially applicable to plants in our area.

Distr. ASSUMPTION; native in tropical America introduced elsewhere.
Notes. Found around settlements sites. Two records from Assumption, *Stoddart* 1076 and *Gwynn & Wood* 1332 (both K).
Vernac. 'poc-poc'.

2. **Passiflora suberosa** L., Sp. Pl.: 958 (1753); F.M.S.: 105 (1877); F.M.C. 143: 47 (1945); F.T.E.A. Passiflor.: 18 (1975).

Herbaceous climber; subglabrous, the lower part of the stems with a thin corky outer layer; stipules entire, narrow and tapering, 6–8 mm long. Leaves very variable in shape, simple, entire, ovate or elliptic oblong, 4–9 × 1.5–5.5 cm or 3-lobed, 3.5–9 × 3.5–9 cm, lobes acute and with a sharp point; petioles 0.7–2 cm long, bearing 2 small, stalked glands near the blade. Flowers small, 1–1.5 cm across, solitary or in pairs. Sepals greenish-yellow, strap-like, fused at the base, reflexed. Petals absent. Corona of two whorls of filaments, white at the apex, purple below; nectary flap membranous. Stamens fused in the lower part to form a tube surrounding the gynophore. Fruit an orbicular berry, 1–1.3 cm in diameter, black when ripe. Fig. 20/3–4.

The above description was prepared from specimens from the Indian Ocean Islands and may not include the range of variation found in this species elsewhere.

Distr. ALDABRA: W, P, M (west), S (west), Esprit; ASSUMPTION; COSMOLEDO; native in tropical America, introduced elsewhere.
Notes. Found on Aldabra near areas of habitation on West Island, rarely in eastern part of South Island, its general distribution at the western end of the atoll suggests possible bird dispersal from an introduction at Settlement. Can flower and fruit throughout the year if sufficient moisture available. Fruit eaten and seed locally dispersed by bulbul and fody, occasionally by coucal. The distribution of the butterfly *Acraea ranavalona* on Aldabra is closely associated with that of *Passiflora suberosa*.

33. CARICACEAE

Soft-wooded trees or shrubs; milky latex present throughout; sparsely branched or unbranched. Leaves large, spirally arranged, in terminal rosettes; stipules absent. Flowers bisexual or unisexual, often several sex-forms on the same plants. Calyx small, 5-lobed. Petals 5, free or united. Stamens usually 10, in 2 whorls, or 1 whorl of 5. Ovary superior (or inferior?), 1-celled, with parietal placentas, ovules many; styles 5, wedge-shaped, spreading. Fruit a large berry or pepo; seeds many.

A small mostly tropical American family with 1 widely distributed species, the papaya, generally naturalized as well as cultivated for its fruit throughout the tropics.

CARICA L.

Trees or shrubs; stems thick, soft, spongy. Leaves large, subpeltate, palmate to digitately 7—9-foliolate; petioles long. Flowers unisexual, occasionally bisexual. Calyx small, 5-lobed. Male flowers: petals united into an elongated tube; stamens 10. Hermaphrodite flowers: resembling male flowers but with ovary. Female flowers: petals free and readily falling, ovary sessile; styles absent, stigmas arising directly from the top of the ovary. Fruit a fleshy berry.

A genus of 45 species.

Carica papaya L., Sp. Pl.: 1036 (1753); F.M.S.: 107 (1877); F.T.E.A. Caric.: 1 (1958).

Stout, sparsely branched trees, up to 10 m high, trees male or female or with various combinations of male, female and bisexual flowers. Leaves 25—75 cm wide, deeply lobed, the lobes deeply and broadly toothed, glabrous; petiole 25—100 cm long, hollow. Male flowers: 2.5 cm long, clustered on long pendulous stalks, calyx cup-shaped, 1 mm long, petals creamy yellow, fused for their length, lobes 5, spreading. Hermaphrodite flowers in clusters with male flowers. Female flowers: 3.5—5 cm long, axillary, shortly stalked, solitary or in few-flowered clusters; calyx cup-shaped, 3—4 mm long; petals free, creamy white and fleshy, linear-oblong; stigmas deeply digitately lobed. Fruit large and heavy, green, turning yellow when ripe, orbicular, ovate-oblong or oblong, up to 30 cm long, flesh orange, edible; seeds many, ovoid, 7 × 4 mm. Fig. 20/5—6.

Distr. ALDABRA: W; tropical America, widely cultivated throughout the tropics.

Notes. Found on Aldabra only at Settlement, where it has been introduced. Flowers during the late wet season, fruits ripen mid-late dry seaon.

Vernac. 'paw-paw' or 'papaya'.

34. CUCURBITACEAE

Mostly postrate or climbing herbs, rarely woody climbers or very rarely small trees; tendrils usually present. Leaves alternate, simple, variously lobed or divided, generally palmately veined; stipules absent. Flowers axillary, solitary or in cymose clusters or variously arranged, unisexual, monoecious or dioecious, 5-merous, regular. Calyx of 5 variously united sepals. Petals united or free. Stamens 5, variously modified and variously united; anthers variously united or free, often twisted and variously dehiscent. Ovary inferior of 3—5 carpels, 1-celled, placentation parietal, or by union of the placentas 3-celled and placentation appearing axile, or entire cavity filled; ovules 1—many; style 1, rarely 3, branched or stigma lobed. Fruit a berry, often very large, tough or soft-skinned, indehiscent or variously dehiscent; seeds 1—many, usually large, sometimes very large.

A medium-sized, largely tropical family, very diverse, with curious evolutionary modifications, difficult to classify; contains many economic species.

1. Petals united; fruit smooth or spiny
  2. Petals entire, yellow
    3. Petals united in their lower part   **1. Cucumis**
    3. Petals united for most of their length   **2. Cucurbita**
  2. Petals fringed, white   **6. Trichosanthes**
1. Petals free; fruit smooth or with wart-like projections
  4. Receptacle tube short and broad; fruit with wart-like projections
  4. Receptacle tube elongated; fruit smooth   **4. Momordica**
    5. Flowers yellow   **5. Peponium**
    5. Flowers white   **3. Lagenaria**

1. CUCUMIS L.

Climbing or trailing herbs; stems annual, rough, hairy. Leaves simple, palmately lobed; tendrils simple. Flowers yellow, monoecious, rarely dioecious. Male flowers solitary or in sparse clusters; receptacle-tube bell-shaped with small, slender lobes; petals united in their lower part; stamens 3, 2 paired and 1 single inserted in the middle of the tube. Female flowers usually solitary; ovary inferior, hairy, perianth as in the male flowers, 3 staminodes often present, stigma 3-lobed. Fruit fleshy, indehiscent, smooth, pubescent or spiny.

23 paleotropical species, mostly in Africa.

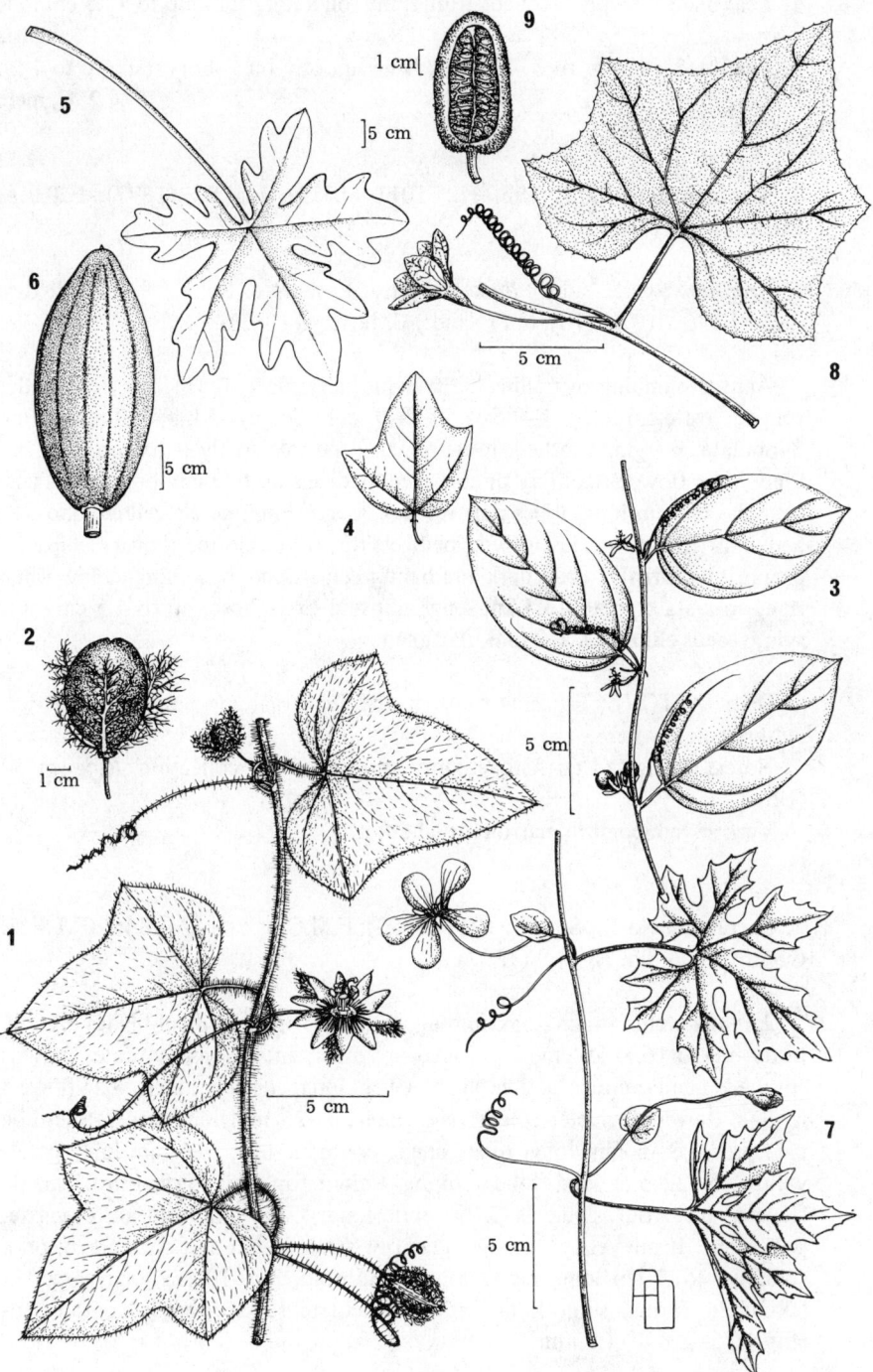

Fig. 20 1, *Passiflora foetida*, habit; 2, fruit. 3, *Passiflora suberosa*, habit;
4, leaf variation. 5, *Carica papaya*, leaf; 6, fruit. 7, *Momordica
charantia*, habit. 8, *Peponium sublitorale*, habit; 9, fruit, sectioned.

1. Leaf-blades deeply 5-lobed; fruit spiny, on a long stalk up to 13.5 cm long
                                                              1. C. anguria
1. Leaf-blades entire to 3–5-lobed; fruit smooth, on a short stalk up to 4 cm
   long   **2. C. melo**

1. Cucumis anguria L., Sp. Pl.: 1011 (1753); F.M.S.: (1877); F.T.E.A.
Cucurb.: 104 (1967).

C. prophetarum L. subsp. *dissectus* sensu Fosberg & Renvoize in Atoll Res.
Bull 136: 107 (1970), not (Naud.) C. Jeffrey (1962).

Annual climbing or trailing herb; stems hairy, up to 2.5 m long. Leaf-blades
rough, ovate-cordate, 4–9.5 × 5.2–9 cm, deeply 5-lobed, lobes often
3-lobulate, margins toothed, roughly hairy; petiole stiffly hairy, up to 12 cm
long. Male flowers: solitary or clustered, pedicels up to 3 cm long; receptacle-
tube 3–3.5 mm long, lobes narrow, 1.5–2 mm long; petals yellow, obovate,
5–6 mm long. Female flowers: pedicels up to 9.5 cm long, ovary ellipsoid,
8 mm long, bristly. Fruit dark and light green striped, becoming yellow when
ripe, on a stalk up to 13.5 cm long, ellipsoid or globose, up to 4.5 cm long,
spiny; seeds elliptic, 6 × 3 mm, flattened.

Distr. ASTOVE; Tanzania south to the Transvaal and South West Africa,
cultivated elsewhere.
Notes. Cultivated on Astove. Only known from 1 collection, *Ridgway* 34
(US).
Vernac. 'concombre marron' or 'gherkin'.

2. Cucumis melo L., Sp. Pl.: 1011 (1753); F.M.C. 185: 147 (1866); F.T.E.A.
Cucurb.: 106, fig. 16/8–9 (1967).

Annual; stems trailing or climbing, up to 1.5 m long. Leaf-blades broadly
ovate, up to 16 × 20 cm, 3–5-lobed or almost entire, base cordate, margins
toothed, membranous; petiole up to 10 cm long, coarsely hairy. Male flowers
in 2–4-flowered clusters; pedicels slender, 3–25 mm long; receptacle-tube
pale green, 3–6 mm long, lobes linear or thread-like, 1–6 mm long; petals
yellow, 5–22 mm long, lobes oblong, united for their lower third. Female
flowers on stout pedicels 3–50 mm long; ovary ellipsoid or elongate,
pubescent. Fruit yellow, orange, green or striped light and dark green, on a
stalk up to 4 cm long, rarely longer, ellipsoid, 3.5–10 cm long, larger in
cultivated forms, smooth or with a reticulate surface; seeds compressed-
elliptic, 5–8 × 1–1.5 mm.

Distr. ASSUMPTION; COSMOLEDO; wild forms in eastern and north-
eastern Africa and in the Indian subcontinent; introduced and cultivated
elsewhere.
Notes. A wild form on Cosmoledo, in low mixed scrub near the beach.
Fosberg 49797 (US), *Renvoize* 1224 (K, US) & 1267 (K).
Vernac. 'melon de France' or 'cantaloupe'.

2. CUCURBITA L.

Climbing or trailing herbaceous plants. Leaves lobed, base cordate; petiole long; tendrils 2—multi-fid. Flowers large, yellow or orange, monoecious. Male flowers: solitary or fascicled, receptacle-tube bell-shaped, 5-lobed; corolla bell-shaped, 5-lobed, lobes recurved at the apex; stamens 3, inserted on the tube. Female flowers: solitary, perianth similar to the male flowers; ovary inferior, oblong. Fruit globose, indehiscent.

25 species including the squashes, pumpkin and marrow, New World in origin, cosmopolitan due to widespread cultivation.

Cucurbita moschata (Duch. ex Lam.) Poir. in Dict. Sci. Nat. 11: 234 (1818); F.T.E.A. Cucurb.: 2 (1967).

C. pepo L. var. *moschata* Duch. ex Lam., Encycl. Méth. Bot. 2: 152 (1786).
C. maxima sensu Fosberg & Renvoize in Atoll Res. Bull. 136: 62 (1970), not Lam. (1786).

Annual; stems long, trailing or climbing, rough to the touch. Leaves large, broadly ovate, up to 13 cm wide, base lobed, margins toothed, herbaceous, scabrid; petiole stout, up to 10 cm or more long, rough to the touch. Male flowers: solitary pedicels slender, up to 6 cm or more long; receptacle-tube green, hairy, lobes linear, 2—3 cm long; petals up to 10 cm long, united for at least half their length. Female flowers: pedicels stout; perianth usually smaller than in male flowers; ovary ovoid. Fruit usually buff, compressed-globose, 30 cm in diameter, flesh yellow to dark orange; seeds dingy white to dark brown, compressed ovate-acute, 20 × 12 mm, margin wavy, hyaline.

Distr. ALDABRA: W; COSMOLEDO: M; native of North America, now widespread throughout the tropics.
Notes. Occurs near areas of settlement and cultivation.
Vernac. 'giraumon' or 'pumpkin'.

3. LAGENARIA Ser.

Climbing or trailing herbs. Leaves simple; petioles with a pair of apical lateral glands; tendrils 2-fid or simple. Flowers large, white, monoecious or dioecious. Male flowers solitary or in spikes; receptacle-tube elongate with small, narrow lobes; petals 5, free, entire; stamens 3, 2 paired and 1 single, inserted on the tube. Female flowers solitary; ovary inferior, hairy; receptacle-tube very short, stigma 3-lobed. Fruit large, fleshy, green, hard-shelled, indehiscent.

A genus of 6 species, mostly in tropical Africa and Madagascar, 1 species pantropical.

Lagenaria siceraria (Molina) Standley in Publ. Field Mus. Nat. Hist. Bot. III: 435 (1930); F.M.C. 185: 103 (1966); F.T.E.A. Cucurb.: 51, fig. 6/9 (1967).

Cucurbita siceraria Molina, Sagg. Chil.: 133 (1782).

Climbing or trailing herb; stems up to 4.5 m long, hairy. Leaves broadly ovate or suborbicular, up to 23 × 23 cm, apex with sharp point, base cordate, entire or slightly 5–9-lobed, margins toothed, softly hairy; petiole up to 12.5 cm long, pubescent, with 2 small apical glands at the junction with the leaf-blade; tendril 2-fid. Monoecious. Male flowers: solitary, pedicels up to 31 cm long; receptacle-tube funnel-shaped, slightly bulbous at the base, pubescent, 1–1.5 cm long, lobes linear or triangular, 3–7 mm long; petals white, opening in the evening, broadly obovate-acute, up to 4.5 cm long. Female flowers: pedicels 6–7 cm long; receptacle-tube 2.5 mm long, lobes acute, 3–3.5 mm long; petals obovate, 3 cm long; ovary ovoid, hairy, 1–1.7 cm long. Fruit green, large, globose or oblong, or in cultivated forms variously shaped, up to 13 cm in diameter, hairy; seeds oblong, compressed, 17 × 7 × 3.2 mm, slightly tapered, slightly 2-horned at the broader end, smooth with 2 flat longitudinal ridges.

Distr. ALDABRA: W; pantropical, but probably introduced outside Africa and Asia.
Notes. Cultivated at Settlement, *Fosberg* 49526 (US).
Vernac. 'bottle gourd'.

4. MOMORDICA L.

Climbing or trailing herbs. Leaves simple or compound; tendrils simple or 2-fid at the apex, usually solitary. Flowers white, cream or yellow, monoecious or dioecious. Male flowers: solitary or clustered; receptacle-tube short and broad, lobes entire; petals 5, free, entire, stamens 2 or 3, 2 paired and 1 single. Female flowers: solitary; ovary inferior, usually ribbed, tuberculate or papillose, stigma 3-lobed. Fruit ovoid, ellipsoid or elongate, fleshy, with wart-like projections, spiny, winged or ridged, indehiscent or dehiscent.

A genus of 42 species in the Old World, mostly in Africa.

Momordica charantia L., Sp. Pl.: 1009 (1753); F.M.C. 185: 30 (1966); F.T.E.A. Cucurb.: 31 (1967).

Annual; climbing or trailing stems up to 5 cm long. Leaves broadly ovate-cordate, up to 10 × 12 cm, deeply 5–7-lobed, margins toothed, membranous, glabrous or pubescent; petiole up to 7 cm long; tendrils simple. Flowers monoecious. Male flowers: solitary, pedicels slender, up to 10 cm long, bearing an orbicular bract half way along its length; receptacle-tube 2–4 mm long, lobes lanceolate, up to 6.5 mm long; petals pale yellow, obovate, up to

22 mm long, 2 with scales at the base; stamens 3. Female flowers: pedicels up to 15 cm long, bearing a small orbicular bract half-way along its length; receptacle lobes linear, 2—5 mm long, reflexed at the tips; petals obovate, up to 12 mm long; ovary ovoid, beaked. Fruit reddish-orange, on a stalk up to 15 cm long, pendulous, ovoid or oblong, up to 11 cm long, with wart-like projections, elastically dehiscent; seeds black with crimson aril, oblong, up to 11 × 6 × 3.5 mm, margins grooved. Fig. 20/7.

Distr. ALDABRA: W; ASSUMPTION; pantropical, but probably introduced in the New World.

Notes. On Aldabra growing in gardens at Settlement, *Fosberg* 49519 (US); on ASSUMPTION growing in the Manager's garden, *Stoddart* 1077 & 1108 (both K, US).

Vernac. 'margoze' or 'pumpkin'.

5. PEPONIUM Engl.

Perennial climbing herbs. Leaves simple, tendrils simple or 2-fid at the apex. Flowers white or yellow, dioecious. Male flowers: solitary or in spikes; receptacle-tube elongate, lobes linear-lanceolate; petals 5, entire, free; stamens 3, inserted on the tube, the anthers fused together. Female flowers: solitary; ovary inferior, ellipsoid, hairy, stigma 3-lobed. Fruit subglobose to ellipsoid and beaked, indehiscent; seeds small, flattended, elliptic.

A genus of 20 species in tropical and southern Africa, Madagascar and the Seychelles.

Peponium sublitorale C. Jeffrey & J.S. Page in K.B. 30: 500, pl. 43/F & 44/F (1975). Type: Aldabra, South Island, Trou Nenez, *Stoddart* 973 (K, holo., US, iso.).

Scrambling or trailing plants, stems 2 m or more long. Leaves broadly ovate, 8—15 × 5—10 cm, base cordate, shallowly 3—5-lobed, membranous, coarsely hairy or rough to the touch; petiole 2.5—5 cm long, rough to the touch; tendrils simple. Male flowers: clustered in short spikes up to 4 cm long, peduncles stout, up to 9 cm long; receptacle-tube 3 cm long, lobes linear, 5 mm long; petals yellow, obovate, 2 cm long. Female flowers: smaller than the male, solitary, peduncles up to 2 cm long; receptacle-tube up to 1 cm long, lobes up to 3 mm long, petals 1.5 cm long; ovary cylindrical, pubescent. Fruit ellipsoid, pubescent, up to 5.5 cm long or more; seeds ovate, flattened, up to 6 × 4.5 × 2 mm (immature). Fig. 20/8—9.

Distr. ALDABRA: S (endemic).

Notes. Known only from the coast of South Island between Trou Nenez and Cinq Cases where it grows on dunes and limestone rock, either prostrate and mat-forming or scrambling over shrubs.

6. TRICHOSANTHES L.

Climbing, herbaceous plants. Leaves entire or 3–9-lobed; tendrils simple or divided. Flowers white, monoecious or dioecious. Male flowers: in spikes, rarely solitary; receptacle-tube cylindrical, enlarged towards the apex, lobes small; petals 5, fused at the base, lobes oblong, fringed; stamens 3, inserted on the tube. Female flowers: solitary, rarely in spikes; perianth similar to the male flowers; ovary inferior, ovoid-spindle-shaped. Fruit elongate or globose, indehiscent.

A genus of 15 Indo-Malesian and Australian species.

Trichosanthes cucumerina L., Sp. Pl.: 1432 (1753); F.T.E.A. Cucurb.: 2 (1967).

T. anguina L., Sp. Pl.: 1008 (1753); F.M.S.: 130 (1877); F.M.C. 185: 161 (1966).

Climbing herbs; stems hairy, angled. Leaves orbicular or oval, up to 12 cm long, 3-lobed, margins toothed; tendrils simple or 2–5-divided. Monoecious. Male flowers: spikes stout, up to 25 cm long; receptacle-tube 2.5–3 cm long; lobes triangular; petals oblong, up to 1 cm long, conspicuously fringed. Female flowers: solitary, pedicels up to 5 cm or more long; ovary elongate, hairy; perianth similar to male flowers. Fruit green and white striped, ovoid to cylindrical, up to 2 m long; seeds embedded in greenish-black pulp, slightly compressed-ovoid, 9–10 × 4–5 mm.

Distr. ALDABRA, W; native from India to Australia, introduced elsewhere. Notes. Cultivated at Settlement on West Island, *Fosberg* 49524 (US). Vernac. 'patole' or 'snake-gourd'.

*35 AIZOACEAE**

Herbs or rarely sub-shrubs; usually fleshy. Leaves opposite to whorled or alternate, simple or reduced to scales; with or without stipules. Flowers solitary or cymose, bisexual, perianth of 1 series or apparently 2 because of petaloid staminodes. Sepals 5–8, partly united. Stamens 5-many, outer ones often sterile and petaloid; filaments free or variously united. Ovary inferior to superior, 1–5 or more cells; ovules many, to rarely 1, placentation axile to basal or parietal; style 1, or stigmas sessile, 2–20, radiating. Fruit a capsule, or leathery and tardily dehiscent, or a berry; seed with strongly curved embryo.

* F.T.E.A. Aizoaceae follows Pax in Engl. & Prantl, Nat. Pflanzenfam. 3, 1B: 33 (1859) and includes the Molluginaceae, which in this account is treated as a separate family.

A medium-small family, mainly South African but with a few widespread genera, especially in the tropics and subtropics.

1. Leaves subequal, broadly linear to oblanceolate, thick; perianth firm, subfleshy, styles 3–4; fruit circumscissile near base, seeds not retained in lid   **1. Sesuvium**
1. Leaves unequal, broad, relatively thin; perianth membranous, styles 1–2; fruit circumscissile near middle, 1 or more seeds retained in lid
                                                                      2. Trianthema

1. SESUVIUM L.

Prostrate to ascending herbs, rarely shrubby. Leaves opposite, fleshy, subsessile, obtuse to subtruncate, somewhat dilated at extreme bases. Flowers solitary, axillary; perianth in 1 whorl. Sepals, 5, arranged on a rather obconical calyx tube, somewhat hooded, margins of lobes everlapping. Stamens 5–60, tending to be coherent or united at base. Ovary 3–5-celled; ovules many. Fruit a capsule; seeds black.

Pantropical and warm-termperate genus with few rather ill-distinguished species, 1 pantropical

Sesuvium portulacastrum (L.) L., Syst. Nat., ed. 10,: 1058 (1759); F.M.S. 108 (1877); Hemsley in B.M.I.K. 1919: 122 (1919); F.T.E.A. Aizoac.: 20, Fig. 7 (1961); Renvoize in P.T.R.S., B. 260: 229 (1971).

Portulaca portulacastrum L., Sp. Pl.: 446 (1753).

Stems prostrate, mat-forming. Leaves broadly linear to spatulate or obovate, to 3 cm long, rounded, very fleshy. Pedicels short. Sepals green outside, pinkish or purplish inside, ovate-oblong, strongly and sharply tipped. Stamens in 1 whorl, coherent below. Seeds black, 1.6 × 1.5 mm. Fig. 21/1–2

Distr. ALDABRA: M, S, Michel, Moustique, lagoon islets; COSMOLEDO: M; ASTOVE; pantropical.
Notes. A plant of maritime shores at and about high water level; often along lagoon margins. Flowers throughout the year. Seed-bearing plants can float for more than 4 weeks and seeds subsequently germinate. On Astove eaten as salad or as a pot herb.
Vernac. 'pourpier'.

2. TRIANTHEMA (Sauvag.) L.

Herbs; usually prostrate, fleshy. Leaves opposite to somewhat alternate, leaf pairs often unequal in size; bases often connected by expanded membranous petioles; a pair of interpetiolar stipules fused to the expanded petiole bases, or petioles absent. Flowers axillary, solitary or cymose. Sepals 5, united below, free from ovary, petaloid. Petals absent. Stamens 5, 10, or more, inserted on calyx tube. Ovary 1–2-celled; ovules 1–many per cell, basally attached; styles 1–2. Fruit a circumscissile capsule, the lid hard, retaining 1 or more seeds at dehiscence, the other remaining in the basal portion; seeds 2–few, embryo strongly curved.

A pantropical genus with few species.

Trianthema portulacastrum L., Sp. Pl. 223 (1753). Alston in Trimen, Handb. 6: 137 (1931); F.T.E.A. Aizoac.: 23 (1961).

Trianthema monogyna L., Mant. 1: 69 (1767); Trimen, Handb. 2: 269 (1894), name illegit.

Prostrate, rather fleshy herb; stems slightly and minutely pubescent, in longitudinal stripes. Leaves unequal, broadly obovate to sub-orbicular, apex rounded to subtruncate; expanded leaf-bases and stipules forming a membranous sheath, stipules triangular-acuminate. Flowers solitary, sessile, partly hidden by sheathing leaf bases. Calyx tube fused to sheath, separating after anthesis; lobes triangular-acuminate, purplish within, bearing a purple awl-shaped appendage outside just below apex. Stamens 10–25. Style 1, awl-shaped. Fruit compressed, 2-lobed, lid firm, carrying with it 1–2 seeds, the membranous lower half retaining 2–8 seeds; seeds black, kidney-shaped, 1.5 × 1.5 mm, dull, rough. Fig. 21/3.

Distr. ALDABRA: Is. Chalen; pantropical.
Notes. Only known from 1 collection, *S. Hnatiuk* 732072, (US) from Chalen, a small rocky islet in the lagoon. Seed possibly carried there by sea-birds.

36. MOLLUGINACEAE

Annual or perennial herbs. Leaves opposite or whorled, rarely alternate, sometimes forming a basal rosette, sometimes subfleshy. Sepals 4–5, free or very slightly united at base. Petals small or absent. Stamens few, hypogynous or almost so. Ovary 3–5-celled; stigmas sessile. Fruit normally a membranous capsule or of (3–) 5 (–6) or 10–15 free carpels; seeds with curved embryo.

A small, mostly tropical family of a few genera; usually united with the Aizoaceae*, but in many respects seems closer to the Caryophyllaceae, from which it differs in the partitioned ovary.

* See Footnote to previous Family, Aizoaceae.

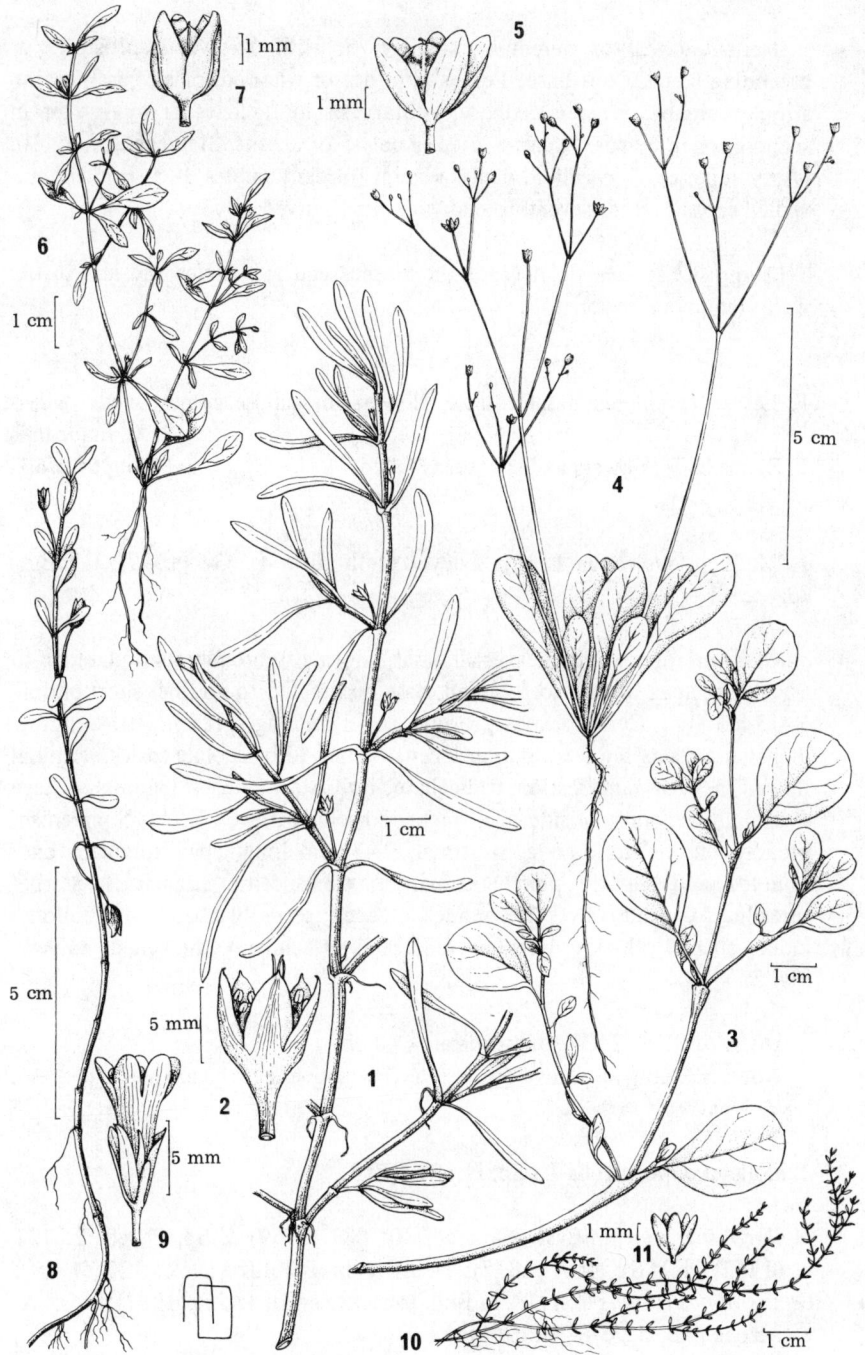

Fig. 21 1, *Sesuvium portulacastrum*, habit; 2, flower, 3, *Trianthema portulacastrum*, habit. 4, *Mollugo nudicaulis*, habit; 5, fruit. 6, *Mollugo oppositifolia*, habit; 7, flower. 8, *Bacopa monnieri*, habit; 9, flower. 10, *Bryodes micrantha*, habit; 11, flower.

MOLLUGO L.

Herbs, annual or perennial; stemless or with 1–several, prostrate or ascending stems from base. Leaves opposite or whorled, or all basal, entire; stipules absent. Flowers axillary, solitary or in fascicles or cymes, or in scapose, open cymes. Sepals 5, free. Petals 5 or absent. Stamens 3, 5, or 10. Ovary superior, 3–5-celled; ovules several to many; styles 3–5. Fruit a thin-walled capsule; seeds several to many, embryo curved.

About 12 species throughout the tropics and subtropics and also in the warm temperate regions.

1. Leaves forming a basal rosette; flowers in scapose cymes; seeds tailless
                                                              1. M. nudicaulis
1. Stems leafy; flowers axillary; seeds tailed   **2. M. oppositifolia**

1. Mollugo nudicaulis Lam., Encycl. Méth. Bot. 4: 234 (1797); F.T.E.A. Aizoac.: 17 (1961).

Glabrous rosette herb. Leaves all basal, obovate to broadly spathulate, up to 5 × 1 cm, apex obtuse to rounded, base contracted to a rather short petiole 5–15 mm long. Cymes scapose, several, to 15 cm long, spreading to ascending, branched; bracts reduced to translucent oblong to lanceolate scales, terminal flower in each ramification pedicellate, ultimate branches tending to have several flowers on one side or a single pedicellate flower. Sepals 5, greenish, broadly elliptic to oblong, subequal, 2–3 mm long, apex rounded, base rounded and slightly pouched, margins translucent. Stigmas 3, sessile, spreading. Capsule valves 3, rounded at apex; seeds black, plump, oblique-kidney-shaped, thickly wart-covered, seed attachment thickened, tailless. Fig. 21/4–5.

Distr. ASSUMPTION; pantropical.
Notes. Possibly adventive on Assumption; growing on guano soil.

2. Mollugo oppositifolia L., Sp. Pl.: 89 (1753).*

Mollugo spergula L., Syst. Nat., ed. 10: 881 (1759) & Sp. Pl. ed. 2: 131
    (1762); F.M.S.: 107 (1877); Hemsley in B.M.I.K. 1919: 122 (1919).
Glinus oppositifolius (L.) DC. in Bull. Herb. Boiss. II, 1: 559 (1901); F.T.E.A.
    Aizoac.: 13, fig. 5/8–9 (1961).

Prostrate herb; stems radiating, to 25 cm long, 4-angled, glabrous or nearly so. Leaves opposite or whorled, narrowly to broadly elliptic, 5–15 mm long, apex acute or slightly acuminate, glabrous to slightly hairy, green to

---

* In F.T.E.A. Aizoac.: 15 (1961) *Mollugo oppositifolia* is treated as *Glinus oppositifolius*.

somewhat glaucous; petiole 0–5 mm long, base expanded. Pedicels axillary, 5–15 mm long, 1–4(–5) at a node, subtended by triangular-lanceolate, scale-like, translucent bracts. Sepals unequal, green, broadly oblong, 3–3.5 mm long, apex slightly hooded, base slightly pouched, margins translucent, spreading at anthesis, closely investing capsule at maturity. Petals absent. Stamens 3, Styles 3, expanded to elliptic, recurved stigmas. Capsule oblong, thin-walled, seeds visible through walls, 3-celled, apex 3-lobed; seeds plump, bright chestnut, obliquely kidney-shaped, 0.5 × 0.3 mm, warty protuberances dark chestnut, seed attachment oblong, fleshy, a quarter the length of the seed, with a prominent, yellow, glistening, coiled appendage longer than the seed. Fig. 21/6–7.

Distr. ALDABRA: ? W, S; pantropical.

Notes. Found mainly at the eastern end of South Island. Few observations but may flower throughout the year if moisture sufficient. Extremely variable in size and in size and shape of leaves. An ephemeral annual in "tortoise pastures" and seasonal pools on platin and in pockets in champignon; grazed by tortoises.

37. RUBIACEAE

Habit various. Leaves opposite or rarely whorled, simple, almost always entire; stipules interpetiolar, very rarely intra-petiolar. Flowers rarely solitary; usually in cymose or thyrsoid inflorescences, less often racemose, rarely spicate; flowers hermaphrodite or variously polygamous or more rarely dioecious, often heterostylous. Calyx mostly united to the ovary, (3–) 4–5 (–8)-toothed or -lobed, sometimes minutely so. Corolla spreading to tubular, usually regular, rarely irregular, (3–)4–5(–11)-lobed. Stamens alternate with corolla lobes, usually inserted on the corolla tube below the sinuses; anthers sessile or on short, rarely long, filaments, dehiscing by longitudinal slits. Ovary usually inferior, 2- or rarely more -celled, ovules 1–many per cell, on axile placentas; style with 2 or rarely more stigmatic branches or lobes, or entire. Fruit a berry, drupe, or capsule, rarely indehiscent and nut-like or a schizocarp; seed with endosperm, testa thin or absent.

A large, mainly tropical family.

1. Plants herbaceous   **3. Hedyotis**
1. Plants woody
  2. Leaves broad, more than half as broad as long, mostly 10 cm wide; corollas over 2 cm long   **2. Guettarda**
  2. Leaves narrower, mostly less than 5 cm wide, 1 cm or less long
    3. Leaves prominently veined beneath, lanceolate; stipules prominently toothed or 3-fid   **7. Triainolepis**
    3. Leaves only moderately veined; stipules entire

  4. Flowers and fruits in terminal cymes
     5. Cymes solitary, open; central flowers sessile or subsessile in forks of
        cyme   **5. Psychotria**
     5. Cymes tending to be congested or less than 3—4 cm long, borne in
        3's or 2's   **6. Tarenna**
  4. Flowers and fruits in lateral or terminal and lateral clusters
     6. Flowers and fruits sessile or subsessile, in close axillary glomerules or
        capitate   **4. Polyspheria**
     6. Flowers and fruits pedicellate, in small cymes or pseudo-umbels
        7. Midrib white or reddish; cymes shortly pedunculate with 2, firm,
           ovate bracts at summit of peduncle, pedicels articulate; corolla
           lobes 4; fruit slightly compressed with a shallow groove on each
           side   **1. Canthium**
        7. Mid rib not differently coloured from blade; corolla lobes 5; fruit
           globose
           8. Cymes strictly axillary, peduncle very short, covered by scale-like
              bracts, pedicels with several, scattered, very small bracts
                                                                 8. Tricalysia
           8. Cymes axillary and in 3's at ends of branchlets, peduncle not
              scaly, pedicel without bractlets except at articulations at summit
              and base   **6. Tarenna**

1. CANTHIUM Lam.

Shrubs or trees. Stipules ovate, apex cuspidate or acuminate.
Inflorescences axillary cymes, often much reduced, even to small fascicles;
flowers 4—5-merous. Corolla-tube short, cylindrical, usually hairy within,
especially at throat, lobes touching in bud, not overlapping, finally spreading.
Stamens inserted in corolla throat; filaments short. Ovary usually 2-celled,
1 pendulous ovule per cell; style 1, stigma cylindrical, 2—4-lobed, included or
exserted. Fruit a drupe with a single, 2-celled or two, 1-celled stones.

A principally African but also Indo-Pacific genus.

Canthium bibracteatum (Baker) Hiern in Fl. Trop. Afr. 3: 145 (1877);
Fosberg in Phytologia 41: 348 (1979).

Plectronia bibracteata Baker, F.M.S.: 146 (1877); Hemsley in B.M.I.K. 1919:
    123 (1919).
Plectronia sp.; Dupont, Report: 37 (1907).

Glabrous shrub, 1—2 (rarely 3—5) m tall; much-branched, branchlets grey,
internodes often short, rather stiff, nodes prominent. Leaves elliptic to
oblong, rarely ovate, 4—8(—9) × 1.5—2.5(—4) cm, apex obtuse to subacute or
slightly acuminate, base rounded to acute, midrib tending to be dull orange
when dry, small sparsely hairy domatia in vein axils beneath; petiole several
mm long; stipules firm, ovate-lanceolate, 3—5 mm long, tardily falling. Cymes
much reduced, umbellate, 5—20-flowered, 10—15 mm long; peduncle short,

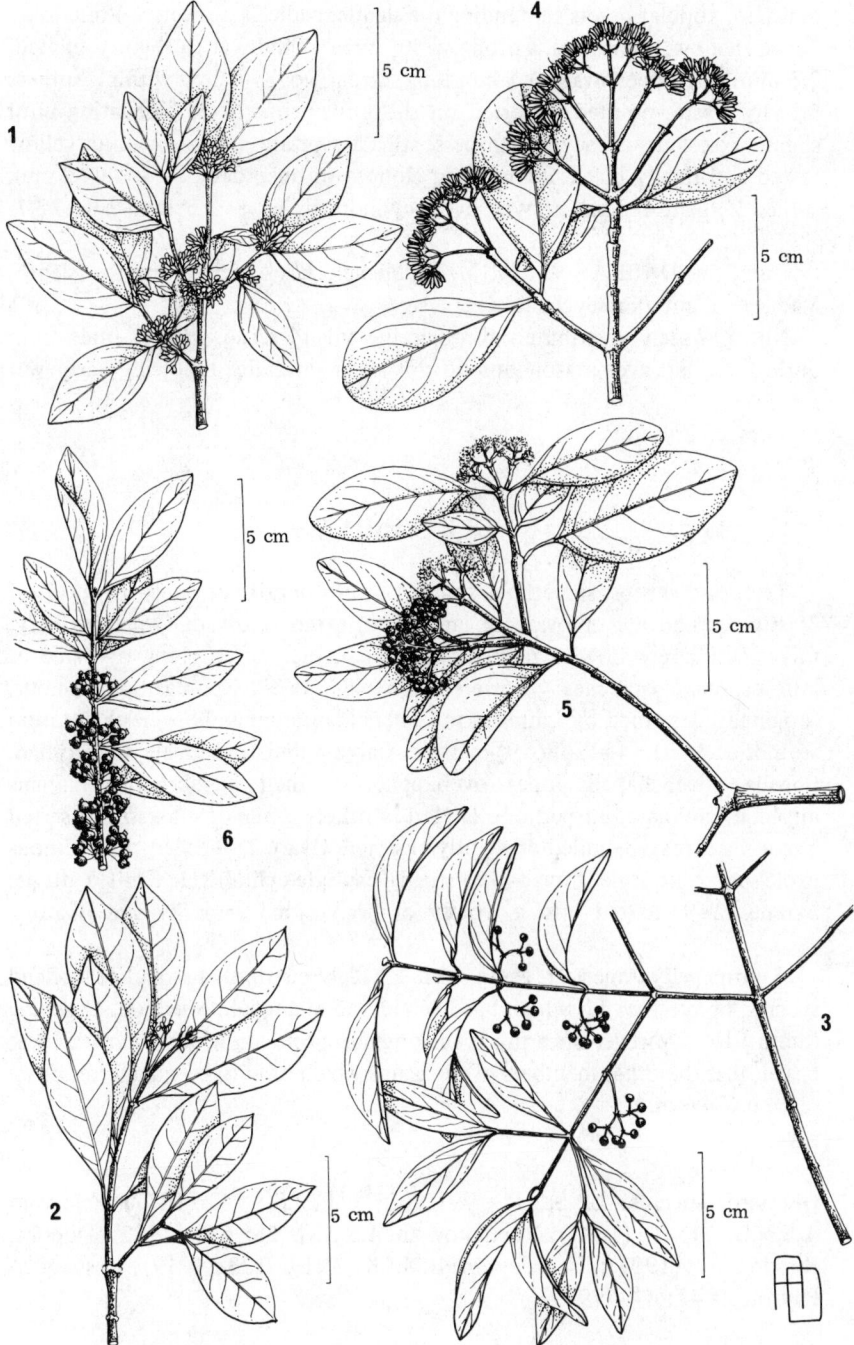

Fig. 22   1, *Canthium bibracteatum*, habit and flowers. 2, *Psychotria pervillei*, habit. 3, *Tarenna supra-axillaris*, habit and fruit. 4, *Tarenna trichantha*, habit and flowers; 5, same, habit and fruits. 6, *Tricalysia sonderana*, habit and fruit.

about a third to half the length of the inflorescence, 2 large, somewhat pouched, stipular bracts subtending the slender pedicels; flowers 3-4 mm long. Calyx shallow, 4-toothed. Corolla white, tube cylindrical to slightly dilated, 2.5 mm long, lobes ovate, 1.5 mm long, throat woolly-haired within. Anthers broadly ovate, pointed, exserted on short filaments. Style elongating until stigma exserted just beyond anthers; stigma capitate. Drupe ripening yellow to red and finally black, compressed globose to obcordate, 3–4 × 4–2 mm; stones 2, light reddish-brown, ovoid-cylindrical, 6 × 2.5 mm. Fig. 22/1.

Distr. ALDABRA: W, P, M, S Michel; E. Africa, Comoro Islands, Madagascar and the Seychelles.

Notes. Widely distributed through the inland scrub. Flower buds form during the late wet season but do not open until the following early wet season.

Vernac. 'bois dur'.

2. GUETTARDA L.

Trees and shrubs, sometimes spiny. Leaves opposite or rarely in whorls of 3; stipules obovate or ovate or lanceolate, often recurved. Cymes axillary, forking or rarely flowers reduced to 2 or 3, or 1, often along one side of inflorescence branches; flowers bisexual, rarely polygamo-dioecious, (originally described by Linnaeus in Genera Plantarum and Species Plantarum as monoecious) (4–)5–8(–9)-merous. Calyx tubular, truncate to toothed. Corolla salver-shaped, lobes overlapping or their membranous margins infolded, undulate, crisped or crenulate. Anthers sessile or subsessile, inserted in corolla throat, included or slightly exserted. Ovary 2–9-celled, 1 pendulous ovule per cell; style thread-like, stigma capitate-cylindrical. Fruit a drupe, pyrenes 2–9, united into a woody or bony stone or corky and floating.

A principally American genus with 1 widely distributed strand or lowland species, *G. speciosa* L., which has been placed in a separate section, *Cadamba* (Sonn.) DC. However since this is the original species on which *Guettarda* was based, it is the other members of the genus which have to be transferred from section *Guettarda*.

Guettarda speciosa L., Sp. Pl.: 991 (1753); F.M.S.: 143 (1877); Schinz in A.S.N.G. 21: 91 (1897); Voeltzkow in A.S.N.G. 26: 552 (1902); Dupont, Report: 37 (1907); Hemsley in B.M.I.K. 1919: 123 (1919); Fosberg in Phytologia 41: 349 (1979).

Shrub to medium-sized tree, up to 8 m tall, very bushy, much branched; branchlet stout, leaf scars conspicuous, horseshoe-shaped, 7 × 6 mm. Leaves opposite and decussate, broadly oblong to more or less ovate or obovate, 10–24 × 7–20 cm, apex obtuse with a slight point, base cordate, thick, papery, midrib and 10–14 pairs of nerves prominent beneath; petiole stout, 1–4 cm long; stipule oblong-ovate, 1 × 1 cm, sheathing, papery, soon falling. Cymes axillary, much-branched, peduncles c 10 cm long, branching, a sessile

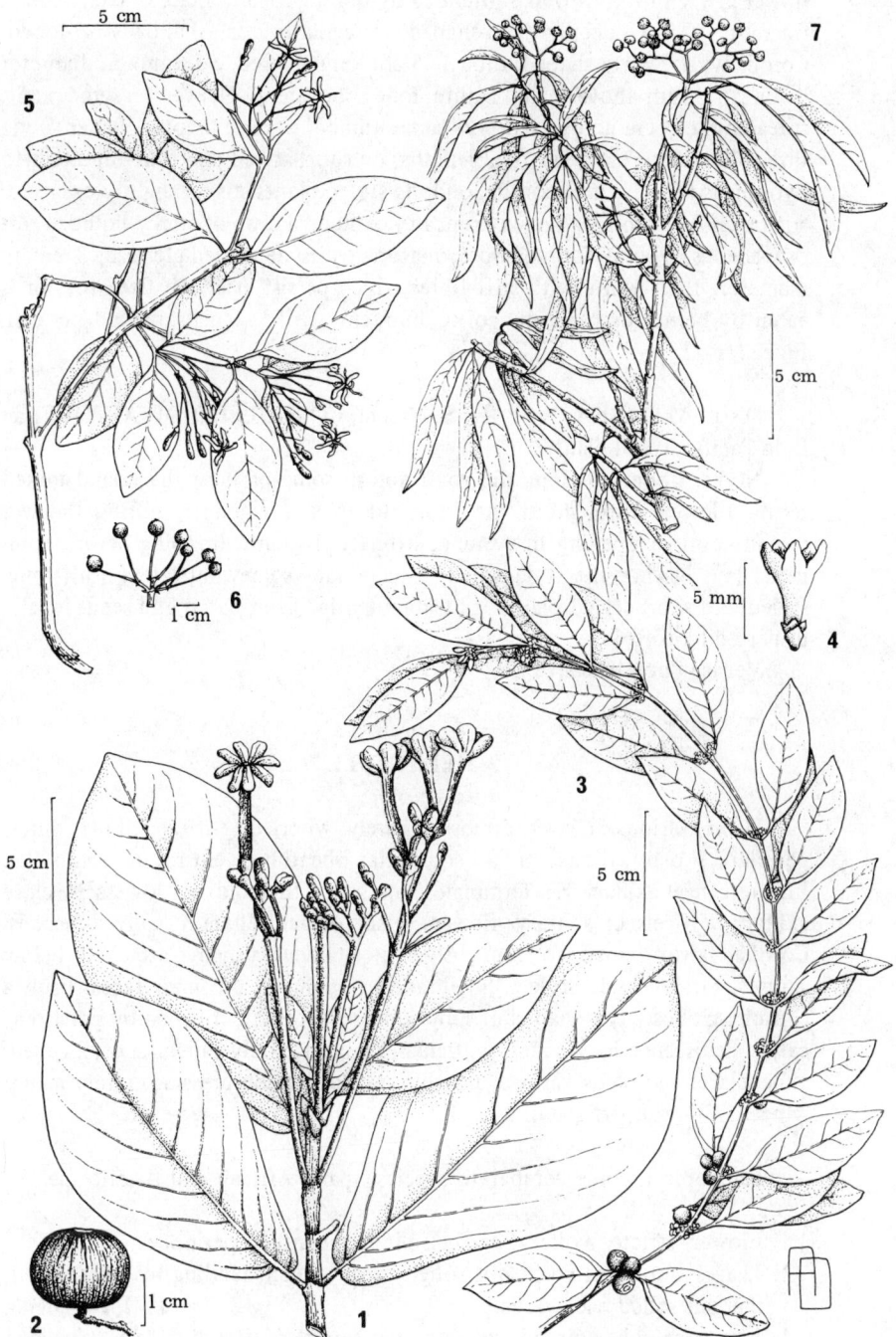

Fig. 23   1, *Guettarda speciosa*, habit; 2, fruit. 3, *Polysphaeria multiflora*,
           habit; 4, flower. 5, *Tarenna verdcourtiana*, habit; 6, fruit.
           7, *Triainolepis africana*, habit.

flower in each fork, flowers subtended by oblong to lanceolate bracts; flowers fragrant. Calyx deeply cup-shaped, irregularly or shallowly 3-lobed. Corolla white, salver-shaped, tube 3—5 cm long, dilated, c 1.5 mm in diameter below to 4 mm above, hairy within, lobes obovate, 6—9, even on same plant, spreading, c 3 cm across. Anthers same number as corolla lobes, linear. Style thread-like, heterostylous, of 2 lengths, on short-styled plants about a half to two-thirds the length of corolla, on long-styled plants stigma slightly exserted; stigma short cylindrical, apex truncate, exuding a drop of sticky liquid. Fruit drupaceous, young fruit globose, mature fruits depressed-globose, 3 cm in diameter, flesh white with stiff fibres which persist after the flesh rots or is eaten by hermit crabs; stone corky, buoyant, with 5—6 cells; seeds 1 per cell. Fig. 23/1—2.

Distr. ALDABRA: W, P, M, S, Michel; COSMOLEDO: M; ASTOVE; an Indo-Pacific strand plant.

Notes. Common along the coast and in some parts of the inland mixed scrub. Flowers throughout the year but most abundantly during the wet season. Corolla opening in evening, strongly fragrant, dropping before noon next day, leaving the style which usually falls somewhat later; both long-styled and short-styled plants fruit abundantly. Fruit eaten and seeds locally dispersed by tortoises.

Vernca. 'bois cassant'.

3. HEDYOTIS L.*

Habit various. Leaves opposite, rarely whorled, entire; stipules inter-petiolar, shortly triangular to somewhat sheathing, entire or comb-like. Inflorescence axillary or terminal, cymose or thyrsoid or flowers in close axillary verticels or solitary; flowers sessile or pedicellate, usually 4-merous. Corolla salver-shaped to funnel-shaped. Stamens usually attached below sinuses, rarely basal. Style 2-fid, branches stigmatic on inner faces. Fruit a capsule, globose, cup-shaped or somewhat compressed, crowned by persistent calyx, dehiscence loculicidal, septicidal, or both, rarely indehiscent; 2-celled; axile placentas; seeds various, usually angular by compression, rarely angles winged, hilum superficial or in a pit.

Pantropical, some temperate species, many African and Asiatic species.

1. Flowers strictly axillary, pedicels predominantly 1 per node.
  2. Leaves obtuse at apex, not rough to the touch, tending to be crowded, margins rolled backwards   **1. H. corallicola**
  2. Leaves sharply pointed, rough to the touch, not tending to be crowded
                                                      2. H. lancifolia var. brevipes
1. Flowers in very loose racemes, as well as in upper axils, usually 2 per node   **3. H. prolifera**

___

* This is treated as the segregate *Oldenlandia* in F.T.E.A. Rub.; for discussion see l.c.: 269 (1976).

1. Hedyotis corallicola Fosberg in K.B. 33: 136 (1978). Type: Cosmoledo, Menai Is., *Frazier* 569 (US. holo.).

Apparently perennial herb, slender, much-branched, prostrate, mat-forming; stems 4-angled, glabrous, marked by rhaphide bundles, rooting at some nodes, producing abundant, congested, rosette-like, leafy, dwarf branches which may later elongate terminally to give rise to further creeping stems, which in turn produce rosettes at their nodes. Leaves stiff, ericoid, oblong or ovate-oblong, 5–8 × 2.5(–3) mm, or smaller, those in the rosettes tending to be somewhat obovate, apex subacute to rounded, base narrowed to a short winged, prominently but minutely bristly haired petiole, apparently somewhat fleshy, margins strongly inrolled; stipular sheaths less than 1 mm long, margins with several, very short, unequal bristles, outer surface of stipules pubescent with minute lanceolate-ovate, flattened, hairs. Pedicels axillary, 1(–2) per node, 3–8 mm long, on elongated stems or in rosettes, somewhat stoutish, glabrous; flowers heterostylous. Calyx lobes somewhat unequal, broad, 1–1.6 × 0.6–0.75 mm, apex subacute to obtuse or slightly pointed, somewhat leaf-like, firm, dorsally keeled. Corolla white, 3–4.5 mm long, deeply 4-lobed, lobes oblong, 2 mm long, apex blunt, inner surface minutely pimply, with a few hairs at base. Short-styled flowers: anthers linear oblong, c 1 mm long, strongly exserted; filaments stout, c 1.5 mm long, glabrous. Long-styled flowers: corolla tube shorter, 1–1.5 mm long; style 3 mm long, 2-fid, stigmas divergent, linear, hairy. Capsule globose to somewhat obovoid or inversely conical, up to 1.5 × 1–2 mm, slightly compressed, thin-walled, crowned with erect, persistent, non-enlarged calyx lobes; seeds almost black, somewhat oblong, c 0.4–0.5 mm across, bluntly angular, surface shallowly netted, somewhat glossy. Fig. 24/1.

Distr. COSMOLEDO: M; ASTOVE; endemic.

2. Hedyotis lancifolia Schum. in Schum. & Thonn., Beskr. Guin. Pl.: 72 (1827).

Oldenlandia lancifolia (Schum.) DC., Prodr. 4: 425 (1830); F.T.E.A. Rub.: 292 (1976).

This species is widespread in Agrica and Madagascar, introduced into tropical America. A number of slightly differing varieties are recognised by Bremekamp in Verh. K. Nederl. Akad. Wet. Afd. Nat. II, 48(2): 230 (1952).

var. **brevipes** (Brem.) Fosberg in K.B. 33: 137 (1978).

Oldenlandia lancifolia var. *brevipes* Brem. in Verh. K. Nederl. Akad. Wet. Afd. Nat. II, 48(2): 234 (1952).

Slender annual or perennial herb, erect to procumbent, stem 4-sided, slightly rough, at least on angles. Leaves narrowly lanceolate, 10–15 × 1.5–2(–3) mm, apex sharply acute to slightly acuminate or spine-tipped, base wedge-shaped, thin, sparsely roughly hairy above; subsessile or petiole short; stipular sheath up to 1 mm long, lobes obtuse to subacute, bristles 5–7, unequal, very slender. Pedicels slender, 1(–2) per node, 3–5(–7) mm long, tending to become reflexed in fruit. Calyx lobes 4, lanceolate, c 1 mm long, acuminate, erect. Corolla pale pink, somewhat funnel-shaped, c 1.5 mm long, tube slightly exceeding lobes, lobes ovate. Anthers linear, exserted, filaments short. Ovary obovoid, c 1 mm long, glabrous; style 2-fid, stigmas linear. Capsule inversely conical-globose, c 2 mm in diameter, crowned by erect, persistent calyx lobes, thin-walled; seeds somewhat oblong, angular, dull brown, net-celled, areolae hexagonal. Fig. 24/2.

Var. *brevipes* was known to Bremekamp only from Ethiopia, but is probably more widespread in Africa. The variety seems to be characterised principally by its short pedicels and by being slightly roughened on stems and pedicels.

Distr. ASTOVE; Ethiopia.

Notes. Occurs on coral sand and coral limestone. Possibly introduced, known only from 3 records, *Ridgway* 55 (US); *Veevers-Carter* 55 (EA) and *Fosberg* 49724 (K, US).

3. Hedyotis prolifera Fosberg in K.B. 33: 138 (1978). Type: Aldabra, West Island, near Settlement, *Renvoize* 2738 (US, holo., K, NY, iso.).

Oldenlandia corymbosa sensu Schinz in A.S.N.G. 21: 91 (1897); Voeltzkow in A.S.N.G. 26: 552 (1902); Dupont, Report: 37 (1907); Hemsley in J. Bot. 54, Suppl. 2: 16 (1916) not L. (1953).
O. sp. '*adhuc indescripta*?'; Hemsley in J. Bot. 54, Suppl. 2: 17 (1916) & in B.M.I.K. 1919: 122 (1919).
O. sp. nov.; Hemsley in J. Bot. Suppl. 2: 16 (1916).
O. sieberi sensu Stoddart & Wright in Atoll Res. Bull. 118: 29 (1967) not Baker (1877).
Hedyotis sp., Fosberg in P.T.R.S. B, 260: 222, 225 (1971).

Slender prostrate or procumbent to ascending or erect (when supported by other plants) herb; root-stock tough, or somewhat thickened; stems to 25(–30) cm long, weakly 4-ridged, subglabrous to minutely hairy. Leaves narrowly linear to linear-lanceolate to narrowly or even rather broadly elliptic to oblong or ovate, to 25(–30) × 1–9 mm, apex acuminate to almost rounded but slightly spine-tipped, base sub-obtuse to acute, margins ciliolate to slightly rough; petiole almost lacking to 3 mm long; stipular sheath dull white, up to 1 mm long, bristles 5–7, strongly unequal. Pedicels in upper axils, (1–)2–4 at a node, axis elongating forming a terminal or rarely axillary raceme, which may proliferate to 5–7 (or more) nodes; pedicels slender to hair-like, straight to arcuate, 3–10 mm long; flowers heterostylous. Calyx

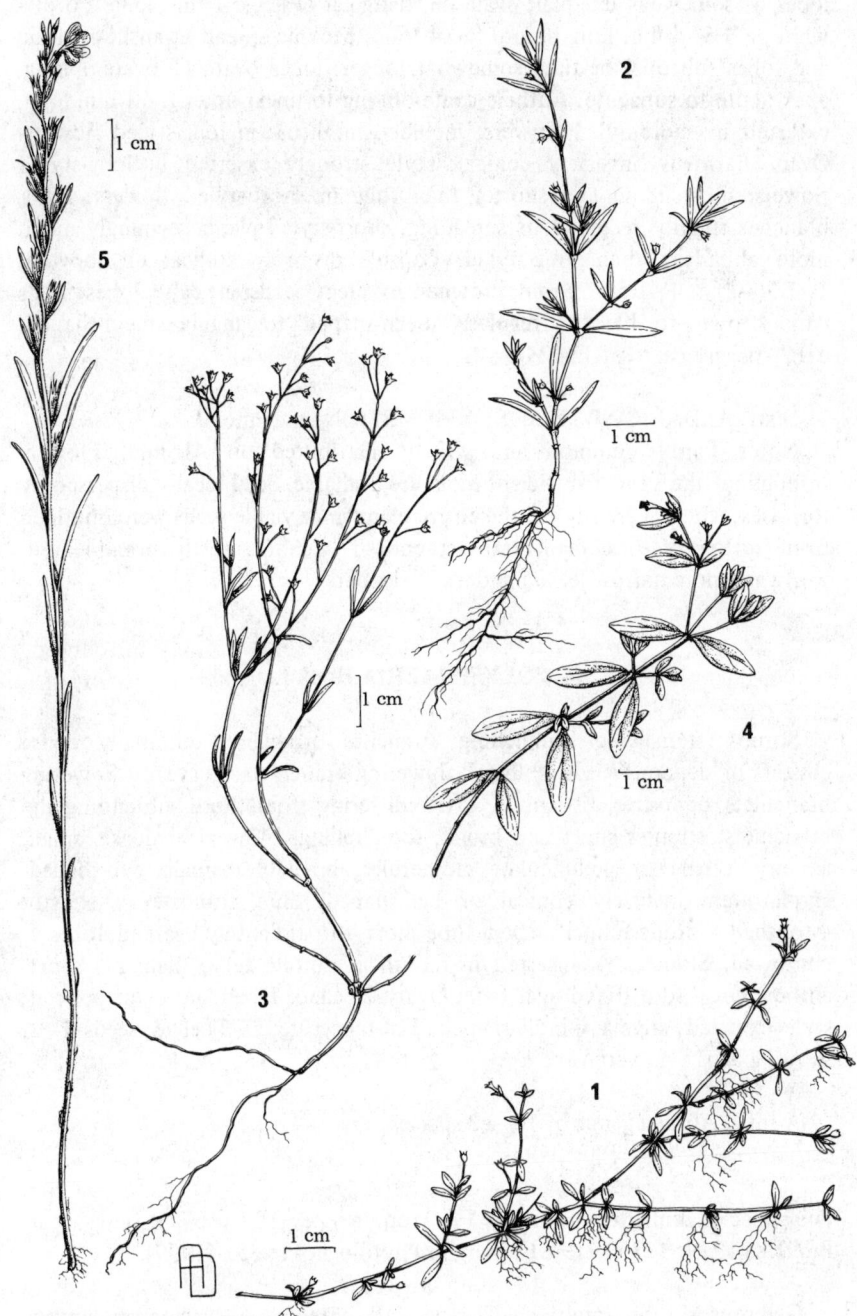

Fig. 24 1, *Hedyotis corallicola*, habit. 2, *Hedyotis lancifolia* var. *brevipes*, habit. 3, *Hedyotis prolifera*, habit. 4, larger leafed variant. 5, *Striga asiatica*, habit.

lobes 4, somewhat unequal, ovate or triangular-ovate, c 1 mm long. Corolla white, c 2.5—4 mm long before the 4 lobes become spread at anthesis, tube and lobes subequal or tube somewhat longer, lobes ovate to ovate-oblong, apex acute to subacute. Anthers ovate-oblong to linear-oblong, c 1 mm long, exserted in short-styled flowers, included in throat in long-styled flowers. Ovary narrowly inversely conical; style strongly exserted in long-styled flowers, included in and shorter than tube in short-styled flowers, 2-fid, branches slightly recurved or spreading, short-styled plants seemingly much more abundant than long-styled. Capsule inversely conical or obovoid, 1—1.5(—2) × 1—1.5(—2) mm, crowned by erect persistent calyx lobes; seeds dark brown to black, irregularly bean-shaped to angular-shield-shaped, c 0.2—0.3 mm across. Fig 24/3—4.

Distr. ALDABRA; W, P, M, S; ASSUMPTION; endemic.
Notes. Fairly common and widely distributed on Aldabra. Flowers throughout the year if sufficient moisture available. Seed locally dispersed by tortoises, who apparently eat the entire plant since viable seeds were obtained from tortoise faeces. Plants at east end of Aldabra mostly broad-leaved, westward more narrow-leaved and more slender.

4. POLYSPHAERIA Hook.f.

Shrubs; stem erect, branching; branches producing terminally dense clusters of slender "supra-axillary" flowering branchlets. Leaves of flowering branchlets opposite, distichous, often differing from those subtending the branchlets; stipules short and broad, soon falling. Flowers in dense, small, axillary, sessile or pedunculate glomerules; bracteoles small, cup-shaped. Hypanthium inversely conical or bell-shaped, limb truncate or shortly 4-toothed. Corolla funnel-shaped, tube short, throat densely bearded, lobes 4, contorted. Stamens 4, inserted in mouth of corolla tube, filaments short, anthers linear, dorsifixed near base. Ovary 2-celled, 1 pendant ovule per cell; style exserted, stigma spindle-shaped. Fruit baccate, 1—2-celled; seeds 1—2, suborbicular, plano-convex.

A small African genus of 10—12 species.

Polysphaeria multiflora Hiern in Fl. Trop. Afr. 3: 127 (1877); Hemsley in B.M.I.K. 1919: 123 (1919); Fosberg in Phytologia 41: 351 (1979).

Glabrous shrubs, rarely small trees, bark shreddy. Leaves conspicuously downward-pointing, elliptic, lanceolate-elliptic, or oblong-elliptic, 3.5—9 × 1.5—3.5 cm, apex acute or acuminate with blunt tip; petiole short; stipules triangular, keeled, soon falling. Flowers in tight, sessile, cymose, axillary glomerules, with stipular scales at base, individual flowers subtended by united bracteoles. Calyx cup-shaped, margin lobed or toothed, glabrous. Corolla white, salver-shaped to funnel-shaped or bell-shaped, 5—6 mm long, 4-lobed, lobes slightly overlapping, not or slightly contorted in bud, becoming recurved after opening, glabrous outside, copiously bearded in throat. Anthers lanceolate, sessile in throat. Ovary 2-celled, ovules 1 per cell; style longer than corolla, minutely pubescent except near base, stigma well-exserted, slightly lobed. Fruit black when ripe, globose, 6–7 mm diameter, fleshy?, stones 2, mottled reddish-brown, hemispheric, 1-seeded. Fig 23/3–4.

Distr. ALDABRA: W, P, M, S, Esprit, Michel; COSMOLEDO: M; ASTOVE; East Africa, Comoros.

Notes. Widely distributed throughout the mixed scrub. Flowers mainly during the mid wet season; fruits remain on the plant for many months. Locally dispersed by birds (seeds from bird regurgitate germinated). Fruit used as a substitute for coffee.

Vernac. 'café'.

5. PSYCHOTRIA L.

Shrubs, rarely trees or lianes. Leaves usually obovate, domatia sometimes present in vein axils; stipules falling or persistent, free or united, often forming a cap (calyptra) enclosing the terminal bud, then usually with 2 or 4 appendages or "ears" at the apex, axils of stipules with a row of erect hair-like glands. Inflorescences terminal or axillary, cymose or thyrsoid, rarely reduced to a fascicle or a single flower; flowers bisexual or dioecious, 4–6-merous. Calyx usually short, lobed, toothed or truncate. Corolla tube cylindrical or slightly dilated, lobes usually as long as to much longer than tube, spreading to reflexed. Stamens inserted in throat just below sinuses. Anthers attached basally or dorsally. Ovary 2-celled, ovules solitary, erect, basifixed; style shorter than to exceeding tube, 2-fid. Fruit a drupe with 2 pyrenes, often dorsally 1–several times keeled, flat on ventral surface, seed filling cavity.

A very large tropical genus, sometimes variously subdivided, with species in all except the driest tropical wooded areas, difficult to classify.

Psychotria pervillei Baker, F.M.S.: 155 (1877); Hemsley in B.M.I.K. 1919: 123 (1919); Fosberg in P.T.R.S. B, 260: 220, 225 (1971); Renvoize in P.T.R.S. B, 260: 231 (1971); Fosberg in Phytologia 41: 352 (1979).

Psychotria sp.; Baker in B.M.I.K. 1894: 198 (1894); Schinz in A.S.N.G. 21: 291 (1897); Voeltzkow in A.S.N.G. 26: 552 (1902); Dupont, Report: 37 (1907); Hemsley in J. Bot. 54, Suppl. 2: 19 (1916) & in B.M.I.K. 1919: 123 (1919).

Shrub to 3 m tall, glabrous. Leaves elliptic to obovate, up to 13 × 4.5 cm, usually much smaller, apex acuminate, base wedge-shaped, dark green; petiole c 1(–1.5) cm long; stipules broadly triangular, c 3 mm long. Cymes 1–3, terminal, sometimes becoming axillary by development of a bud at same node; peduncles slender, 3–5 cm long, branchlets minutely pubescent, bractlets minute, ciliolate, ultimate triads of cyme with centre flower sub-sessile; flowers fragrant. Calyx lobes 5, low-triangular. Corolla white, tube 3–4 mm long, lobes oblong-ovate, c 1.5–2 mm long, apex slightly hooked, inner surface densely minutely pubescent. Anther tips exserted. Style with 2-fid stigma exserted. Fruits fleshy, bluish-grey or pale purple, globose or depressed globose, slightly compressed, 3–4 × 3–4 mm, pyrenes 2, hemispherical, dorsally ribbed. Fig. 22/2.

Vegetatively could be confused with *Pandaca mauritiana* (Family 44/3), which has a milky sap, leathery leaves, which are less sharply wedge-shaped at the base, and lack stipules.

Distr. ALDABRA, S (east) Michel; Seychelles.

Notes. Very rare on Aldabra, near Takamaka Grove; populations recently affected by attacks of a coccid. From the few observations available flowers during the wet season.

Vernac. 'bois couleuvre'.

6. TARENNA Gaertn.

Shrubs, rarely trees. Leaves simple, opposite; petiole usually present; stipules ovate, falling. Inflorescence terminal or becoming lateral, cymose; flowers usually (rarely 4–)5–6-merous. Calyx usually lobed. Corolla salver-shaped to somewhat funnel-shaped, lobes spreading, overlapping in bud. Stamens inserted in corolla throat below sinuses, filaments short, anthers linear. Ovary 2–4-celled, placentas fleshy, ovules 1–several or more; style elongating, stigma becoming strongly exserted, club-shaped to linear. Fruit fleshy, endocarp thin; seeds tending to be subglobose with a cavity on one side.

A large African-Indo-Pacific genus. Aldabra records of *Pavetta* species belong here.

1. Inflorescence and flowers except for corolla limb, glabrous
  2. Cymes soon becoming axillary; corolla 5-lobed   **1. T. supra-axillaris**
  2. Cymes remaining terminal; corolla 4-lobed   **3. T. verdcourtiana**
1. Inflorescence and flowers notably pubescent; cymes remaining terminal;
    leaves usually at least slightly hairy beneath   **2. T. trichantha**

1. Tarenna supra-axillaris (Hemsley) Bremekamp in Fedde, Repert. Sp. Nov. 37: 206 (1934); Fosberg in Phytologia 41: 354 (1979). Type: Aldabra, Ile Esprit, *Fryer* s.n. (K, holo.).

Pavetta supra-axillaris Hemsley in J. Bot. 54, Suppl. 2: 19 (1916) & in B.M.I.K. 1919: 123 (1919).

Shrub to 3 m tall; glabrous except for the inflorescence; branchlets slender, pale, circular in cross-section, diverging widely from branches. Leaves elliptic or lanceolate-elliptic to narrowly ovate or lanceolate-ovate, often somewhat sickle-shaped, often somewhat folded, up to 10 × 3 cm but mostly much smaller, apex acute to usually acuminate, point usually blunt, base acute, somewhat attenuate, thinly papery; petiole slender, 4–8 mm long; stipules ovate, somewhat acuminate, rounded at tip, somewhat sheathing at base but soon separated by growth of lateral branchlets or inflorescences. Cymes up to 2–3 cm long, borne in pairs at terminal nodes, but soon becoming lateral by elongation of stem; peduncle slender, curved so that the cyme is usually pendent, branching rather congested, a cup-like whorl of 4, more or less united, scale-like bractlets subtending each ramification and each flower bud, distal parts and sometimes entire inflorescence liberally covered by a some-what granular resinous secretion, possibly from axils of stipular bractlets; flowers 5-merous. Calyx lobes suborbicular, densely white-ciliate, overlapping at base, minutely pubescent inside. Corolla greenish-white, tube glabrous, slightly dilated upward, limb club-shaped in bud, strongly pimply externally, minutely pubescent toward tips, lobes narrowly oblong-elliptic, c 1.5 mm long, strongly but minutely ciliolate, bearded at base. Anthers exserted, linear, 3.5 mm long. Ovary 2-celled, 1 ovule per cell; style at maturity long-exserted, stigma narrowly spindle-shaped or paddle-shaped (perhaps by collapse), tardily becoming apically 2-fid. Fruit black, globose, 2.5–3.0 mm in diameter, crowned with a conical ring of slightly toughened, overlapping calyx lobes, fleshy; endocarp sclerified, 1-celled; seed 1, chestnut-brown, globose, with a deep somewhat irregular cavity in one side. Fig. 22/3.

This species, apparently endemic to Aldabra Atoll, is somewhat anomalous in *Tarenna*, as its paired cymes are borne at a terminal node and soon become apparently truly axillary. The origin and development of these inflorescences needs further investigation in the field, and living plants studied over a period of time.

The species may be, as Hemsley suggested, close to *T. nigrescens* (Hook.f.) Hiern (*Coptosperma nigrescens* Hook.f.), from Madagascar, Comoro Is. and Mozambique, but not to the Aldabra species that Hemsley referred to *T. nigrescens*, which is *T. verdcourtiana* Fosberg.

Distr. ALDABRA: W, P, M, S, Esprit; endemic.
Notes. Occurs in coastal and inland scrub. Flowers during the wet season; buds form in the late wet season and remain dormant throughout the dry season.

2. **Tarenna trichantha** (Baker) Bremekamp in Fedde, Repert, Sp. Nov. 37: 207 (1934); Fosberg in Phytologia 41: 355 (1979. Type: Aldabra, *Abbott* s.n. (K, holo., US, iso).

Pavetta trichantha Baker in B.M.I.K. 1894: 148 (1894); Schinz in A.S.N.G. 21: 91 (1897); Voeltzkow in A.S.N.G. 26: 552 (1902); Dupont, Report: 37 (1907); Hemsley in B.M.I.K. 1919: 123 (1919); Fosberg in P.T.R.S. B, 260: 218, 225 (1971); Renvoize in P.T.R.S. B, 260: 231 (1971).

Rutidea coriacea sensu Hemsley in B.M.I.K. 1919: 124 (1919), not Baker (1877).

Shrub or small tree, to 4 m tall; wood hard; branches pale greyish-brown, young growth usually more or less pubescent. Leaves broadly obovate to oval or broadly elliptic, up to 10 × 5.5 cm, apex rounded or obtuse, base broadly wedge-shaped to rounded, upper surface minutely pubescent or rough to glabrous, lower surface shortly hairy to almost glabrous; petiole slender, mostly 0.5–1 cm long; stipules triangular-ovate, 1.5 mm long, acuminate, keeled. Cymes dense, corymbiform, tending to be very rounded or hemi-spherical, borne singly or in 3's at terminal nodes of branchlets, conspicuously pubescent, much branched; bracteoles at ramifications strap-shaped with broad base, strap-shaped or awl-shaped; flowers 5-merous, fragant. Calyx teeth 5, triangular, densely white woolly-haired. Corolla white, salver-shaped, tube cylindrical, c 5 mm long, appressed-puberulent, somewhat dilated at top, limb club-shaped in bud, densely and minutely appressed-pubescent outside, glabrous inside, lobes 5, broadly oblong, c 1.5 × 1 mm, apex rounded or obtuse, reflexed, glabrous inside, throat not bearded. Anthers ovate-oblong, c 2 mm long, erect, apex shortly and bluntly tipped, base sagittate. Ovary inferior, densely white woolly-haired; style exserted 3–5 mm, included portion thinly and minutely pubescent, exserted part glabrous, distal part fluted, somewhat spindle-shaped. Fruit black, globose, 3–5 mm diameter, sparsely and minutely appressed-pubescent, crowned by minute, persistent calyx-lobes surrounding a thickened ring-like disc, endocarp very thin, paper-like, brittle; seeds 2, 3 or 4, even on same plant, glossy dark brown, somewhat compressed-globose, c 2 mm across, deep linear scar on one edge. Fig. 22/4–5.

This species can be recognised by its densely congested, whitish- or grey-pubescent cymes. Vegetative characters are very variable, especially leaf-shape and hairiness; glabrous forms occur along the south coast of Aldabra. A specimen from Assumption, *Frazier* 608 (US), has the corollas and hypanthia almost glabrous.

Distr. ALDABRA: W, P, M, S, Esprit, Michel; ASSUMPTION; COSMOLEDO; ASTOVE; coasts of Kenya, Tanzania, Mozambique and Comoros.
Notes. A frequent and widespread constituent of coastal and inland scrub. Deciduous and semi-deciduous during periods of drought. Flowers mainly during the wet season; the fruit remains on the trees for several years. Fruit eaten and locally dispersed by doves and blue pigeon.

3. Tarenna verdcourtiana Fosberg in Phytologia 41: 357 (1979). Type: Aldabra, West Island, *Wood* 1631 (US, holo.).

Tarenna nigrescens sensu Hemsley in J. Bot. 54, Suppl. 2: 17 (1961) & in B.M.I.K. 1919: 123 (1919); Renvoize in P.T.R.S. B, 260: 231 (1971); not (Hook.f.) Hiern (1877).

Glabrous shrub, to 4 m tall, usually much smaller; branchlets grey, slender, diverging at about 45°–50° from larger branches, branching may be slightly supra-axillary, basal internode may be much reduced, with only a pair of stipules at its summit. Leaves elliptic to slightly obovate, to elliptic-lanceolate or rarely very broadly elliptic, mostly 4–7 × 1.5 × 2(–4.3) cm, apex acute to somewhat acuminate, base acute to obtuse, firm-papery to almost leathery; petiole 5–8 mm or shorter; stipules ovate, usually long-acuminate, 2 mm long. Cymes 3 or 1 by suppression, terminal, 2–4 cm long; pedicels up to 1(–1.5) cm long, with 2 scale-like bracts near base; flowers fragrant.. Calyx bell-shaped, 1.5 mm long, shallowly 4-lobed, lobes obtuse to rounded, minutely jagged or ciliolate. Corolla white, tube c 4–5 mm long, somewhat dilated upward, glabrous without, throat lightly bearded, limb bluntly spindle-shaped in bud, lobes 4, oblong, c 7 mm long, apex rounded, spreading to somewhat reflexed, margins rolled back. Anthers broadly linear, c 6 mm long, exserted from sinuses, strongly curved after dehiscence. Style eventually longer than corolla tube, slightly pubescent, stigma linear, flattened or slightly spindle-shaped, c 7 mm long, somewhat papillose-puberulent. Fruit black, globose, 5–6 mm in diameter; seeds 1–4, arranged radially, with convex surface dull brown, gently wrinkled, a deep pit on inner angle. Fig. 23/5–6.

Distr. ALDABRA: W, P, M, S (west and central); ASSUMPTION; endemic.

Notes. A constituent of the coastal mixed scrub. Flowers during the early wet season; buds formed during the late wet season remain dormant throughout the dry season.

7. TRIAINOLEPIS Hook.f.

Shrubs or small trees. Leaves opposite, usually conspicuously nerved; stipules usually with 3 apical points. Cymes terminal, small. Calyx bell-shaped, teeth 5–7, unequal. Corolla salver-shaped, hairy outside, tube glabrous inside, bearded in throat, lobes 5, spreading, glabrous inside, touching, not overlapping in bud. Anthers oblong, dorsifixed; filaments short. Ovary 5–10-celled, ovules erect, 1–2 per cell. Fruit globose, fleshy, with 5–7, bony, united, 1-seeded pyrenes.

Small, principally Madagascan genus with 1–2 African species.

Triainolepis africana Hook. f. in Gen. Pl. 2:126 (1873); F.T.E.A. Rub.:149 (1976).

subsp. **hildebrandtii** (Vatke) Verdc. in K.B. 30:282 (1975); F.T.E.A. Rub.:150, fig. 13 (1976); Fosberg in Phytologia 41:359 (1979).

Triainolepis hildebrandtii Vatke in Oestr. Bot. Zeit. 25:230 (1975); Hemsley in B.M.I.K. 1919:123 (1919).

Psathura fryeri Hemsley in J. Bot. 54, Suppl. 2:20 (1916). Type: Aldabra, *Fryer* 44 (K, holo.).

Trianolepis fryeri (Hemsley) Bremekamp in Proc. K. Nederl. Akad. Wet. C,
59:12 (1956); Fosberg in P.T.R.S. B,260:218,225 (1971); Renvoize in
P.T.R.S. B,260:231 (1971).

Shrub 1–4 m tall; branchlets glabrous, youngest growth light brown to
greenish, orange or pinkish, lightly hairy. Leaves elliptic-lanceolate to
lanceolate, up to 13 × 4 cm, usually much smaller, apex narrowly acuminate,
base acute, somewhat decurrent on petiole, nerves c 10 pairs, distinct,
pubescent beneath, with small domatia in nerve axils; petiole 5–10 mm long,
slender; stipules short, hairy, somewhat sheathing, each side with a low
obtuse lobe with 3–5 linear processes on margin. Cymes 3, rarely 1 or 5, at
terminal node, sometimes 1 in each axil at the next node below, slender,
loosely branched, usually 2.5–4(–5) cm long, sparsely to densely minutely
pubescent, bractlet linear, very small, at each ramification. Flowers fragrant;
hypanthium hemispherical, longitudinally wrinkled and minutely mealy-
pubescent. Calyx broadly cylindrical or somewhat bell-shaped, wrinkled outside, minutely pubescent, sinuses each with a conspicuous gland and a dense
tuft of hair, lobes 5–6, unequal, triangular to almost strap-shaped with broad
base, apex blunt, hairy, especially inside and on margins. Corolla white,
densely woolly-pubescent outside, less so on lobes, tube cylindrical,
7–8.5 mm long, upper 1–1.5 mm abruptly strongly dilated, limb truncate in
bud, lobes 5, linear-oblong, spreading to recurved, with a strong hook-like
appendage subapically within, throat and bases of lobes bearded. Anthers 5,
exserted, lanceolate, erect. Ovary 7–10-celled; style thread-like, glabrous,
7–10-branched at apex. Fruit white to pink, globose or depressed globose,
drupe c 8.5 mm in diameter, furrowed outside when dry, endocarps united
into a furrowed bony putamen 6 mm or less in diameter; seed developed in
some cells, others abortive. Fig. 23/7.

This species is variable, depending on the conditions, such as available
moisture and quality of soil. In dry conditions its leaves point strongly downward. *T. fryeri* was described as an Aldabra endemic and has commonly been
treated as such. However, it differs from *T. hildebrandtii* only in obscure
details of the fruit. The Aldabra population, itself varies more in other
characters, from plant to plant, than it does from the Madagascar and
Comoro population called *T. hildebrandtii* Vatke, reduced to a subspecies of
T. africana by Verdcourt.

Distr. ALDABRA: W, P, M, S, Esprit, Michel; ASSUMPTION;
COSMOLEDO; Kenya, Tanzania, Zanzibar, Pemba, Mozambique, Malawi,
Zambia Madagascar and Comoro Is.
Notes. A common constituent of mixed scrub and on platin. Flowers
during the wet season; may be leafless during droughts. Fruits eaten and
locally dispersed by blue pigeons and doves.
Vernac. 'coeur de boeuf'.

8. TRICALYSIA A. Rich. ex DC.

Shrubs or small trees. Leaves elliptic or lanceolate; stipules awl-shaped. Inflorescences few-flowered, small, axillary, bracteate cymes; flowers 4–8-merous. Calyx toothed or rarely 2-lipped. Corolla lobes overlapping or contorted in bud. Anthers linear, exserted on short filaments from corolla throat. Ovary 2-celled, ovules 2 or more per cell, paired or in 2 vertical rows on a fleshy placenta; style exserted, stigma 2-lobed. Fruit globose, fleshy; seeds 1–9 (or 12?), more or less rounded or irregularly obtusely angular, deeply sunken in the remains of the fleshy placenta.

A small African-Indo-Pacific genus.

Tricalysia sonderana Hiern in Fl. Trop. Afr. 3: 119 (1877); Fosberg in P.T.R.S. B, 260: 218, 225 (1971); Renvoize in P.T.R.S. B, 260: 231 (1971); Fosberg in Phytologia 41: 360 (1979).

T. cuneifolia Baker in B.M.I.K. 1894: 148 (1894); Schinz in A.S.N.G. 21: 91 (1897); Voeltzkow in A.S.N.G. 26: 552 (1902); Dupont, Report: 37 (1907); Hemsley in J. Bot. 54, Suppl. 2: 18 (1916) & in B.M.I.K. 1919: 123 (1919). Type: Aldabra, *Abbott* s.n. (K, holo., US, iso.).

Shrub or small tree, 1–5 m tall; branchlets pale yellowish-grey. Leaves ovate to elliptic or narrowly elliptic, up to 10 × 4 cm, usually smaller, apex acute to slightly acuminate, base acute to somewhat abruptly contacted, glabrous; petiole 1–3 mm long; stipules ovate, sheathing, abruptly acuminate, tip spine-like, tending to diverge somewhat from stem; cymes very condensed, once or twice branched, or reduced to single triads, scale-like bracts conspicuous; pedicels varying greatly in length, flowers 5-merous, hypanthium glabrous, very slightly contracted at base of calyx. Calyx very broadly bell-shaped to saucer-shaped, 2–2.5 mm across, shallowly 5-lobed, lobes shallow-triangular. Corolla white, c 10 mm long just before opening, tube and limb subequal, tube dilated upward, limb conical in bud, lobes broadly oblong, somewhat exceeding tube, basal part of lobes strongly overlapping, upper part somewhat less so, becoming reflexed, bases of lobes and throat usually copiously and conspicuously bearded. Anthers completely exserted and reflexed, linear, 4 mm long overall, cells 3 mm, apical 1 mm of the connective enlarged into an oblong fleshy appendage. Ovary 2–(3–) celled, up to 6 ovules per cell; style glabrous, exserted beyond the beard, apex divided into 2 strongly flattened, lanceolate, somewhat recurved branches. Fruits black, globose, 6–7 mm in diameter; seeds 6–9(–12?), dark chestnut brown, irregularly compressed, subangular, attached on one angle, with an irregular cavity on one side near attachment. Fig. 22/6.

One flowering specimen, *Wood* 1614 (K, US) has the bases of the corolla lobes and throat only slightly bearded, but otherwise does not differ significantly.

The inclusion of its seeds embedded in the developed fleshy placentae suggests that the genus may belong in the *Gardeniae* Cham. & Schlecht. rather than in the *Ixoriae* (Benth.) Hook., where it seems to be placed in recent classifications of Rubiaceae.

Distr. ALDABRA: W, P, M, S (east), Esprit, Michel, Egret; ASSUMPTION; East Africa, southern Africa, Madagascar and Comoro Is.

Notes. Found in mixed scrub and in small areas or clumps of scrub on platin. May lose most or all its leaves during droughts. Flowers mainly in the wet season but unseasonal rain can stimulate massive flowering. Fruit eaten and seed dispersed by blue pigeon.

38. COMPOSITAE

Habit various, mostly herbs or shrubs; often aromatic. Leaves alternate or opposite, simple, often deeply cut or pinnately divided, rarely compound; stipules absent. Flowers (florets) condensed into heads (capitula) on a flat to hemispherical or conical receptacle, surrounded by an involucre of 1 or more whorls of usually highly modified bracts; florets of 1 or 2 kinds in each capitulum, outer ray florets and inner disk florets or all ray or disk florets. Calyx epigynous, reduced to a crown of scales, spines or hairs, or absent. Corolla tubular with 4 or 5 lobes (disk florets) or limb expanded and showy, split down one side of the rube (ray florets) or rarely 2-lobed. Stamens 5, filaments free; anthers united into a tube. Ovary 1-celled, ovule erect; style 2-fid at apex into 2 stigmas. Fruit a nut (achene), usually crowned by persistent modified calyx (pappus); indehiscent; seeds 1, basal, free from wall.

A very large cosmopolitan family of c 900 genera with over 13,000 species.

1. Shrubs or herbs; flowers mauve or white   **6. Vernonia**
1. Herbs; flowers yellow or yellow and white
  2. Capitula with ray florets only   **2. Launaea**
  2. Capitula with disk and ray florets
    3. Achenes of 1 type
     4. Capitula solitary, on long peduncle; achenes with feathery pappus
                                        5. Tridax
     4. Capitula several, in branched inflorescence; achenes without pappus or pappus reduced to 2–3 bristles
      5. Leaves all simple; achenes without pappus   **3. Melanthera**
      5. Leaves simple above, compound below; achenes with 2–3 barbed bristles   **1. Bidens**
    3. Achenes of 2 types, 3-angled, with 2–3 horn-like bristles, or flattened and winged and with 2 horn-like bristles or spines   **4. Synedrella**

1. BIDENS L.

Annual or perennial herbs. Leaves opposite, entire or variously dissected or compound; petiole present or absent. Capitula solitary, neuter florets ligulate, hermaphrodite florets tubular. Involucre of 2–3 rows of bracts, outer herbaceous, inner usually with papery margins. Ray florets white, pink, deep red, orange or yellow. Achenes compressed or 3–4 (–5)- angled; pappus absent or reduced to 2–5 barbed bristles.

A cosmopolitan genus of about 230 species.

Bidens pilosa L., Sp. Pl.: 832 (1753); F.M.S.: 169 (1877); F.M.C. 189: 663, fig. 121/7–14 (1963).

Annual, up to 1 m high; stems square, sparsely hairy to glabrous. Leaves compound, 3–5-foliolate or upper leaves simple, leaflets 3–5, ovate to oblong-ovate, 3.5–10 × 1.5–3.5 cm, apex acute, base truncate, margins ciliate, variously toothed; petiole 2–5 cm long. Capitula bell-shaped, up to 1 cm across; peduncles 2.5–6.5 cm long, arranged in sparse, lax, open cymes. Outer involucral bracts 7–10, oblong-linear, 5 mm long; inner bracts linear-oblong, larger. Ray florets white, disc florets yellow, tubular. Achenes black, oblong, 5–10 mm long, 3-sided, 4–6-ribbed; bristles 2–3, 2–4 mm long, barbed.

The above description applies to plants in the Madagascan region only.

Distr. ASTOVE; a pantropical weed.
Notes. Known from 3 collections, *Ridgeway* 96 (US); *Veevers-Carter* 96 (EA) and *Fosberg & Frazier* 49750 (K, US), in coconut plantation.
Vernac. 'la ville-bague' or 'blackjack'.

2. LAUNAEA Cass.

Glabrous herbs. Flowers yellow, all ligulate; receptacle flat. Involucre of many rows of overlapping bracts, usually with membranous margins; inner bracts subequal, outer bracts shorter. Ligule truncate, 5-toothed, Achenes narrow, 4–5-ribbed; pappus copiously bristly with fine, simple hairs.

A genus of 20 species distributed from the Mediterranean to East Asia, tropical and southern Africa and Madagascar.

1. Erect, annual herb; leaves lanceolate   **1. L. intybacea**
1. Low, stoloniferous herb; leaves spathulate   **2. L. sarmentosa**

1. Launaea intybacea (Jacq.) P. Beauv. in Bull. Soc. Bot. Genève II, 2: 114 (1910).*

Lactuca intybacea Jacq., Ic. Pl. Rar. 1: 16, fig. 162 (1781); Fosberg in K.B. 29: 259 (1974).
Sonchus oleraceus sensu Hemsley in B.M.I.K. 1919: 124 (1919), not L. (1753).

Erect annual herb, up to 2 m high; taproot stout, vertical, with milky sap. Leaves sessile, lanceolate, up to 20 cm long, deeply dissected, margins toothed. Inflorescence 10–50 cm long, branches ascending bearing clusters of up to 6 capitula on bracteate peduncles 2–3 mm long; capitula narrowly oblong, 1–1.2 cm long. Inner involucral bracts ± 8, linear, 10 mm long. Florets light yellow. Achene 3–4 mm long, strongly warty, 4-ribbed; pappus 6–8 mm long.

Distr. ALDABRA: W, S (west); ASSUMPTION; COSMOLEDO: M; ASTOVE; a native of the Carribean region, now widespread through the tropics.
Notes. Occurs around areas of settlement. Flowers during the late wet season and early dry season. Leaves eaten in salads.
Vernac. 'lasteron'.

2. Launaea sarmentosa (Willd.) Alston in Trimen, Handbook Fl. Ceylon 6: 173 (1931).

Prenanthes sarmentosa Willd., Phytogr.: 10. t.6, fig. 2 (1794).
Launaea bellidifolia Cass. in Dict. Sci. Nat. 25: 321 (1822); Hemsley in B.M.I.K. 1919: 129 (1919); F.M.C. 189 Fig. 164/4–7 (1963).
L. pinnatifida Cass. in Ann. Sci. Nat. 23: 85 (1831); F.M.C. 189: 883, fig. 164/810 (1963).
Microrhynchus sarmentosa (Willd.) DC., Prodr. 7: 181 (1838); F.M.S.: 181 (1877).

Small, slender, perennial herbs; stems slender, simple, prostrate, rooting at the nodes. Leaves basal, spathulate or narrowly obovate, up to 15 cm long, usually much less, margins entire, toothed or sinuate, rarely so deeply divided as to appear almost pinnate. Capitula 1.5–2 cm long, usually solitary on a bracteate peduncle 1–2 cm long. Inner involucral bracts ± 8, linear or narrowly oblong, 10–15 mm long. Florets yellow. Achene 3.5–5 mm long, 4-ribbed; pappus 6–10 mm long. Fig. 25/1.

Distr. ALDABRA: S; ASSUMPTION; COSMOLEDO: W, M; an Indo-Pacific strand plant.
Notes. A rather rare plant growing in loose sand on dunes or beaches above high water mark. Flowers throughout the year; florets close mid-morning.

---

* Maintained in *Launaea* by S.A.R.

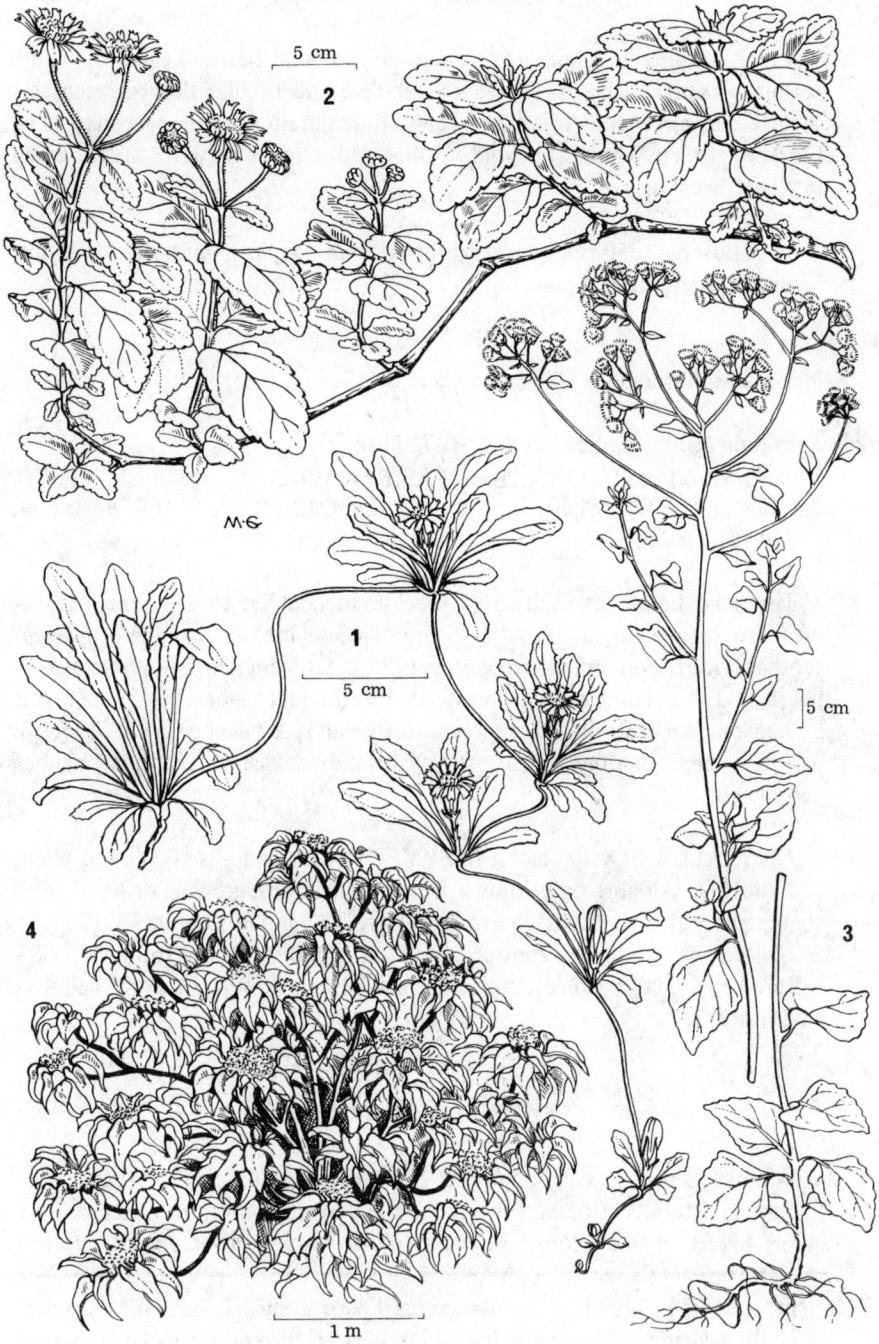

Fig. 25 1, *Launaea sarmentosa*, habit. 2, *Melanthera biflora*, habit.
3, *Vernonia cinerea*, habit. 4, *Vernonia grandis*, habit.

3. MELANTHERA Rohr

Erect, trailing or climbing, annual or perennial herbs. Leaves opposite; petiole present or absent. Capitula of ligulate and tubular florets; receptacle slightly convex. Involucre of 2–3 rows of membranous or herbaceous bracts. Florets yellow. Achenes obovoid, 3–4-angled; pappus of 0–12 stiff, barbed or ciliate bristles.

A genus of c 50 species distributed from tropical America to Africa, Madagascar and India.

Melanthera biflora (L.) Wild in Kirkia 5: 4 (1965)*.

Verbesina biflora L., Sp. Pl. ed. 2: 1272 (1762).
Wollastonia biflora (L.) DC., Prodr. 5: 546 (1836).
Wedelia biflora (L.) Wight, Contrib. Bot. Ind.: 18 (1837); F.M.C. 189: 644, fig. 118 (1963).

Perennial herbs; stems trailing, much-branched, up to 1 m long. Leaves ovate to lanceolate, 2–10 × 1–6 cm, apex acute to acuminate, margins toothed, rough on both surfaces; petiole 0.5–3 cm long. Capitula hemispherical, 1–1.5 cm across, in few-headed, terminal clusters. Involucral bracts herbaceous, scabrid. Florets yellow, shortly hairy. Achenes obovoid, 3-angled, 3.5 mm long, glabrous except for slight pubescence at the apex; pappus absent. Fig. 25/2.

Distr. ALDABRA: S (east); ASSUMPTION; an Indo-Pacific strand plant.
Notes. On Aldabra only known from 2 collections, growing on very rough champignon at Cinq Cases, *Renvoize* 880 (K, US) and *Fosberg* 48902 (K, US); 2 collections from Assumption *Stoddart* 1060 & 1105 (both K, US). Inflorescence may become extremely swollen and succulent due to the saline environment.

4. SYNEDRELLA Gaertn.

Annual herbs. Leaves opposite, toothed. Capitula small, consisting of ligulate and tubular florets. Involucre of few bracts, outer bracts herbaceous, inner bracts membranous. Florets yellow. Achenes of ligulate florets

---

* In an as yet unpublished discussion Fosberg shows that, if the genera related to *Verbesina* L., including *Melanthera* Rohr, are not to be combined in a single, large, ill-defined genus *Verbesina* L., the genus *Wollastonia* DC. ex Doone should be restored for the species originally described as *Verbesina biflora* L., here called *Melanthera biflora* (L.) Wild, and its close relatives. *Wollastonia* is closest to, if it does not actually include, the Hawaiian genus *Lipochaeta* DC. rather than to *Melanthera*.

compressed, 2-winged, achenes of tubular florets compressed or 3-angled, not winged.

A genus of c 50 species in tropical America, Africa, India and Madagascar.

Synedrella nodiflora (L.) Gaertn., Fruct. 2: 456, t.171, fig. 7 (1791).

Verbesina nodiflora L., Cent. Pl. 1: 28 (1755).

Pubescent herbs 13–30 cm high. Leaves opposite, elliptic-lanceolate, 3–7 × 1–4 cm, apex acute, base wedge-shaped, margins toothed; petiole 0.5–2 cm long. Capitula solitary or clustered, sessile or subsessile, terminal and axillary, oblong-ovoid, 1 cm long. Involucral bracts broad, outer bracts 2, green, 7.5–10 mm long, inner papery. Florets few, c 15, yellow. Achenes of tubular florets 2–3-angled, 5 mm long, bearing, 2–3, apical, spine-like, bristles, 3 mm long; achenes of lingulate florets oblong, with 2 pappus spines, flattened, 4 mm long, margins winged, wings frilled, often reduced to a series of slender segments 2–3 mm long. Fig. 28/1–2.

Distr. ALDABRA: W; a native of tropical America now widespread through the tropics.

Notes. An abundant weed at Settlement on Aldabra, only known from 1 collection, *Wickens* 3512 (K), apparently a recent introduction. Observations few but apparently flowers during the mid dry season.

5. TRIDAX L.

Annual or perennial herbs. Leaves opposite, toothed. Capitula solitary, composed of ligulate male florets and tubular hermaphrodite florets; receptacle flat or convex. Involucre of 2 rows of bracts, outer bracts short, broad and herbaceous, inner bracts narrower and membranous. Florets white and yellow. Achenes oblong or obovoid; pappus of long or short, feathery bristles.

A genus of 26 species in Mexico and tropical America.

Tridax procumbens L., Sp. Pl: 900 (1953); F.M.S. 170 (1877); F.M.C. 189: 666, fig. 115/29–32 (1963).

Pubescent herbs with erect or ascending stems up to 20 cm high, excluding the inflorescence. Leaves ovate or elliptic, 2–6 × 1–3 cm. Capitula solitary, ovoid, 1 cm long, borne on long, erect peduncles, 15–30 cm long. Outer bracts ovate-acuminate, half the length of the florets or less; inner bracts oblong-obtuse. Ligulate florets white; tubular florets yellow. Achenes oblong, 3 mm long; pappus feathery, 6 mm long. Fig. 28/3–4.

Distr. ALDABRA: W; native of tropical America, now a pantropical weed. Notes. Apparently a recent introduction, only known from 2 records, *Huatiuk* 731757 and *Merton* 7065 (both K) from Settlement on Aldabra. Vernac. 'herbe caille'.

6. VERNONIA Schreb.

Trees, shrubs or herbs. Leaves alternate, rarely basal. Capitula solitary or usually in a branching inflorescence; flowers tubular, receptacle honey-combed. Involucre of 4–10 rows of overlapping bracts. Florets purple, mauve or white. Achenes columnar or obovoid; pappus in 2 rows, outer of short scales or rarely bristles, inner of many, long, serrated bristles.

A large genus of over 1000 species, widespread throughout the world.

1. Small erect annual; florets mauve   **1. V. cinerea**
1. Stout shrub up to 4 m high; florets mauve or white   **2. V. grandis**

1. Vernonia cinerea (L.) Less. in Linnaea 4: 291 (1829); F.M.S.: 161 (1877); F.M.C. 189: 18, fig. 3/11 (1960).

Conyza cinerea L., Sp. Pl.: 862 (1753).

Erect, sparsely pubescent, annual herbs, 20–75 cm high. Leaves ovate, obovate or lanceolate, 2–5 × 0.5–3 cm, apex acute, base tapering, margins serrate to subentire; petiole 1–2 cm long. Capitula small, clustered in a spreading, much branched, terminal inflorescence. Involucre bracts in 5 series, linear-lanceolate, 2–4 mm long, apex acute or acuminate. Florets mauve. Achenes oblong or ellipsoid, 1–2 mm long, hairy, topped by a corona of short upward ciliate hairs; pappus white, of 1 row of upward ciliate bristles, 3 mm long. Fig. 25/3.

Distr. ALDABRA, W; ASSUMPTION; ASTOVE. pantropical weed.
Notes. Found near areas of habitation; little collected: Aldabra, *Renvoize* 773; Assumption, *Stoddart* 1040 *Frazier* 24; Astove, *Renvoize* 1210 and *Stoddart & Poore* 1276, 1301 (all K).
Vernac. 'herbe quérit vite' or 'herbe de flacque'.

2. Vernonia grandis (DC.) J. Humb., Fl. Madag. 189: 44, fig. 8 (1960).

Decaneurum grande DC., Prodr. 5: 67 (1836).
Vernonia aldabrensis Hemsley in J. Bot. 54, Suppl. 2: 20 (1916) & in B.M.I.K. 1919: 124 (1919). Type: Aldabra, *Fryer* 41 (K, holo.).
Psiadia sp., Hemsley in B.M.I.K. 1919: 124 (1919).

Spreading shrub or small tree, 2–4 m high; young branches pubescent, becoming glabrous with age. Leaves lanceolate, 10–17 × 4–10 cm, apex and base tapering, margins entire, both upper and lower surfaces covered with small dot-like glands, subglabrous to sparsely pubescent with pubescent nerves; petiole 1–2.5 cm long. Capitula numerous, ovoid, 4–5 mm across, clustered in large, much-branched, terminal inflorescences, 10–20 cm across. Involucral bracts in 4–5 rows, lanceolate-acuminate or the uppermost bracts acute, margins ciliate, apex glandular. Florets white or pale mauve, fragrant. Achenes narrowly obovoid, 2–3 mm long, ribbed, glandular; pappus of 1 row of upward ciliate or spinous bristles 5 mm long. Fig. 25/4.

This species is very similar to *V. colorata* (Willd.) Drake in mainland Africa. The principal differences appear to be in the leaves, which in *V. colorata* are densely pubescent beneath and generally less tapered at the base and apex. Some specimens of *V. colorata* however approach *V. grandis* in leaf characters. Differences between these species do undoubtedly exist but at what level they should be recognized cannot be decided until the genus has been studied in depth. Humbert in Flora of Madagascar, l.c. divides *V. grandis* into several varieties distinguished principally on the shape of the bracts. The precise segregation however is not clear and the species is here recognised in the broad sense.

Distr. ALDABRA: W, P, M, S, Michel; ASSUMPTION; COSMOLEDO: M; ASTOVE; Madagascar and Comoro Is.
Notes. Occurs mainly in *Pemphis* scrub. Flowers mainly during the wet season, but sporadically during the dry season if sufficient moisture available.

39. GOODENIACEAE

Herbs, shrubs, or small trees. Leaves alternate or rarely opposite or basal, simple; stipules absent. Inflorescence cymose, or racemose, or flowers solitary, axillary; flowers bisexual, usually 5-merous. Calyx lobed or cup-shaped. Corolla irregular, limb 5-lobed or 2-lobed, lobes with margins touching or rolled inwards. Stamens 5, alternate with corolla lobes, filaments free, epigynous or attached to base of corolla, anthers free or coherent to form a cylinder around style. Ovary usually inferior, 1–2-celled, placentation basal or axile, ovules 1, 2, or many, ascending; style 1, stigma simple or lobed, subtended by a cup-like flap. Fruit a capsule, berry, drupe or nut.

A small Australian family with a few Pacific island species and 2 widespread strand species.

SCAEVOLA L.

Shrubs or rarely small trees. Leaves usually alternate, usually with white hairs in axils. Flowers in axillary cymes or solitary, rarely apparently in racemes. Calyx truncate, cup-shaped, more rarely 5-lobed. Corolla with tube split down dorsal side, usually to base, lobes with membranous margins, inrolled in bud, lobes spreading in a fan-like arrangement. Ovary 2-celled, ovules 1 per cell; style usually curved at apex. Fruit a drupe with a bony stone which in some species is surrounded by softer, aerogenous or corky tissue, 2-celled and with 2, small, empty cavities; seeds 1–2.

Principally Australian, with a few insular species and two widespread strand species.

Scaevola taccada (Gaertn.) Roxb., Hort. Beng. 15 (1814).*

Lobelia taccada Gaertn., Fruct. Sem. 1: 119, t. 24, fig. 5 (1788).
Scaevola koenigii Vahl, Symb. Bot. 3: 36 (1794); F.M.S.: 182 (1877); Baker
    in B.M.I.K. 1894: 148 (1894); Schinz in A.S.N.G. 21: 91 (1897);
    Voeltzkow in A.S.N.G. 26: 552 (1902); Dupont, Report: 37 (1907);
    Hemsley in B.M.I.K. 1919: 124 (1919).

Rounded, mound-like shrubs, or when well-developed, small trees; branching characterized by the growth of a branchlet from the lower side of the terminal branchlet and extending beyond it. Leaves usually in conspicuous rosettes at ends of branchlets, obovate, 15–21 × 7–9 cm, apex rounded, base wedge-shaped, decurrent into the petiole, venation except the midrib obscure; petiole 5 mm long, tuft of long white hairs in leaf axils. Cymes axillary, slender. Corolla white or purple, tube 15 mm long, split down one side, white hairs at the throat, lobes spreading, 13 × 6 mm. Style curving upwards through the corolla slit; stigma wedge-shaped, pointing towards the throat. Drupe white or purplish, ovoid or subglobose, 1.3 cm in diameter, soft, tasteless; stone ellipsoid, ribbed, outer layer aerogenous or corky. Fig. 26/1–3.

Distr. ALDABRA: W, P, M, S; ASSUMPTION; COSMOLEDO: W, M; ASTOVE; an Indo-Pacific strand plant.

Notes. Occasional in coastal areas on champignon and sand, more common in interior on platin where it is locally a component of mixed scrub, very locally forming pure stands. Flowers throughout the year but mainly during the wet season. Viable seed found in beach drift; seeds can float for at least 8 weeks and subsequently germinate. Drupe eaten and seeds locally dispersed by tortoise, pigeon, turtle-dove and bulbul.

Vernac. 'bois manioc' or 'veloutier manioc'.

---

* According to F.T.E.A. Good.: 1 (1978) the correct name of this species should be *Scaevola sericea* Vahl [Symb. Bot. 2: 37 (1791)]. The relevant additional synonymy is as follows: *Scaevola taccada* Roxb., Hort. Beng.: 15 (1814) (not based on *Lobelia taccada* Gaertn. (1788)); *S. taccada* (Gaertn.) Roxb., Fl. Ind. 2: 146 (1824), name illegit. (not *S. taccada* Roxb. (1814)).

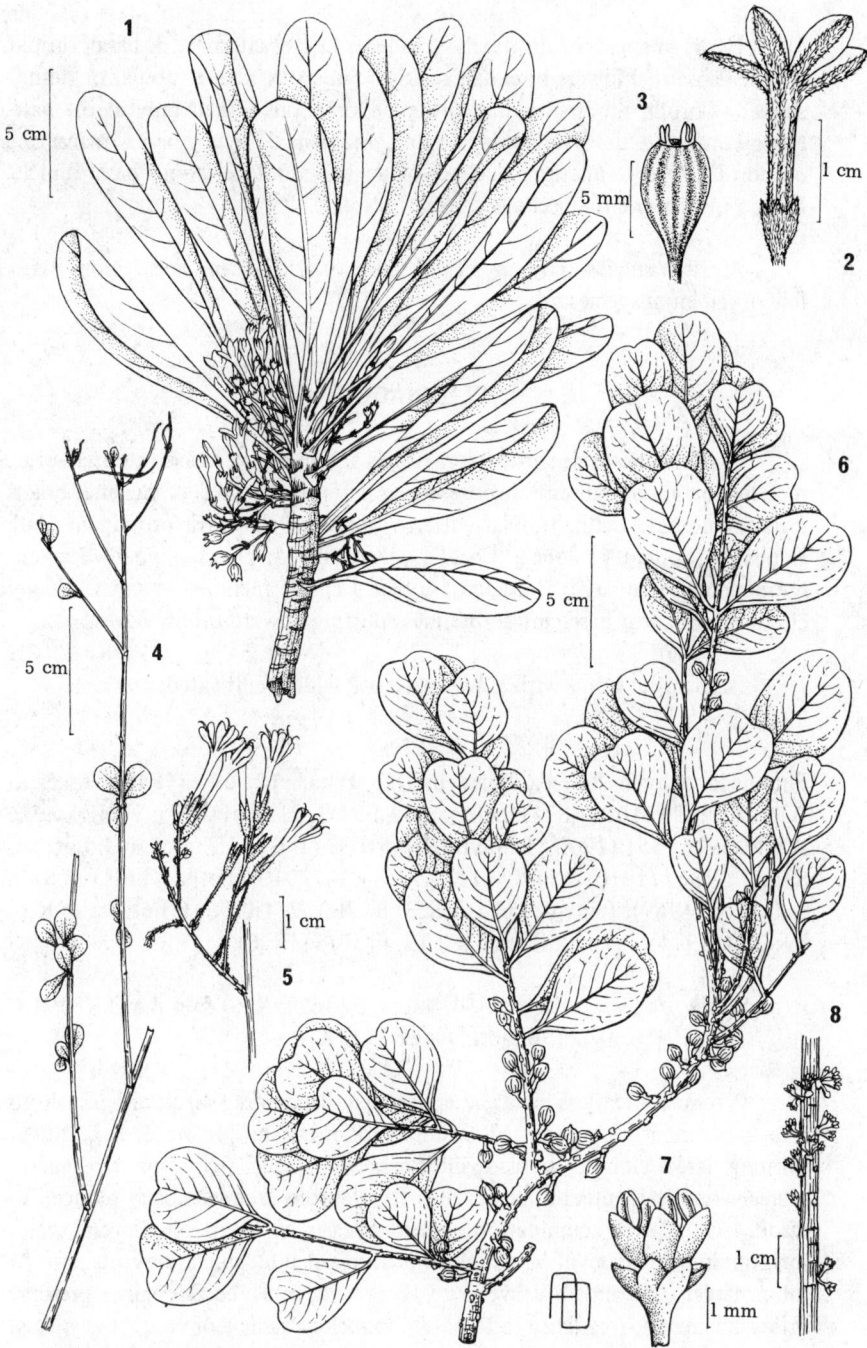

Fig. 26 1, *Scaevola taccada*, habit; 2, flower viewed from the entire side;
3, fruit. 4, *Plumbago aphylla*, habit of leafy form; 5, flowers.
6, *Sideroxylon inerme* subsp. *cryptophlebia*, habit; 7, flower;
8, flowering section of stem.

40. PLUMBAGINACEAE

Herbs or small shrubs, rarely scandent. Leaves alternate or basal, simple; stipules absent. Flowers bisexual, regular, 5-merous. Calyx tubular or funnel-shaped. Corolla tubular or funnel-shaped or petals joined only at the base, lobes contorted and overlapping in bud. Stamens 5, hypogynous or inserted on corolla. Ovary superior, 1-celled, ovule 1, pendulous from a basal funicle. Fruit a utricle or circumscissile capsule.

A rather small family, mostly of saline or arid or semi-arid regions, with a few ornamental species.

PLUMBAGO L.

Perennial herbs or shrubs, sometimes subscandent. Leaves alternate, rarely reduced or almost absent. Inflorescence a terminal spike or raceme; bracts usually present. Calyx tubular, often ribbed, usually with prominent stalk glands, 5-toothed or lobed. Corolla salver-shaped, broadly lobed. Stamens hypogynous, free; style 1, stigma 5-lobed. Capsule included in calyx, usually circumscissile near base, and sometimes splitting upward into 5 valves.

Pantropical genus with a few species, 2 widely cultivated.

Plumbago aphylla Boj. ex Boiss. in DC., Prodr. 12: 694 (1848); Baker in B.M.I.K. 1894: 148 (1894); Schinz in A.S.N.G. 21: 88 (1897); Voeltzkow in A.S.N.G. 26: 551 (1902); Dupont, Report: 37 (1907); Hemsley in J. Bot. 54, Suppl. 2: 362 (1916) & in B.M.I.K. 1919: 124 (1919); Fosberg in P.T.R.S. B, 260: 216 (1969); Renvoize in P.T.R.S. B, 260: 231 (1969); Fosberg in K.B. 29: 259 (1974); F.T.E.A. Plumbag.: 12, fig. 3/8 (1976).

P. parvifolia Hemsley in J. Bot. 54, Suppl. 2: 362 (1916) & in B.M.I.K. 1919: 124 (1919). Type: Aldabra, *Fryer* 115 (K, holo.).

Practically leafless, pale glaucous green, broom-like shrub or herb, up to 1 m tall; root-stock woody, sometimes prolonged and prostrate, 1 cm thick, forming large clones; stems cylindrical but deeply furrowed, alternately sparsely branched, erect branches in tufts, glabrous. Leaves mostly reduced to black, triangular or acuminate, toughened scales, c. 1 mm long, several scales surrounding the base of a branch or a dormant bud; small, grey obovate or suborbicular expanded leaves c 10 × 4 mm occasionally present. Inflorescence an irregularly, alternately branched panicle of congested spikes, with scale-like bracts, smaller branches with very sticky, stalked glands, these more abundant distally and on spikes and bracts, ultimate branches and spikes with short, dense, white hairs; flowers very easily disarticulating from spikes. Calyx tubular, 7–8 mm long, pubescent, abundantly glandular, lobes tardily separating to about half this length. Corolla white, thinly parchment-like, tube 13−15 mm long, lobes wedge-shaped-obcordate, 4−6 mm long, with a projecting central nerve several mm long in the notch. Stamens and style included. Capsule spindle-shaped, 8 × 2 mm, strongly toughened, anomalous in the genus in not being clearly circumscissile but the valves separating from the base. Fig. 26/4−5.

Further investigation, using living plants, of the nature of the capsule and its dehiscence would be desirable.

Distr. ALDABRA: W, P, M, S, Esprit, Michel; ASSUMPTION; COSMOLEDO: W; ASTOVE; Tanzania; Madagascar.
Notes. Locally dominant in coastal and inland scrub. Flowers throughout the year if sufficient moisture available, but mainly during the wet season. Fruit readily adhere to passing birds, animals and man.
Vernac. 'herbe paille en queue'.

41. SAPOTACEAE

Trees and shrubs, generally with milky sap. Leaves usually alternate, simple, entire, usually leathery; stipules rarely present. Flowers axillary, solitary or in fascicles or cymes, bisexual, regular. Sepals 4−12, or shortly united, lobes in 1 or 2 whorls or spirally arranged. Corolla bell-shaped to shortly tubular, 4−5 or more-lobed, lobes overlapping in bud. Stamens in 2−3 whorls, outer whorl often reduced to staminodes, inserted on the corolla. Ovary superior, 4−5 or more-celled, 1 ovule per cell; style 1, stigma minute. Fruit a berry; seeds 1 or more, large, hard, with conspicuous attachment scar, its size and placement characteristic of the genera.

A medium-sized mostly tropical family, very difficult taxonomically.

SIDEROXYLON L.

Trees or shrubs. Leaves thick, leathery; stipules usually absent. Flowers densely clustered in the leaf axils or on the lower, leafless parts of the branches, pedicelled. Sepals 5, free or fused at the base, lobes ovate or rounded. Corolla lobes 5, the tube often of equal length to the lobes. Stamens 5, as long as or longer than the corolla; staminodes petaloid or strap-like, wavy or fringed on the margin. Ovary subglobose or conical. Fruit a subglobose berry, bearing the short persistent style; seeds 1.

A genus of c 100 tropical species, mostly in South America, the number depending on the circumscription of the genus adopted.

Sideroxylon inerme L., Sp. Pl.: 192 (1753); Hemsley in J. Bot. 54 Suppl. 2: 23 (1916) & in B.M.I.K. 1919: 125 (1919); F.T.E.A. Sapot.: 33 (1968).

Large shrubs or small trees, up to 5 m high. Leaves elliptic to obovate or suborbicular, 2−10 × 2−6 cm, apex obtuse to notched, base tapering; petiole 1−3 mm long. Flowers tiny, pedicels up to 7 mm long. Sepals ovate, up to 2.5 mm long. Corolla greenish-white, tube up to 1.5 mm long, lobes up to 2.5 mm long. Ovary subglobose; style short, up to 1.5 mm long. Fruit green, becoming black when mature, globose, 6−15 mm in diameter; seeds fleshy, 1, depressed-subglobose, 4−9 mm in diameter, smooth, with 1−5 longitudinal ridges and 1−4 small pits at the base.

Distr. (of species as a whole) Somalia to the Cape and the Aldabra Group

subsp. **cryptophlebia** (Baker) J. H. Hemsley in K. B. 20: 477 (1966).

Myrsine cryptophlebia Baker in B.M.I.K. 1894: 149 (1894); Schinz in A.S.N.G. 21: 88 (1897); Voeltzkow in A.S.N.G. 26: 551 (1902); Dupont, Report: 38 (1907): Hemsley in B.M.I.K. 1919: 125 (1919). Type: Aldabra, *Abbott* s.n. (K. holo).

Leaves shortly and broadly obovate to suborbicular, 2.5−8 × 3−5.5 cm, abruptly tapering at the base, glabrous or subglabrous, greyish green. Flowers slightly smaller than in the typical subspecies; pedicels up to 3 mm long, elongating in fruit to 4 mm. Sepals 2 mm long, persistent in fruit. Corolla 3 mm long, seeds up to 4.5 mm in diameter, ridges obscure. Fig. 26/6−8.

This subspecies is distinguished from subsp. *inerme* and subsp. *diospyroides* (Baker) J. H. Hemsley from the African mainland by suborbicular or more obovate and slightly smaller leaves, smaller flowers on shorter pedicels and smaller seeds.

May be confused during the dry season with *Margaritaria anomala* (Family 62/3) which has a distinctly netted appearance to the undersurface of the leaf, with veins at c. 80° to the midrib instead of 45°·

Distr. ALDABRA: W, P, M, S, Esprit, Michel; ASSUMPTION; COSMOLEDO: W, M; ASTOVE; endemic.

Notes. A frequent constituent of the mixed scrub. Locally heavily infested by woolly coccids. Flowers mainly during the wet season but unseasonal rains can also stimulate flowering; most leaves may fall during periods of drought. Fruit eaten by the blue pigeon.

Vernac. 'bois zak' or 'bois de natte'.

42. OLEACEAE

Trees or shrubs, sometimes trailers or scramblers. Leaves opposite, rarely alternate or subalternate, simple or compound; stipules absent. Flowers regular, in usually few-flowered fascicles, cymes or panicles. Sepals free or united, 4–5-lobed. Corolla 4–5-lobed, rarely free, margins touching or overlapping. Stamens 2. Ovary superior, 2-celled; style 1, stigma minute. Fruit a capsule berry or drupe; seeds 1–4.

A rather small, widely distributed family, containing the domestic olive and a number of ornamentals, including the lilacs, privets, and jasmines.

JASMINUM L.

Shrubs or climbers. Leaves opposite, simple or compound; petiole usually present. Flowers bisexual, solitary or in terminal or axillary clusters. Sepals united, lobes 4–10, short or elongate. Petals often scented, white, yellow or tinged with pink, longer than the sepals, united to form a tube, lobes 4–10. Stamens 2, inserted high up inside the corolla tube; filaments short. Ovary 2-celled, 2 ovules per cell; style slender. Fruit a simple or 2-lobed berry; seeds 1 per lobe.

A genus of 290 species, distributed throughout the tropical and warm temperate regions of the Old World.

Jasminum elegans Knobl. in Engl., Bot. Jahrb. 17: 538 (1893); F.M.C. 166: 83, fig. 17 (1952).

J. mauritianum sensu Baker in B.M.I.K. 1894: 149 (1894); Schinz in A.S.N.G. 21: 88 (1897); Voeltzkow in A.S.N.G. 26: 551 (1902); Dupont, Report: 39 (1907), not Bojer ex DC. (1844).
J. aldabranum Gilg & Schellenb. in Engl., Bot. Jahrb. 51: 86 (1913). Type: Aldabra, *Abbott* s.n. (B, holo. †, K, US, iso.).
J. aldabrense Hemsley in J. Bot. 54, Suppl. 2: 21 (1919) & in B.M.I.K. 1919: 125 (1919). Types: Aldabra, *Abbott* s.n. (K, syn., US, isosyn.); *Thomasset* 247 (K, syn.); *Fryer* 6 & 21 (both K, syn.).

Small, evergreen, somewhat woody shrubs, up to 1 m high; stems scrambling, pubescent. Leaves 3-foliate, central leaflet usually much larger than the lateral leaflets, narrowly to broadly ovate, 2.5–5.5 × 2–3.5 cm, apex obtuse, base sub-wedge-shaped, margins entire, glossy green above, glabrous except for a pubescent midrib; petiolule of central leaflet 5–10 mm long, pubescent, of lateral leaflets 1–2 mm long; petiole 5–10 mm long, pubescent. Flowers in terminal or axillary clusters. Sepals united at the base into a small cup 2–3 mm long, lobes 5, 1–7.5 mm long, pubescent. Petals white, faintly scented, united in the lower part into a tube 12–20 mm long, lobes 5, spreading or reflexed, narrowly ovate, 6–10 mm long. Anthers linear, 3 mm long, situated at the mouth of the tube. Style slender, as long as the corolla tube. Fruit a purple-black, globose berry, 10–12 mm in diameter, glabrous; seeds purple-black, globose, 6 mm in diameter. Fig. 27/1–2.

Distr. ALDABRA: W, P, M, S, Esprit, Michel; NW. Madagascar.

Notes. Occurs frequently throughout Aldabra, growing on coral limestone rock as an undershrub in the mixed scrub. Flowers mainly during the late wet season, also sparsely throughout the year.

Vernac. 'jasmin'.

43. SALVADORACEAE

Trees or shrubs, unarmed or with axillary spines. Leaves opposite, simple, entire, often thick and leathery; stipules minute or absent. Flowers bisexual or dioecious, in dense terminal or axillary clusters. Calyx-lobes 2–4. Petals 4, free or united at the base. Stamens 4, attached to the corolla just below the lobes or at the base of the petals and alternate with them, filaments free or united at the base; glands sometimes present and alternating with the stamens. Ovary superior, 1–2-celled, ovules 1–2, erect; style short. Fruit a berry or drupe, seed 1–2.

A small palaeotropical family of 3 genera.

1. Shrubs, armed with axillary spines; petals free; fruit a berry   **1. Azima**
1. Low-shrubs, unarmed; petals united at the base; fruit a drupe
                                                                  2. Salvadora

1. AZIMA Lam.

Dioecious, much-branched, often spiny shrubs. Flowers small, clustered in the leaf axils. Calyx 4-toothed or 2-lobed. Petals 4, free, lanceolate. Male flowers: stamens 4, free, slightly longer than the petals and alternating with them; anthers oblong, almost as long as the filaments; ovary rudimentary. Female flowers: staminodes 4, small, alternating with the petals; anthers rudimentary; ovary ovoid, 2-celled, 1 or 2 basal ovules per cell; stigma sessile, bilobed, pubescent. Fruit a 1 or 2 seeded berry.

A genus of 4 species distributed from tropical and southern Africa and Madagascar to India, Malaya and the Philippines.

Azima tetracantha Lam., Encycl. Méth. Bot. 1: 343 (1783) & Illustr. t. 807 (1799); Hemsley in B.M.I.K. 1919: 125 (1919); F.M.C. 118: 3, fig. 1. (1968).

Densely branching shrubs, up to 3 m high; branches pale, glabrous, striated, bearing at each node 4 spines 0.5–3 cm long, spines rarely absent. Leaves ovate or elliptic, 3–6 × 0.7–3 cm, apex spine-tipped, glabrous, leathery, somewhat shiny above; petioles short, 2–3 mm long; stipules spinose.

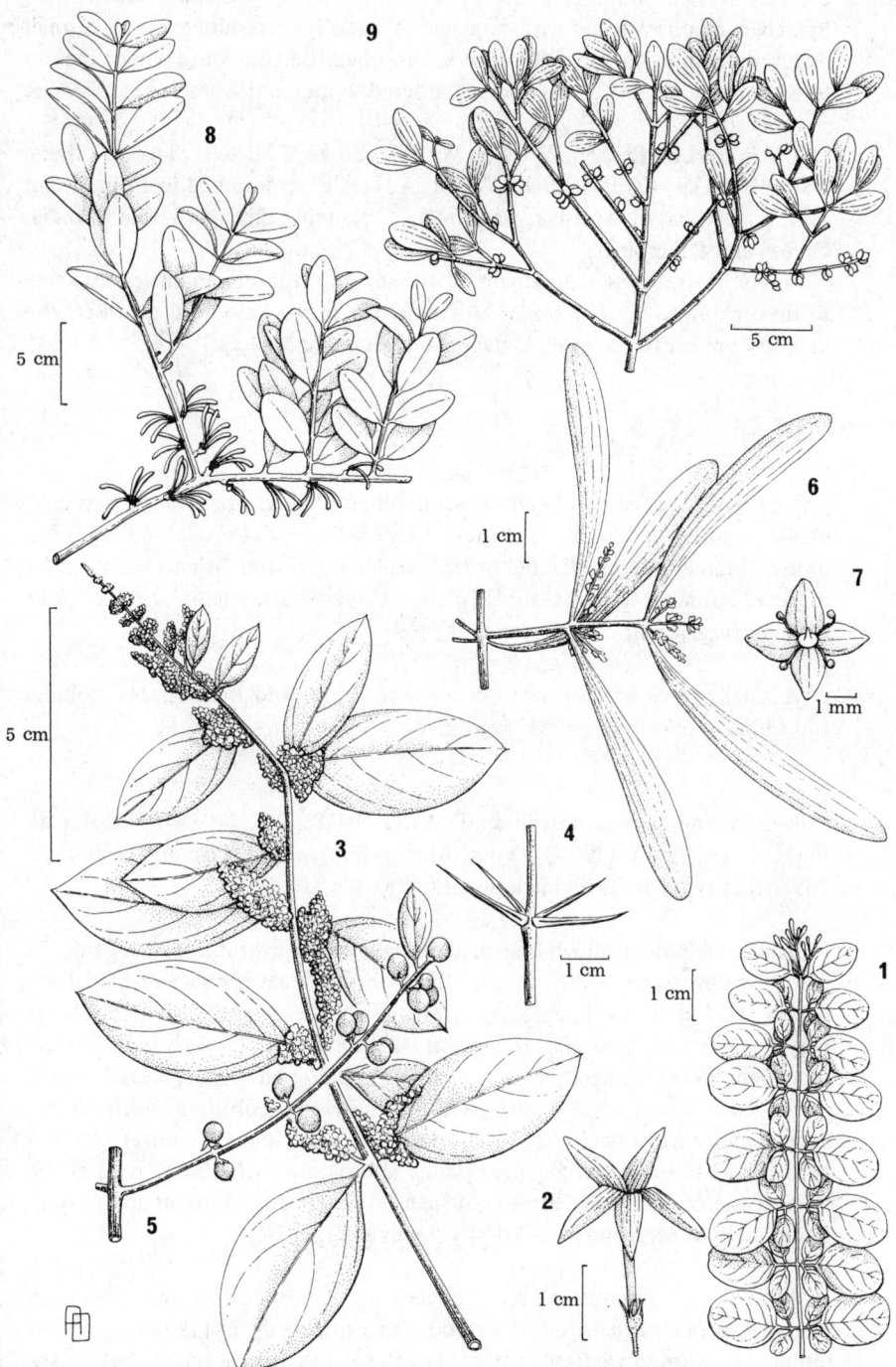

Fig. 27   1, *Jasminum elegans*, leaves; 2, flower. 3, *Azima tetracantha*,
flowering branch; 4, spines; 5, fruits. 6, *Salvadora angustifolia*,
leaves and flowers; 7, flower. 8, *Bakerella clavata* var. *aldabrensis*,
habit. 9, *Viscum triflorum*, habit.

Flowers 3 mm long, greenish yellow, in opposite pairs or clustered on branched or unbranched axillary racemes, each flower subtended by a small pubescent, spine-tipped bract. Calyx 4-toothed. Berries white when ripe, up to 8 mm in diameter; seeds 2, black, discoid, 6 mm in diameter. Fig. 27/3–5.

Distr. ALDABRA: W, P, M, S, Esprit, Michel, lagoon islets; ASSUMPTION; COSMOLEDO: W, M; ASTOVE; widespread in tropical and southern Africa, Madagascar; Comoro Is., extending through Arabia to India, Ceylon and Philippines.

Notes. A frequent constituent of the mixed scrub communities. Appears to flower during the dry season. A few female plants have been recorded, the majority appear to be male. A favourite food of the tortoise.

2. SALVADORA L.

Unarmed trees or shrubs, often scrambling. Flowers small, hermaphrodite, sessile or pedicelled, in dense clusters. Calyx 4-toothed. Corolla 4-lobed, lobes obtuse. Stamens 4, inserted below the corolla lobes, alternating with 4 glands, or glands absent. Ovary 1-celled, ovule 1; style short, stigma cap-like. Fruit globose, fleshy, with a hard stone.

A small genus of 4 species occurring in Africa and from Arabia to India and China.

Salvadora angustifolia Turrill in B.M.I.K. 1918: 202 (1918); Hemsley in B.M.I.K. 1919: 125 (1919). Types: Aldabra, *Dupont* 15; Cosmoledo, *Dupont* 289; other syntypes from Madagascar (all K, syn.).

Low, scrambling, much-branched, tough, woody shrubs or small tree, to 3 m high; bark greyish-brown, smooth, glabrous. Leaves oblanceolate, strap-like, 5–10 × 0.5–1 cm, apex acute or obtuse, base tapering to a very short petiole, 3-nerved, glabrous, grey-green, slightly fleshy, Flowers (not seen on specimens from our area) sessile, in opposite pairs, on short spikes, clustered in the axils of the leaves. Calyx green, 1.5 mm long. Corolla greenish-yellow, tube 3 mm long, lobes 1 mm long, often reflexed. Stamens shorter than the corolla lobes; anthers triangular, glands alternating with the stamens at the base of the corolla tube. Fruit (not seen on specimens from our area) ovoid, 5 mm in diameter, (immature) fleshy; seeds 1. Fig. 27/6–7.

The above description only applies to specimens in our area and Madagascar (var. *angustifolia*). Verdoorn in Bothalia 7: 14 (1958) segregated under var. *australis* (Schweik) Verdoorn the South African plants, which are small trees or shrubs with pubescent leaves and branchlets.

Distr. (var. *angustifolia*) ALDABRA: S (west); COSMOLEDO; Madagascar.
Notes. Found on Aldabra only at Anse Tambalico, *Hnatiuk* 731263 (K). Does not appear to be native. No fruiting or flowering material collected;

more material desired. Dupont records that the leaves and twigs of this species are boiled together to make a drug. Verdoorn (1963) in Fl. Southern Africa records that leaf infusions are used for sore eyes by natives east of Lebombo Mountains.

Vernac. 'tampalocoque' or 'tambalico'.

44. APOCYNACEAE

Trees, shrubs, climbers and herbs; usually with milky sap. Leaves opposite or whorled, rarely alternate, simple, entire; stipules usually absent. Flowers in cymes, usually with bracts or solitary in leaf axils. Calyx usually 5 free sepals or at least deeply lobed. Corolla various in form, lobes 5, usually overlapping and contorted in bud. Stamens 5, inserted at top of corolla tube, alternate with lobes; anthers sessile or nearly so, united in a closed ring around the stigma, 2-celled. Ovaries of 2, rarely 3–5, separate carpels united at apex by a single style and surrounded at base by a varied arrangement of glands or nectaries, each ovary 1-celled, ovules usually several to many; style 1, stigma (or clavuncle) thickened, variously shaped. Fruit usually 2 dry or fleshy follicles or rarely drupes or berries, sometimes 1 by abortion, or rarely a capsule; seeds few to many, often bearing a tuft of hairs or winged.

A fairly large, mostly tropical family.

1. Herbs; flowers pink, purple or white   **2. Catharanthus**
1. Trees or shrubs; flowers usually white
  2. Leaves ovate, glossy above, up to 3 cm long   **1. Carissa**
  2. Leaves elliptic, dull above, usually over 5 cm long   **3. Pandaca**

1. CARISSA L.

Shrubs or trees, often spiny. Leaves opposite, rarely whorled, leathery, persistent. Flowers in axillary cymes. Calyx minute, deeply 5-toothed. Corolla white or pinkish, tube slender, hairy within, lobes 5, oblong-ovate, acute, twisted to the right. Stamens 5, inserted half-way down the corolla tube, filaments short, anthers oblong. Ovary 2-celled; style shorter than the corolla tube; stigma 2-fid. Fruit a large or small berry.

c 40 species in the Old World tropics.

Carissa edulis Vahl, Symb. Bot. 1: 22 (1790); F.M.C. 169: 23 (1976).

C. sechellensis Baker, F.M.S.: 222 (1877).

C. edulis Vahl var. *sechellensis* (Baker) Pichon in Mém. Inst. Sci. Madag. Sér. B, 2: 33, fig. 1/1 (1949).
C. edulis Vahl subsp. *madagascariensis* Pichon var. *sechellensis* (Baker) Pichon in Mém. Inst. Sci. Madag. Sér. B, 2: 134 (1949).

Small, erect, glabrous, evergreen tree, 6–8 m high; trunk grey, deeply grooved, branches slender, spines absent. Leaves opposite, broadly ovate to sub-circular, 2–5 × 1.5–3 cm, apex blunt or acute, base rounded to broadly wedge-shaped, glossy dark green above, dull light green beneath; petiole 1–2 mm long. Flowers sweetly scented, in axillary cymes; pedicels 5–7 mm long. Calyx 1.5–2 mm long, deeply 5-toothed, margins of teeth pubescent. Corolla white, tube 6–8 mm long, lobes 3.5–5 mm long, spreading. Fruit not seen. Fig. 28/4–6.

The above description applies to Aldabra material only, elsewhere this species varies considerably in presence and absence of spines, pubescence, shape and size of leaves, density of the inflorescence, and length of corolla tube. Although many varieties are recognized by some authors we have not attempted to equate our plants with any of them. The genus is awaiting revision.

Distr. ALDABRA: S (east); tropical Africa, Madagascar and the Seychelles to tropical Asia.
Notes. Only known for the Bassin Frigate area, *Gibson* 7, 17, 18 (all K) apparently widely scattered but rare, in the depths of groves within approximately 3/4 km ^{2} immediately south and east of Bassin Frigate.
Vernac. 'bois sandal' or 'sandal'.

2. CATHARANTHUS G. Don

Erect herbs, or somewhat woody. Leaves opposite, entire. Flowers axillary. Calyx 5-lobed. Corolla salver-shaped, tube slender, apex club-shaped and bearded within, throat contracted, lobes 5, obovate. Anthers 5, sessile. Ovaries 2, elongate, with 2 similar glands at the base; style 1, stigma capitate, enclosed by anthers, bearded at top, with a membranous collar below. Fruit of 2, cylindrical follicles, with a double false septum; seeds 16–20, small, without a tuft of hairs.

A mainly Madagascan genus of 8 species; 1 cultivated as an ornamental and is now widespread and naturalised throughout the tropics.

Catharanthus roseus (L.) Don, Gen. Hist. 4: 95 (1838); F.M.C. 169: 152 (1976).

Vinca rosea L., Sp. Pl.: 305 (1753); F.M.S.: 224 (1877); Baker in B.M.I.K. 1894: 49 (1894); Schinz in A.S.N.G. 21: 89 (1897); Voeltzkow in A.S.N.G. 26: 551 (1907); Dupont, Report: 39 (1907).

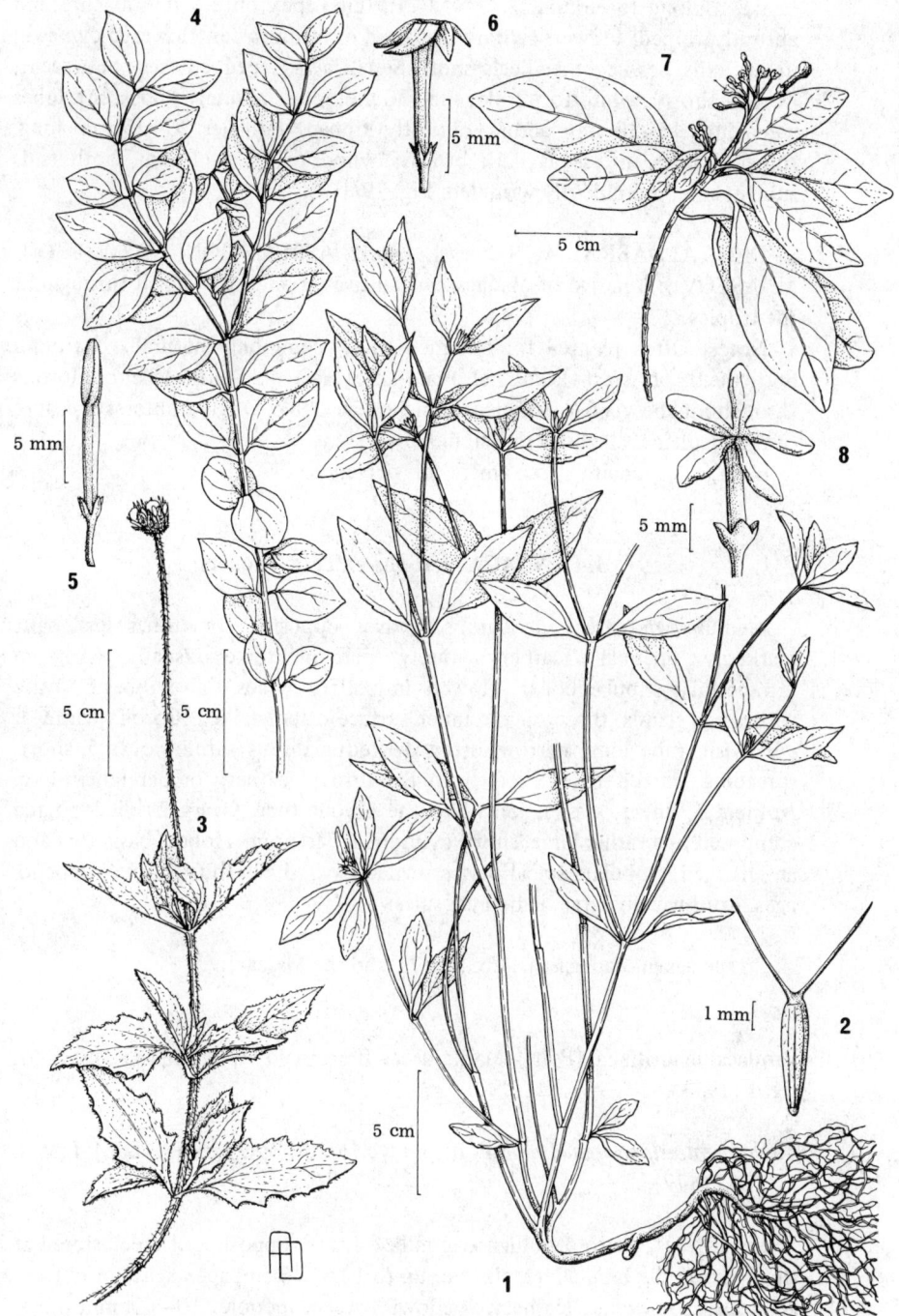

Fig. 28 1, *Synedrella nodiflora*, habit; 2, fruit. 3, *Tridax procumbens*,
       habit. 4, *Carissa edulis*, habit; 5, flower in bud; 6, flower open.
       7, *Pandaca mauritiana*, habit; 8, flower.

Erect, sparsely branched herbs to 60 cm tall, minutely or thinly pubescent. Leaves oblong to elliptic, 2–3 × 1–1.5 cm, apex obtuse to subacute and abruptly tipped. Flowers solitary or paired or in short, few-flowered cymes in upper axils, sessile or pedicels short. Sepals awl-shaped, sparsely pubescent. Corolla showy, white to purple, tube 30 mm long, minutely pubescent, lobes spreading, 15–30 mm across, abruptly tipped. Follicles 20–25 mm long, ribbed, pubescent; seeds dark brown, cylindrical, 2 × 1 mm, both ends subtruncate, prominently wrinkled. Fig. 29/1–2.

Distr. ALDABRA: W, S (west), Esprit; ASSUMPTION; COSMOLEDO: M; ASTOVE; a native of Madagascar, cultivated and naturalized throughout the tropics.

Notes. Often planted for ornament, now commonly naturalized around Settlement on West Island, also at Anse Mais on South Island. Flowers throughout the year if sufficient moisture available. Of great interest because of the medicinal alkaloids it contains.

Vernac. 'saponaire', 'rose amère' or 'pervenche'.

3. PANDACA Noronha ex Petit-Thouars

Medium-sized trees or shrubs. Leaves opposite, sometimes leaf pairs markedly unequal, leathery, rarely parchment-like; usually with an intrapetiolar stipular collar. Flowers in axillary cymes. Calyx lobes 5, ovate with scaly glands, those on the inner surface often stalked. Corolla white or yellowish, tube long, narrow, often twisted, glabrous within, lobes 5, short, spreading, curved to the left from the throat, leathery or parchment-like. Anthers 5, linear, sessile, joined to the corolla tube. Ovary 2-celled; stigma composed of a rather large, short, cylindrical part from a lobed, basal ring and an enlarged, globular, apical part crowned by 2, small, short, conical appendages. Fruit usually large, paired capsules.

26 species in Madagascar, Comoro Is. and the Mascarenes.

Pandaca mauritiana (Poir.) Markgraf & Boiteau in Adansonia II, 13: 244, fig. 1 (1973).

Tabernaemontana mauritiana Poir., Encycl. Méth. Bot. 7: 530 (1806); F.M.S. 224 (1877).

Small trees, up to 5 m high; sap milky. Leaves opposite, often clustered at the apex of the branchlets, elliptic, up to 12 × 7.5 cm, apex acute or obtuse, glabrous, somewhat leathery, yellowish-green; petiole 10–12 mm long; stipules united, forming a narrow sheath and cuff. Flowers in sub-terminal cymes. Calyx green, cup-shaped, 3–4 mm long, lobes fused for half of their length. Corolla creamy-white, tube 1 cm long, lobes oblong, 1 cm long, spreading, slightly twisted. Anthers inserted near the apex of the corolla tube, 2.5 mm long, apex acuminate. Fruit, twin elliptic capsules, each 4.5 × 2.5 cm; seeds numerous, light brown, ovoid-wedge-shaped 8 × 5 mm. Fig. 28/7–8.

Vegetatively could be confused with *Psychotria pervillei* (Family 37/5) whose leaves are parchment-like, not leathery, more sharply wedge-shaped at the base and with more sharply defined venation, the presence of stipules and lacking milky sap.

Dist. ALDABRA: S (east); Mauritius, Réunion.

Notes. A rare constituent of mixed scrub south and west of landing strip trace to the south of Anse Cèdres. Only known from 1 collection, *Gibson* 4 (K).

Vernac. 'bois de lait'.

45. ASCLEPIADACEAE

Herbs, shrubs or climbers; sap usually milky. Leaves simple, almost always opposite or whorled; stipules absent. Flowers 5-merous, tending to be in umbellate cymes. Sepals almost free. Corolla united. Stamens 5, fused to the single, fleshy stigma, often surrounded by an elaborate corona; pollen usually fused into waxy masses (pollinia) which are transmitted by insects to the stigma of another flower. Carpels usually 2, free below but held together by the common stigma; ovules many. Fruit of 2, separate follicles; seeds many, usually with a tuft of hairs at one end.

A large, mainly tropical family.

Traditional methods of distinguishing genera in this family are based principally on variation of the corona and pollinia, 2 characters which are difficult to describe except in a way which can only be understood by specialists of this group. No attempt therefore has been made to distinguish clearly between the generic descriptions and the reader is referred to the illustrations for a clearer picture of the differences between them.

1. Plants leafless or with scale leaves up to 1 mm long; stems trailing,
   glaucous, somewhat fleshy   **3. Sarcostemma**
1. Plants with leaves
  2. Semi-erect shrubs; flowers with a conspicuous corona of thread-like
     filaments   **1. Pentopetia**
  2. Climbing or scrambling herbs, sometimes woody
    3. Inflorescences much-branched cymes, up to 10 cm long; petals waxy
       white, turning yellow, 1.5–2 mm long; leaves wedge-shaped at the
       base   **4. Secamone**
    3. Inflorescence an umbellate-cyme (occasionally slightly branched in
       Tylophora)
      4. Leaves elliptic to oblong, usually 1–3 × 0.5–1.5 cm, petiole 0.5–1 cm
         long; inflorescence small, 3–5-flowered, petals white, flecked with
         purple streaks within, 6–10 mm long   **2. Pleurostelma**

4. Leaves ovate, elliptic or oblong, 3–6 × 2–4.5 cm, petiole 1–1.8 cm long; inflorescence large, 10 or more-flowered; petals green, 4–5 mm long   **5. Tylophora**

1. PENTOPETIA Decne.

Small shrubs or trees. Flowers in axillary umbellate cymes. Sepals 5, almost free. Petals 5, united at the base. Stamens united into a column surrounding the stigma and bearing at the base a corona of 5 slender lobes; pollen in groups of 4, not pollinia.

A genus of 10 species, confined to Madagascar and the Comoros.

Pentopetia androsaemifolia Decne. in DC., Prodr. 8: 500 (1844).

Much-branched woody shrub. Leaves opposite, ovate to elliptic, (1–)2–4 (–8) × (1–)1.5–2(–5) cm, apex obtuse, acute or shallowly indented, leathery or parchment-like, glabrous or sparsely hairy, rarely densely hairy, discolourous. Flowers in umbellate cymes; peduncle 3–10 mm long; pedicels 5–15 mm long. Sepals ovate, 2–3 mm long. Petals ? white (colour not recorded) 4–8 mm long, free for three quarters their length. Corona of 5 slender filaments, 6–10 mm long. Fruit a slender spindle-shaped pod, 12 × 0.5 cm; seeds oblong, 10 × 2 mm long, with an apical tuft of silky hairs 3.5 mm long. Fig. 30/1–4.

Distr. ASSUMPTION; Madagascar, Comoro Is.
Notes. Found near Settlement, *Frazier* 34 (K); probably introduced as an ornamental.

2. PLEUROSTELMA Baill.

Perennial scrambling herbs. Flowers borne in axillary umbellate cymes. Sepals 5, almost free. Petals 5, united at the base. Stamens united to form a column around the stigma and bearing a corona of 5 minute lobes; anthers with well-developed lateral wings.

A monotypic genus.

Pleurostelma cernuum (Decne.) Bullock in K.B. 10: 612 (1956).

Astephanus cernuus Decne. in Ann. Sci. Nat. II, 9: 342 (1838).
A. arenarius Decne. in DC., Prodr. 8: 507 (1844); Baker in B.M.I.K. 1894: 149 (1894); Schinz in A.S.N.G. 21: 89 (1897); Voeltzkow in A.S.N.G. 26: 551 (1902); Dupont, Report: 39 (1907).
Microstephanus cernuus (Decne.) N.E. Br. in B.M.I.K. 1895: 249 (1895); Hemsley in J. Bot. 54, Suppl 2: 21 (1916) & in B.M.I.K. 1919: 125 (1919).

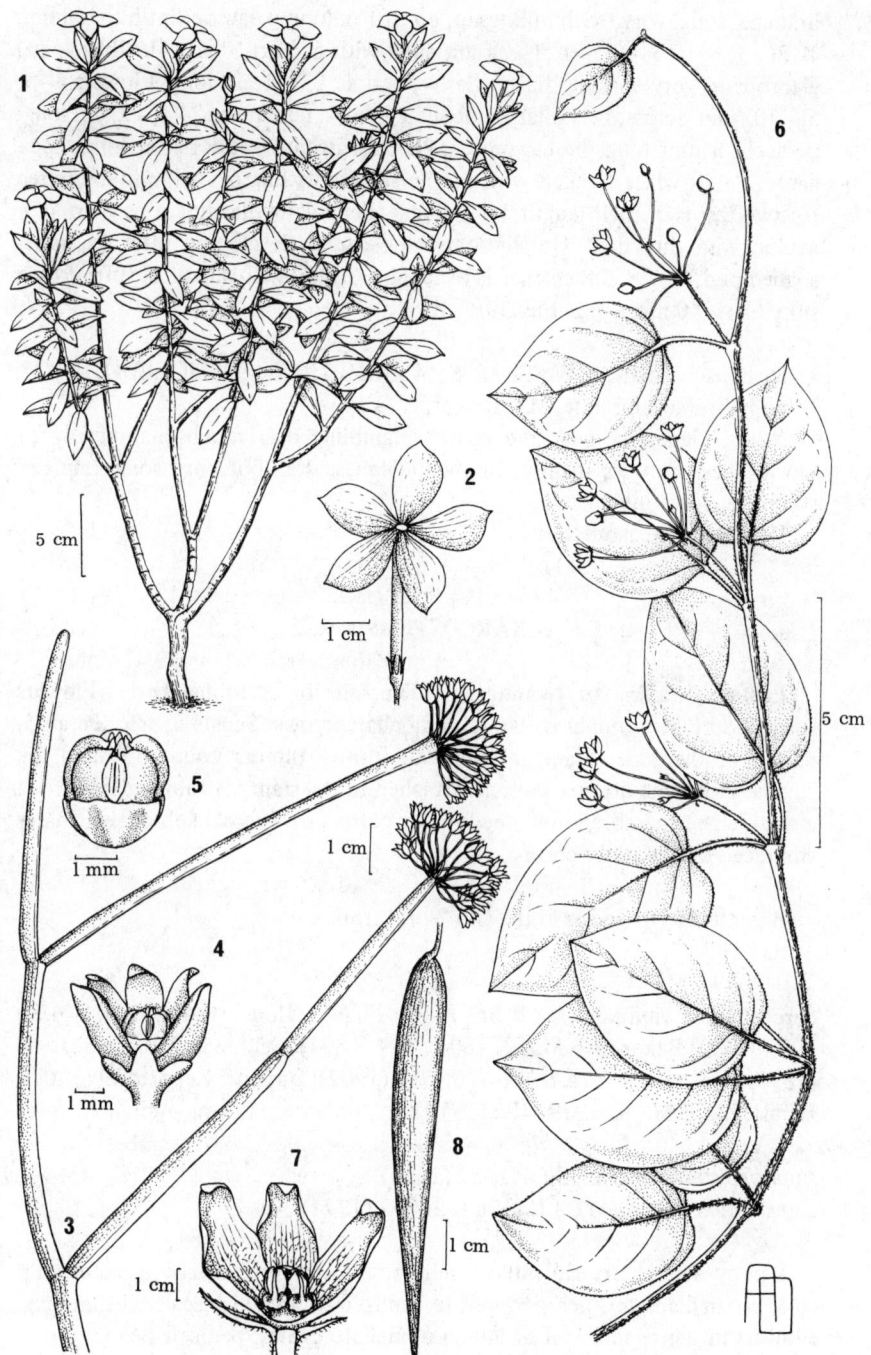

Fig. 29   1, *Catharanthus roseus*, habit; 2, flower. 3, *Sarcostemma viminale*,
habit; 4, flower; 5, detail of corona. 6, *Tylophora indica*, habit;
7, flower; 8, fruit.

Low scrambling, loosely mat-forming or climbing herbs; rootstock woody; branches long, wiry, with milky sap, up to 1 m long. Leaves elliptic to oblong, 1–3(–5) × (0.3–)0.5–1.5(–2) cm, apex with a short, sharp, flexible point, glabrous to very sparsely hairy, fleshy; petiole 0.5–1 cm long. Flowers 3–5, 5(–10) mm across, in axillary umbellate cymes; peduncles 2–5(–7) cm long; pedicels 5 mm long, bracts present. Sepals green, ovate, 1 mm long, apex acute. Petals white flecked with purple streaks within; 6(–10) mm long, free for two-thirds of their length, lobes linear-triangular, spreading, tips twisted in a clockwise direction. Usually only one carpel developing into a spindle-shaped pod, 4–6 × 0.8 cm; seeds ovate, 4 × 2 mm, with an apical tuft of long silky hairs, 40 mm long. Fig. 30/5–8.

Distr. ALDABRA: W, P, M, S; ASSUMPTION; COSMOLEDO: W, M; Ethiopia to Mozambique, Madagascar.

Notes. Generally near the coast, scrambling over rough champignon or low bushes. Flowers mainly during the wet season, but unseasonal rain can stimulate flowering.

Vernac. 'liane caoutchouc'.

3. SARCOSTEMMA R.Br.

Leafless, trailing or twining, succulent shrubs or robust herbs. Flowers small, borne in terminal or lateral, umbellate cymes. Sepals 5, free. Petals 5, united at the base. Stamens united to form a tubular column around the stigma. A double corona present, attached to the staminal tube; outer corona a cup-shaped membrane or lobed; inner corona of 5, erect lobes, their bases embraced by the outer corona.

A genus of 10 species in the Old World tropics.

Sarcostemma viminale (L.) R.Br., Prodr. Fl. Nov. Holl. : 464 (1810); F.M.S.: 227 (1877); Baker in B.M.I.K. 1894: 149 (1894); Schinz in A.S.N.G. 21: 89 (1897); Voeltzkow in A.S.N.G. 26: 551 (1902); Dupont, Report: 39 (1907); Hemsley in B.M.I.K. 1919: 125 (1919).

Euphorbia viminalis L., Sp. Pl. 452 (1753).
Cynanchum viminale (L.) L., Mant. 2: 392 (1771).

Woody scrambler; sap milky; stems round, glabrous, fleshy, glaucous, up to 5 mm in diameter. Leaves opposite, reduced to brownish scales, 1 mm long. Flowers in dense terminal or lateral umbellate cymes; pedicels slender, up to 1 cm long. Sepals greenish-yellow, ovate, 1 mm long, apex acute. Petals yellow to almost white, lobes ovate to oblong, 3.5–8 mm long, apex obtuse or acute. Corona and staminal tube forming a lobed column enclosing the 2 carpels. Usually only 1 carpel developing into a linear or spindle-shaped pod up to 12 × 1 cm; seeds flattened, ovate, 8 × 3 mm, bearing an apical tuft of silky hairs 20 mm long. Fig. 29/3–5.

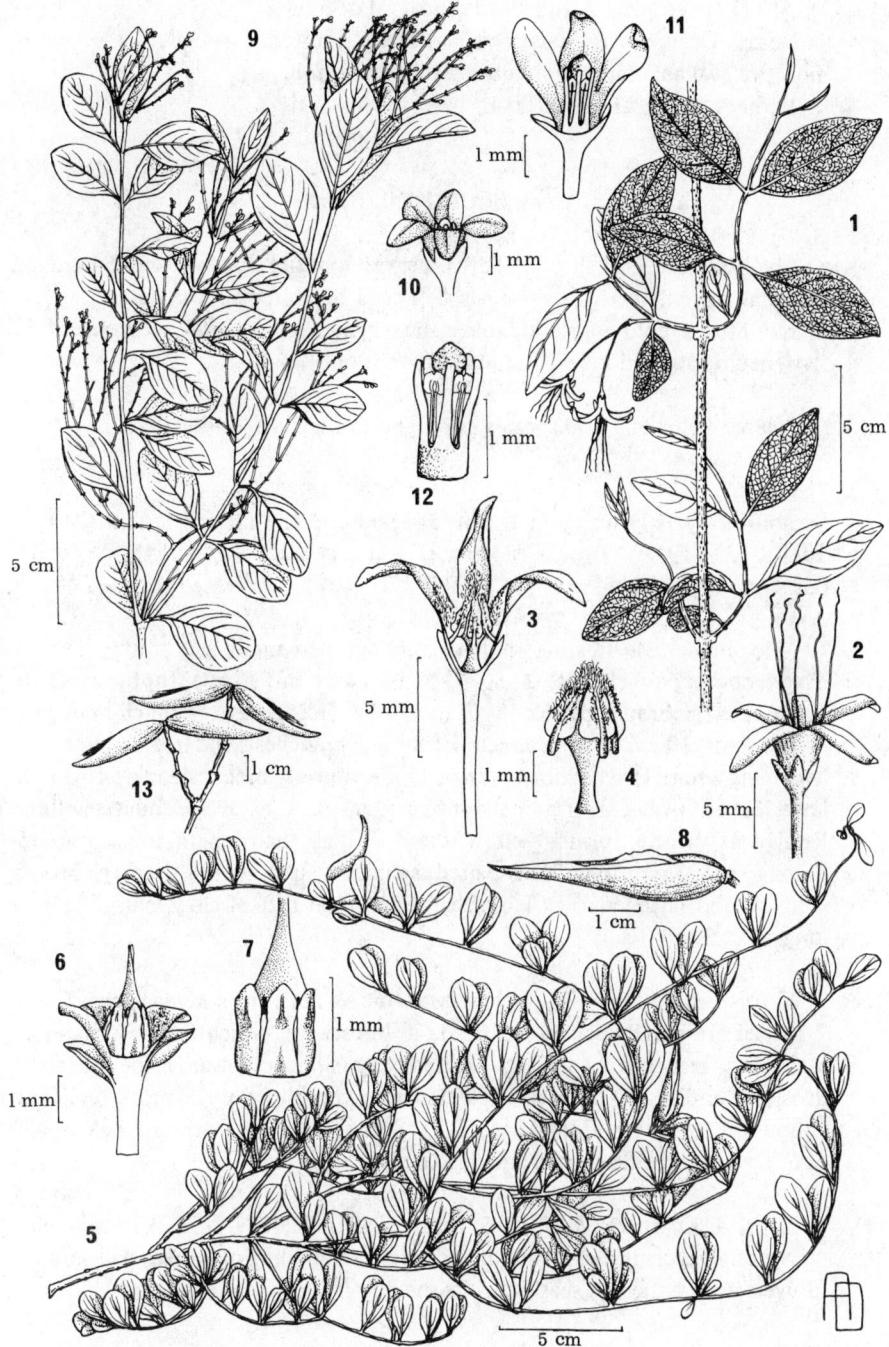

Fig. 30 1, *Pentopetia androsaemifolia*, habit; 2, flower; 3, flower section;
4, detail of androgynecium. 5, *Pleurostelma cernuum*, habit;
6, section of flower; 7, detail of androgynecium; 8, fruit.
9, *Secamone fryeri*, habit; 10, flower; 11, flower section; 12, detail
of androgynecium; 13, fruit.

Distr. ALDABRA: W, P, M (west), S, Esprit; ASSUMPTION; COSMOLEDO: M; ASTOVE; tropical Africa, Madagascar, Mascarenes.

Notes. Occurs in the inland mixed shrub. Flowers throughout the year if moisture available but mainly during the wet season.

Vernac. 'lianes sans feuilles' or 'liane calé'.

4. SECAMONE R.Br.

Climbing shrubs or shrubby herbs. Leaves opposite. Flowers small, borne in terminal or axillary cymes. Sepals 5. Petals 5, united at the base. Stamens united into a column surrounding the stigma and bearing a corona of 5, flattened or pointed lobes; pollinia in groups of 4.

A large genus of c 100 species, throughout the Old World tropics.

Secamone fryeri Hemsley in J. Bot. 54, Suppl. 2: 22 (1916) and in B.M.I.K. 1919: 125 (1919). Types: Aldabra, *Thomasset* 225, *Dupont* 47, *Fryer* 61 (all K, syn.).

Much-branched, twining and scrambling, perennial; sap milky. Leaves ovate, obovate or elliptic, 2–5(–7) × 1–3(–5) cm, apex abruptly acute to acuminate, glabrous; petiole 5–7 mm long. Inflorescences much-branched cymes up to 10 cm long, peduncle 1 cm long, branches slender, ascending and becoming progressively shorter towards the flowers, bracts slender, 1–8 mm long. Sepals ovate, 1–1.5 mm long, a small dark basal appendage within. Petals waxy white, turning yellow, fused for half their length, lobes ovate to oblong, 1.5–2 mm long. Pods paired, spindle-shaped, 4–8 × 1 cm, glabrous; seeds ovate, flattened, 5 × 1 mm, with an apical tuft of silky hairs 3–5 mm long. Fig. 30/9–13.

S. *fryeri* is similar to S. *pachystigma* Jum. & Perr. from Madagascar. These 2 species are readily separated by the inflorescence, which in the former has ascending branches becoming progressively shorter towards the ultimate divisions, and in the latter species has either very short or spreading branches (Choux in Ann. Mus. Col. Mars. III, 11: 386 (1914). Furthermore the capsule of S. *fryeri* is spindle-shaped and in S. *pachystigma* ovoid.

Distr. ALDABRA: W, P, M, S, Esprit; ASSUMPTION; ASTOVE; endemic.

Notes. Occurs in the inland mixed scrub, scrambling over trees and shrubs. Flowers during the wet season; fruits shed by late dry season.

5. TYLOPHORA R.Br.

Trailing or climbing herbs. Leaves opposite. Flowers small, in dense or sparse, umbellate cymes. Sepals 5, united at the base. Petals 5, united at the base. Stamens fused into a column surrounding the stigma and bearing a corona of 5 flat fleshy lobes.

A genus of 50 species found throughout the tropics of the Old World.

Tylophora indica (Burm.f.) Merrill in Phillip. J. Sci. 19: 373 (1921).

Cynanchum indicum Burm.f., Fl. Ind.: 70 (1768).
Tylophora asthmatica Wight & Arn. in Wright, Contrib.: 51 (1834); F.M.S.: 229 (1877); Dupont, Report: 39 (1907); Hemsley in B.M.I.K. 1919: 125 (1919).*

Trailing or climbing perennial herbs; milky sap absent; stems slender sub-glabrous to pubescent. Leaves ovate, elliptic or oblong, 3–6 × 2–4.5 cm, apex with a short sharp, flexible point, base truncate, glabrous, somewhat leathery, fleshy; petioles 1–1.8 cm long, pubescent, veins and petioles often tinged purple. Flowers in axillary umbellate cymes; peduncles 1–2 cm long, pedicels slender, up to 3 cm long. Sepals linear, 1.5–2 mm long, pubescent. Petals green, oblong, 4–5 mm long, apex acute, glabrous. Pods paired, spindle-shaped 4–9 × 1 cm; seed flattened, ovate, 6 × 3 mm with an apical tuft of hairs 2.5 mm long. Fig. 29/6–8.

Distr. ALDABRA: S; tropical Asia.
Notes. Only known from the south coast at Le Renfin, *Merton* 7101 (K) and Entreboi, *Hnatiuk* 731137 (K), growing on coastal champignon.

46. BORAGINACEAE

Herb, shrubs or trees; often rough to the touch or stiffly hairy. Leaves usually alternate, simple, usually entire or sometimes serrate or toothed; stipules absent. Inflorescences usually cymose or cymose-paniculate; cymes often 1 or 2-ranked, rarely forming compact heads or elongate cymes; flowers usually bisexual, regular, rarely irregular. Sepals 5, free or partially united. Corolla variously shaped, (4–)5(–7)-lobed. Stamens usually 5, inserted on the corolla and alternating with the lobes. Ovary superior, 2- or rarely 4-celled, ovules 1 or 2 per cell; style 1, arising from the base of the ovary or terminal. Fruit 4, 1-seeded nutlets, or a 1–4-seeded nut or drupe.

A fairly large almost cosmopolitan family, strongly represented in arid and semi-arid lands and in the tropics, where there are many woody members.

1. Plants densely silky-haired; leaves thick, grey   **3. Tournefortia**
1. Plants glabrous or nearly so; leaves thin, green

---

* Hemsley loc. cit. cites *Dupont* 8, a plant which was collected on Mahé, Seychelles.

2. Petioles 4—5 or more cm long; flowers large, orange; fruits 1—2 cm or more across **1. Cordia**
2. Petioles less than 2 cm long; flowers small, white; fruits less than 1 cm across **2. Ehretia**

1. CORDIA L.

Shrubs or trees. Cymes axillary, more or less paniculate and corymbiform. Calyx lobes united in bud, opening regularly or irregularly. Ovary entire, 4-celled; style twice forked. Drupe with 1 hard pyrene, partly or mostly enveloped by calyx, pyrene 1—4-seeded.

A large tropical genus.

Cordia subcordata Lam.,   Encycl. Méth. Bot. 7: 41 (1806); F.M.S.: 200 (1877); Baker in B.M.I.K. 1894: 149 (1894); Schinz in A.S.N.G. 21: 90 (1897); Voeltzkow in A.S.N.G. 26: 552 (1902); Hemsley in B.M.I.K. 1919: 128 (1919); Fosberg in P.T.R.S. B,260: 219; 224 (1971); Renvoize in P.T.R.S. B,260 230 (1971).

Small tree or shrub, up to 10 m high. Leaves ovate to broadly ovate, up to 22 × 17 cm, apex usually slightly acuminate, base obtuse to slightly sub-cordate, margins entire; petioles 5—11 cm long. Cymes axillary or apparently terminal, up to 10 × 14 cm, usually smaller, loosely branched; pedicels 5—10 mm long. Calyx cylindrical, 12—18 × 6—8 mm, lobes 5, obtuse, spreading. Corolla deep orange, funnel-shaped, c 4 cm long, lobes obtuse to rounded, spreading, very thin. Heterostylous; long-styled plants with style about 3 cm long; short-styled plants with style 2—2.5 cm long. Fruit completely enclosed in enlarged calyx, drupe hard, globose, to 2.5 cm diameter, beaked, 4-celled. Fig. 31/1—3.

Distr. ALDABRA: W,P,M, S, Esprit; ASSUMPTION; COSMOLEDO: W,M; ASTOVE; an Indo-Pacific strand plant.
Notes. Locally common, usually in sandy places, rarely on bare limestone; absent from the southern and eastern beaches of Aldabra. Flowers through-out the year if sufficient moisture available. Sea dispersed; viable seeds found amongst beach drift.
Vernac. 'porché', 'porcher' or 'bois porcher'.

2. EHRETIA P.Br.

Trees and shrubs. Leaves usually rather rough, margins entire or somewhat toothed. Cymes axillary or terminal, often corymbiform or paniculate; flowers white, usually small. Corolla tube cylindrical, lobes spreading. Stamens exserted. Ovary entire; style 2-fid in upper part, stigmas small, terminal on style branches. Fruit a globose drupe; pyrenes 2—4.

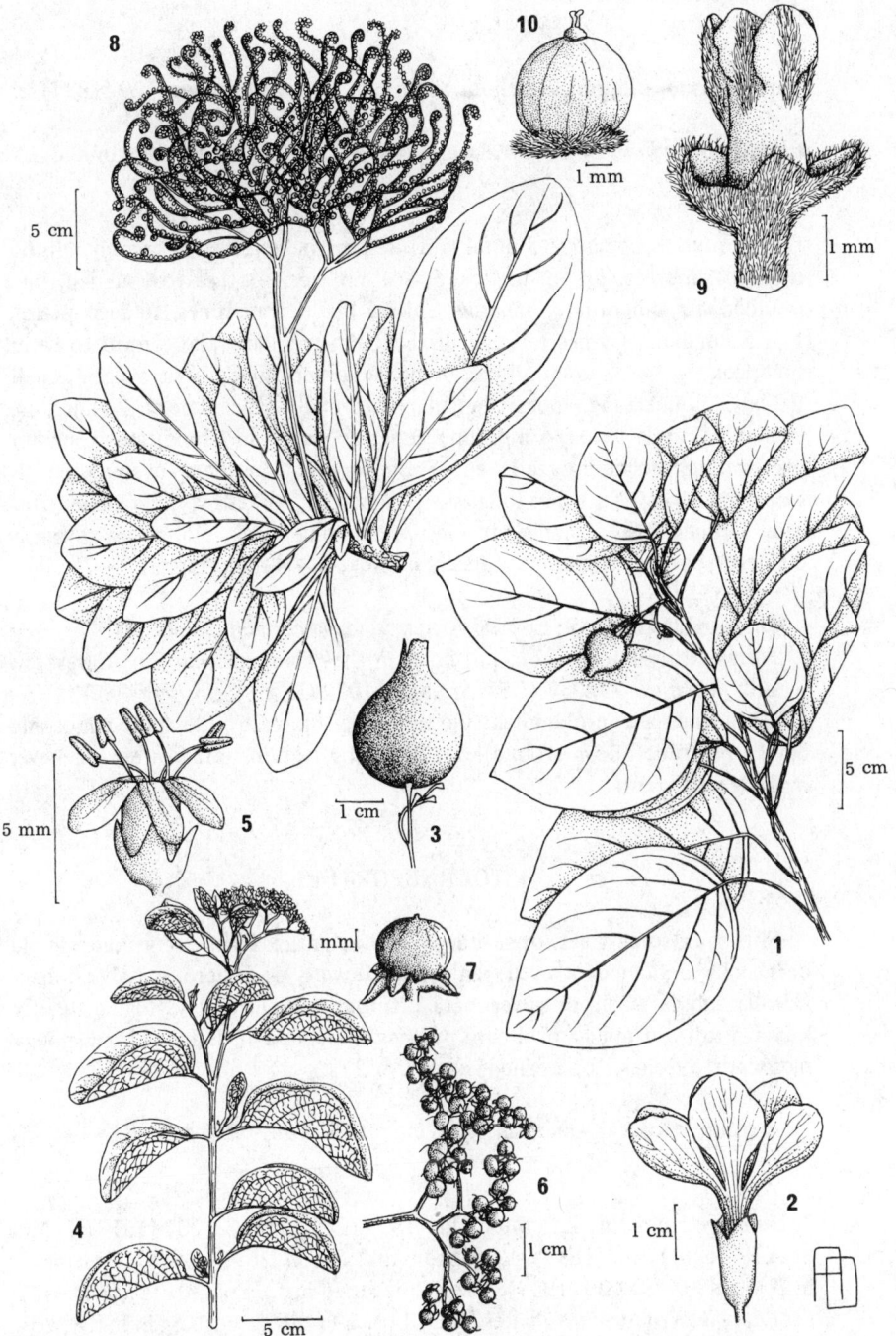

Fig. 31 1, *Cordia subcordata*, habit; 2, flower; 3, fruit. 4, *Ehretia cymosa*, habit; 5, flower; 6, fruiting inflorescence; 7, fruit. 8, *Tournefortia argentea*, habit; 9, flower; 10 fruit.

A pantropical genus.

Ehretia cymosa Thonn. in Schum. & Thonn., Beskr. Guin. Pl.: 129 (1827).

Ehretia corymbosa Bojer ex A.DC., Prodr. 9: 505 (1845); Fosberg in K.B. 29:
260 (1974).

Glabrous shrub, up to 3 m tall; diffusely branched. Leaves broadly elliptic,
oval, to suborbicular, up to 7 × 5(–6) cm, apex obtuse to rounded, base
rounded to subcordate, margins entire, blade sometimes folded; petiole
1–1.5 cm long. Cymes terminal or becoming axillary, flat-topped to hemispherical, 3–4 × 4–6 cm, much branched, bracts none; flowers many, small.
Calyx 2.5 mm long, lobes triangular to broadly lanceolate, c 2 mm long,
ciliolate. Corolla white, 4 mm long, lobes obovate 3 mm long, apex rounded,
overlapping, becoming reflexed. Stamens erect, anthers ovate, strongly
exserted. Style 2.5–3 mm long, stiffly exserted, forked, style branches truncate. Drupes fleshy, orange, turning red to blue-black, globose, c 6 mm in
diameter, pyrenes trigonous, dorsally convex, 4 × 4 mm. Fig. 31/4–7.

Distr. ALDABRA: S; tropical Africa, Madagascar and Comoro Is.
Notes. Only known from pot holes in the champignon at Cinq Cases; few
records, *Hnatiuk* 732057 (US); *Stoddart* 1022 (K, US); *Fosberg* 49172 (US).
The branches are intolerant of the salt spray blown by the trade winds and
barely protrude above ground level. Few observations but flowers whenever
moisture sufficient.

3. TOURNEFORTIA L.

Shrubs or small trees, or scandent. Leaves entire. Cymes scorpioid, simple
or usually branched or even paniculate; flowers 4–5-merous. Calyx lobed.
Corolla tubular, bell- or salver-shaped, throat without scales. Stigma slightly
lobed. Fruit a drupe; 2 or 4 hard pyrenes embedded in fleshy or hard spongy
mesocarp, pyrenes 1-or 2-celled; seeds 1 or 2.

A mainly tropical American genus close to *Heliotropium* L.

Tournefortia argentea L.f., Suppl. Pl.: 133 (1781); F.M.S.: 201 (1877); Schinz
in A.S.N.G. 21: 90 (1897); Voeltzkow in A.S.N.G. 26: 552 (1902); Hemsley
in B.M.I.K. 1919: 126 (1919); Vezey-Fitzgerald in J. Ecol. 30: 11, 14 (1942);
Stoddart & Wright in Atoll Res. Bull. 118: 29 (1967); Fosberg in P.T.R.S. B,
260: 219, 224 (1971); Renvoize: P.T.R.S. B, 260: 230 (1971).

Shrub or small tree, crown dome-shaped, to 9 m high. Leaves obovate, up
to 30 × 11 cm, apex rounded to obtuse or subacute, base narrowed to a
decurrent petiole 1 or 2 cm long, blade fleshy, grey-green, densely silky-haired.
Cymes 1-sided, terminal, 20–30 × 15–20 cm, peduncle long and stout,

densely silky-haired; flowers very fragrant. Calyx cylindrical, 3 mm long, deeply lobed. Corolla white, tube shorter than the calyx, lobes exserted, 7 mm across, anthers included. Style fleshy, globose, lobes 2. Fruit globose, 4–6 mm across, mesocarps of a very coarse, firm, aerogenous tissue (not corky as often described), pyrenes 2, flattened on inner side. Fig. 31/8–10.

Distr. ALDABRA: W, M, S, Michel; ASSUMPTION; COSMOLEDO: W, M; ASTOVE; an Indo-Pacific strand plant.

Notes. Usually found on dunes or other sandy places, also on coastal champignon. Flowers throughout the year if sufficient moisture available. Fruits are a common component of beach drift.

Vernac. 'bois tabac' or 'veloutier à tabac fleurs'.

47. CONVOLVULACEAE

Herbaceous or woody twiners or creepers, rarely erect herbs; frequently with milky sap but not copiously so. Leaves alternate, frequently cordate, hastate or sagittate, simple but occasionally deeply or palmately divided, rarely pinnately divided; stipules absent. Sepals 5. Corolla funnel- or salver-shaped, usually not deeply divided. Stamens 5, attached to corolla near base, or part-way up tube. Ovary superior, 1–4(–5)-celled, ovules 4–6, basal; style usually 1, rarely 2–3, slender, branched; stigmas 1–4, thickened. Fruit a 2, 3 or 4-celled, dehiscent or indehiscent capsule, rarely fleshy or corky; seeds 1–2 per cell.

An important tropical family with a few temperate members; includes many widespread pioneers and weedy species, as well as tropical forest lianas.

1. Small wiry herb, prostrate to ascending, not extensively creeping or
   twining; leaves less than 2 cm long, silky-haired   **1. Evolvulus**
1. Creepers or twiners; leaves over 2.5 cm long, usually cordate or subcordate
   or apex 2-lobed   **2. Ipomoea**

1. EVOLVULUS L.

Prostrate to ascending herbs. Leaves alternate, simple. Flowers 1–few, axillary; peduncle, if 1-flowered, with a pair of bracts part way up. Corolla saucer-shaped to very broadly bell-shaped, entire or almost so. Stamens 5, included, or at least shorter than corolla; anthers straight. Styles 2, forked; stigmas elongate. Fruit a dehiscent capsule, 2-celled; seeds 4, black, smooth.

A pantropical genus, principally American but with 2 wide-ranging species.

Evolvulus alsinoides (L.) L., Sp. Pl. ed. 2: 392 (1762); Baker in B.M.I.K. 1894: 149 (1894); Schinz in A.S.N.G. 21: 89 (1897); Voeltzkow in A.S.N.G. 26: 551 (1902); Dupont, Report: 38 (1907); Hemsley in B.M.I.K. 1919: 126 (1919); F.T.E.A. Conv.: 18 (1963).

Convolvulus alsinoides L., Sp. Pl.: 157 (1753).

A wiry annual or perennial herb; stems many, pubescent, ascending toA w spreading, sparsely branched. Leaves ovate to ovate-lanceolate, 5–8 × 2–6 mm, apex acute, densely silky-haired, sessile. Flowers 1(–5) in the upper leaf axils; peduncle slender, 6–15 mm long, pedicels 2–5 mm long. Sepals ovate-lanceolate, 1.5–2 mm long. Corolla bluish-purple with white centre, saucer-shaped, 8–9 mm across, very delicate. Capsule globose, 3–4 mm across, 4-valved; seeds 4, pale brown and black, ovoid, 1.7 × 1 mm. Fig. 32/1–3.

Distr. ALDABRA: W, M, S, Esprit; ASSUMPTION; COSMOLEDO: W, M; ASTOVE; pantropical.
Notes. Common along paths in habitation sites, rare or occasional else-where. Flowers throughout the year if sufficient moisture available.

2. IPOMOEA L.

Twining herbaceous to woody vines, rarely erect herbs or shrubs. Leaves alternate, simple, usually cordate, entire to deeply divided, petiolate. Flowers axillary, solitary or in few- to many-flowered cymes. Corolla narrowly funnel- or bell-shaped, regular to somewhat irregular, entire to somewhat 5-lobed, folded fan-like in bud, median lines heavy, intervening areas membranous. Anthers included or exserted, straight. Style 1, slender, elongate, stigmas globose, 1–3. Fruit a capsule. 3–10-valved or irregularly dehiscent.

A large pantropical and warm-temperate genus, found on almost all Indian Ocean islands including those in the Aldabra group; many of the species wide-spread and weedy, found on beaches and in disturbed places, a few found in lowland forests.

1. Leaves ovate to oblong, apex shallowly notched   **4. I. pes-caprae**
1. Leaves cordate or dissected, apex not shallowly notched
  2. A large twiner, stems several mm thick; leaves orbicular; flowers large, white   **2. I. macrantha**
  2. Slender twiners or creepers, if twining stems 1–2 mm thick; flowers 4 cm or less long
    3. Plants strongly twining; leaves neither angled nor dissected, thin, pubescent; flowers white or cream with dark centre   **3. I. obscura**
    3. Plants creeping but scarcely twining; leaves often dissected, at least angled, glabrous; flowers violet or lilac   **1. I. batatas**

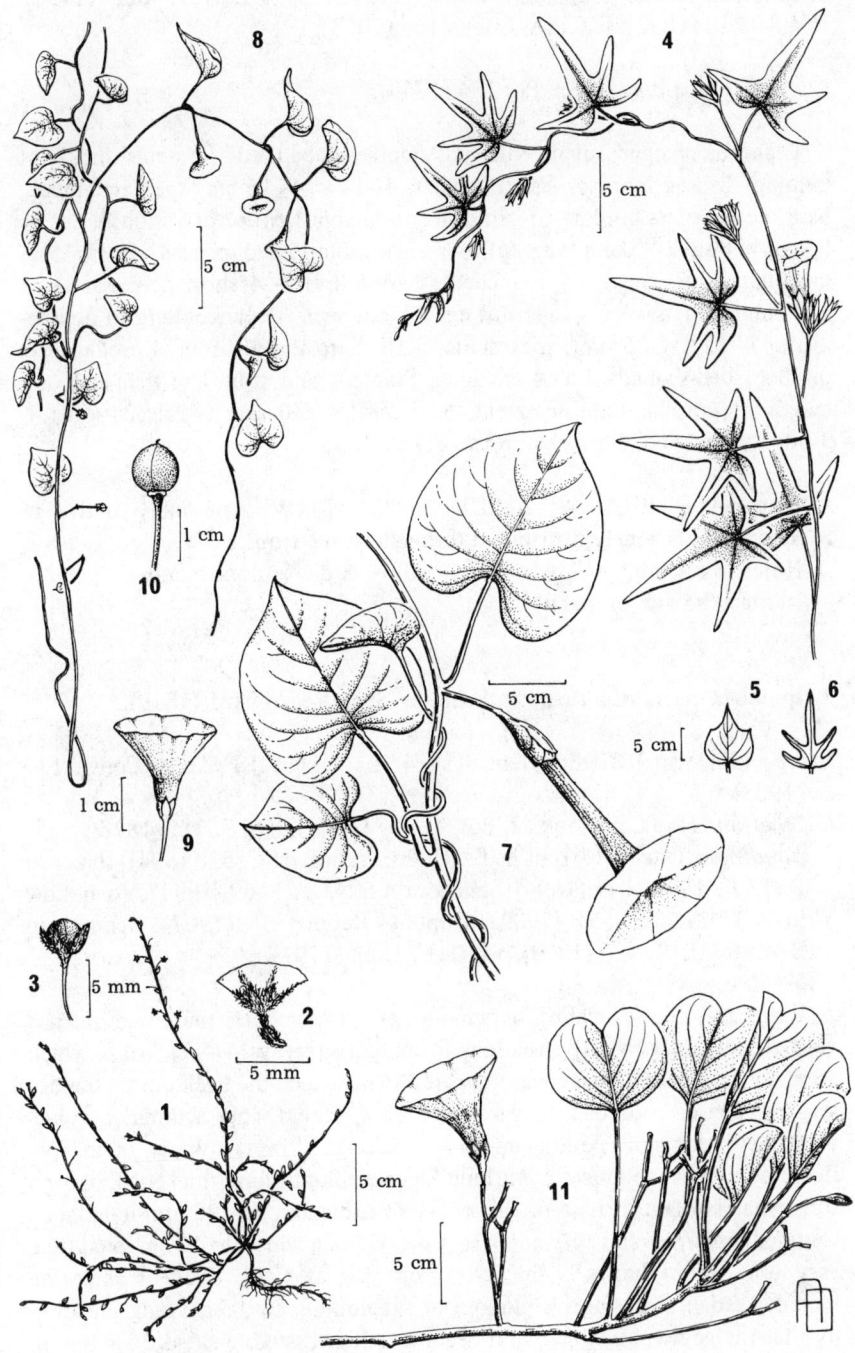

Fig. 32 1, *Evolvulus alsinoides*, habit; 2, flower; 3, fruit. 4, *Ipomoea batatas*, habit; 5 & 6, leaf variation. 7, *Ipomoea macrantha*, habit. 8, *Ipomoea obscura*, habit; 9, flower; 10, fruit. 11, *Ipomoea pes-caprae*, habit.

1. Ipomoea batatas (L.) Lam., Illustr. Encycl. Méth. Bot. 1: 465 (1793); F.M.S.: 210 (1877); F.T.E.A. Conv.: 114 (1963).

Convolvulus batatas L., Sp. Pl.: 154 (1753).

Glabrous creeper; tubers enlarged, spindle-shaped, edible; stems thick, not twining. Leaves more or less triangular, 4—14 × 4—16 cm, apex acuminate, base more or less cordate to sub-truncate, margins entire or variously cut or lobed; petiole 4—20 cm long. Inflorescence axillary, peduncles 3—18 cm long, umbellately or irregularly branched at apex into 1—4 short pedicels, 3—12 mm long, bracts very small; flowers seldom seen. Sepals oblong to oblong-ovate, 7—12 × 3—5 mm, apex acute, with sharp tip, glabrous. Corolla violet or lilac, bell-shaped, 3.5—4 cm long. Stamens and style less than half the length of corolla. Capsule ovoid, 5—6 mm in diameter, 4-valved; seeds 4, 3-sided, ovoid, 4 × 3 mm, glabrous. Fig. 32/4—6.

Distr. ALDABRA: W; ASSUMPTION; ASTOVE; probably native in S. America now widely distributed throughout the tropics.
Notes. Cultivated and persistent in fallow and abandoned land.
Vernac. 'batate' or 'patate'.

2. Ipomoea macrantha Roem. & Schultes, Syst. Veg. 4: 251 (1819).

I. tuba (Schlecht.) G.Don, Gen. Syst. 4: 271 (1837); F.T.E.A. Conv.: 137 (1963).*
I. glaberrima Hook. in Hook., J. Bot. 1: 357 (1834); F.M.S. 211 (1877).
I. grandiflora (Jacq.) H. Hall in Engl., Bot. Jahrb. 18: 153 (1894); Baker in B.M.I.K. 1894: 149 (1894); Schinz in A.S.N.G. 21: 89 (1897); Voeltzkow in A.S.N.G. 26: 551 (1902); Dupont, Report: 38 (1907); Hemsley in B.M.I.K. 1919: 126 (1919), not (L.f.) Lam. (1797).

Extensive, coarse, glabrous, twining herb, lower parts thickened but not very woody, in very dry situations forming a short, very thick trunk which annually produces herbaceous, elongated stems which die back during the dry seasons. Leaves cordate-orbicular, up to 15 × 15 cm, apex acuminate, somewhat fleshy or not; petiole up to 6 cm long. Flowers solitary or in few-flowered cymes; peduncle 6 cm long; pedicels 2 cm long, thickened, or even somewhat swollen at base of calyx. Sepals orbicular, 15—25 mm long, apex rounded, enlarged and very thick in fruit. Corolla white, to 10 cm long, tube very long, limb flaring 8 cm across, opening at night. Stamens and style included, stigmas 2. Capsule globose or subglobose, to 2.5 cm long, enclosed by the enlarged sepals which later become reflexed; seeds 3-sided, 10 × 8 mm, black-pubescent with longer hairs along the edges. Fig. 32/7.

---

* The earlier name *I. macrantha* was discovered by Gunn in Brittonia 24: 158 (1972) after the publication of F.T.E.A. Recently Manitz in Feddes, Report 88: 265 (1977) has suggested that the earliest name is *I. violacea* L.

Distr. ALDABRA: W,P,M,S, Esprit, Michel; ASSUMPTION; COSMOLEDO: W; ASTOVE; pantropical strand plant.

Notes. Usually found near areas of habitation, climbing over shrubs and low trees fringing the shore. Flowers throughout the year, but mainly during the wet season. Viable seeds found amongst beach drift.

Vernac. 'batatran blanc' or 'patatran blanc'.

3. Ipomoea obscura (L). Ker-Gawl. in Bot. Reg. 3: t. 239 (1817); F.M.S.:209 (1877); F.T.E.A. Conv.: 116 (1963).

Convolvulus obscurus L., Sp. Pl.: 159 (1753).

Slender twiner, almost glabrous to noticeably hairy. Leaves broadly cordate, up to 7 × 7 cm, apex acuminate, basal sinus broad, rounded, margins sub-entire to finely crenate, more or less ciliate; petioles slender, 2–6.5 cm long. Flowers 1–several, in pedunculate cymes, 1 flower per cyme opening at a time; peduncles 2–5(–10) cm long, pedicels 1–1.5 cm long, somewhat dilated upward. Sepals elliptic or elliptic-ovate to oval or, in fruit, orbicular, 4–6 mm long, inner sepal broader, sharp tipped. Corolla white or cream to sulphur-yellow, with dark purple centre, bell- to funnel-shaped, flaring, 1.5–2 cm long, 1.5–2.5 cm across. Stamens and style included. Capsule sub-globose, to 1 cm long, beaked, exserted from calyx, firm; seeds dark brown, ovoid 2.5 × 3.5 mm, plump, silky-pubescent. Fig. 32/8–10.

Distr. ALDABRA: W; widespread through the Old World tropics.

Notes. A recent introduction, found in scrub communities around Settlement on West Island; *Renvoize* 846 (K,US), *Hnatiuk* 732021 (US). Observations few, but apparently flowers throughout the year if sufficient moisture available.

4. Ipomoea pes-caprae (L.) R.Br. in Tuckey, Narr. Exp. Zaire: 477 (1818); F.M.S.: 211 (1877) partly; F.T.E.A. Conv.: 121 (1963); Fosberg in K.B. 29: 260 (1974).

subsp. **brasiliensis** (L.) van Ooststr. in Blumea 3: 533 (1940); F.T.E.A. Conv.: 121 (1963).

Convolvulus brasiliensis L., Sp. Pl.: 159 (1753).
Ipomoea pes-caprae sensu F.M.S.: 211 (1877) partly; Schinz in A.S.N.G. 21: 89 (1897); Voeltzkow in A.S.N.G. 26: 551 (1902); Dupont, Report: 38 (1907), not (L.) Br. (1818).
I. biloba Forssk., Fl. Aegypt.-Arab.: 44 (1775); Hemsley in B.M.I.K. 1919: 126 (1919).*

---

* *I. biloba*, according to F.T.E.A. Conv.: 122, probably belongs to subsp. *brasiliensis* rather than subsp. *pes-caprae* but the type material is too poor for a firm decision.

A prostrate, coarse, glabrous creeper, capable of forming dense mats; Leaves oblong or oval to ovate, up to 13 × 13 cm, apex normally 2-lobed or at least prominently indented, base rounded or somewhat cordate, leathery; petiole up to 14 cm long, peduncles erect, stout, with 1-several pedicellate flowers in a cyme. Sepals elliptic to orbicular up to c 1 cm long, outer sepals narrower, apex obtuse, with a small prominent tip, glabrous. Corolla rose-purple, darker in centre, funnel- to bell-shaped, 3–5 cm long. Stamens and style included; stigmas 2. Capsule globose, 12–15 mm high; seeds 3-sided, ovoid, 6–10 × 6–10 mm, densely brown-pubescent, dull yellowish when the pubescence is worn off. Fig. 32/11.

Ipomoea pes-caprae L. subsp. *pes-caprae*, which has deeply 2-lobed leaves, is found throughout the Indian Ocean region, usually on sea-beaches and strand habitats, but is apparently not found in the Aldabra Group, where it is replaced by the subsp. *brasiliensis*.

Distr. ALDABRA: W, P, M, S; ASSUMPTION; COSMOLEDO: W, M; ASTOVE; a pantropical strand species.

Notes. Occurs on beaches just above high-tide mark. Flowers throughout the year provided sufficient moisture available, but mainly during the wet season. Seeds sea-dispersed; viable seed found amongst beach drift. Plants eaten by tortoise.

Vernac. 'batatran rouge', 'patatran rouge','batata à durand' , 'batatrant' or 'batate ronde'.

48. SOLANACEAE

Herbs, vines, shrubs, or more rarely, small trees. Leaves alternate, simple or rarely compound; stipules absent. Inflorescence racemose, spicate, paniculate, cymose, fasciculate or flowers solitary. Calyx 3–6-lobed. Corolla usually 5-lobed. Stamens 5, inserted at base of corolla or part-way up corolla tube, anthers 2-celled. Ovary 2-(rarely more)-celled, ovules numerous, placentation axile. Fruit a many-seeded berry or capsule.

A large, principally tropical family; many genera contain poisonous or medicinal substances. Many genera are cultivated as food-, condiment- or drug-plants, or as ornamentals.

1. Flowers nearly saucer-shaped; fruit berry-like
  2. Anthers short, not converging; calyx teeth minute   **1. Capsicum**
  2. Anthers linear or linear-oblong, converging; calyx teeth well-developed, triangular, ovate or lanceolate   **4. Solanum**
1. Flowers tubular ot funnel-shaped; fruit capsular
  3. Corolla over 10 cm long, strongly flaring, capsule spiny, thick-walled
                                                                  2. Datura

3. Corolla about 5 cm long, limb narrow; capsule not spiny, thin-walled
3. Nicotiana

1. CAPSICUM L.

Herbs or shrubs. Leaves alternate, simple, entire. Flowers solitary or several together on pedicels, terminal but becoming axillary; pedicels tending to be curved so flowers drooping; flowers bisexual. Calyx bell-shaped, margins truncate but with 5, separated, tiny teeth. Corolla almost saucer-shaped, limb deeply 5-lobed, lobes with margins touching, not overlapping. Stamens 5, inserted at top of corolla tube, anthers not converging, oblong, cells 2, longitudinally dehiscent. Ovary 2—4-celled, ovules many; style 1, usually slender, sometimes club-shaped. Fruit a berry, usually red, usually containing capsicin; seeds medium-small, black, flattened.

A small tropical American genus, several species of which are widely distributed in cultivation and naturalised. The fruits of some species contain a pungent substance, capsicin, valued in cooking as a spice.

1. Herbaceous; fruit broad, pleasantly strong-flavoured but scarcely pungent
1. C. annuum
1. Woody; fruit narrow, very pungent to taste   **2. C. frutescens**

1. Capsicum annuum L., Sp. Pl.: 189 (1753).

Low, branched, glabrous herb. Leaves broadly ovate, up to 8 × 5—5.5 cm or more, apex strongly acuminate, base truncate to subcordate, decurrent, thin; petiole up to 3 cm long. Flowers greenish, c 1 cm across. Fruit red, ovoid, apex blunt, large, fleshy.

Distr. ALDABRA: W; widely cultivated throughout the tropics.
Notes. Cultivated at Settlement, 1 collection only, *Fosberg* 49425 (US).
Vernac. 'piment', 'tili', 'chillies', 'red' or 'sweet pepper'.

2. Capsicum frutescens L., Sp. Pl.: 189 (1753).

Small glabrous shrub; stems tending to be zigzag. Leaves ovate, up to 7 × 3 cm or more, apex bluntly acuminate, base rounded to somewhat sub-cordate. Flowers white, abruptly drooping, c 8 mm across. Fruit narrow, lanceolate in outline, up to 20 × 5—7 mm, pointed, pedicels straight, very hot to taste. Fig. 33/1.

Distr. ALDABRA: W; widely cultivated and naturalised throughout the tropics.

Notes. Found growing in champignon near Settlement; 1 collection, *Stoddart* 963 (K, US).

Vernac. 'piment arbrisseau', 'bird chillies', 'piment' or 'tili'.

2. DATURA L.

Coarse herbs, shrubs or small trees (sect. *Brugmansia* (Pers.) Bernhardi); branching falsely dichotomous. Leaves usually shallowly lobed. Flowers terminal, often in the fork formed by two branches, large. Calyx dehiscing by a line near base. Corolla trumpet-shaped, folded in bud, lobes 5, usually sharply pointed, tips recurved. Stamens 5, filaments joined to inside of corolla tube, anthers linear, included. Ovary 4-celled below, 2-celled above, ovules many; style slender, stigma capitate, included. Capsule large, on an erect or recurved pedicel, spiny or partially or wholly smooth, 4-valved, or irregularly dehiscent, or indehiscent, sitting on a persistent, woody, disc-like calyx remnant; seeds numerous, slimy when moistened.

A widespread genus, a few cultivated species, all parts toxic, some species used as hallucinagens.

Datura metel L., Sp. Pl.: 179 (1753).

Coarse herb or shrub to 2(−3) m tall, young growth minutely pubescent, becoming glabrous. Leaves ovate, up to 15 × 10 cm, rarely more, with 2−3 obtuse lobes separated by shallow, rounded sinuses, midribs minutely pubescent, becoming glabrous; petiole up to 7 cm long. Flowers solitary, erect, pedicel 5 mm long. Calyx cylindrical, 6−9 cm long, lobes 5, acute or acuminate, c. 1 cm long. Corolla white to dull purple, trumpet-shaped, 11−18 cm long, flaring to 6−8 cm across, tapering point of lobes 10−15 mm long, recurved. Capsule reddish-brown, globose, c 2.5 cm long, covered with short, stout spines, breaking irregularly, thick-walled; seeds whitish, asymmetrically obovate, 4−5 × 4−5 mm, barely stalked, outer margin strongly thickened, with 2−4 irregular, prominent, rounded ridges encircling it. Fig. 33/2−4.

Distr. ALDABRA: W; ASSUMPTION; ASTOVE; widespread through the tropics of the Old World.

Notes. Planted and naturalised around settlement sites. Dried corollas are reported to be smoked as a cure for coughing and asthma.

Vernac. 'feuille du Diable' or 'fleur poison'.

3. NICOTIANA L.

Erect herbs or shrubs; often glandular. Leaves alternate. Inflorescence a panicle, crowded. Calyx tubular to bell-shaped, united almost to summit, 5-lobed. Corolla tubular to salver-shaped, throat somewhat expanded, lobes

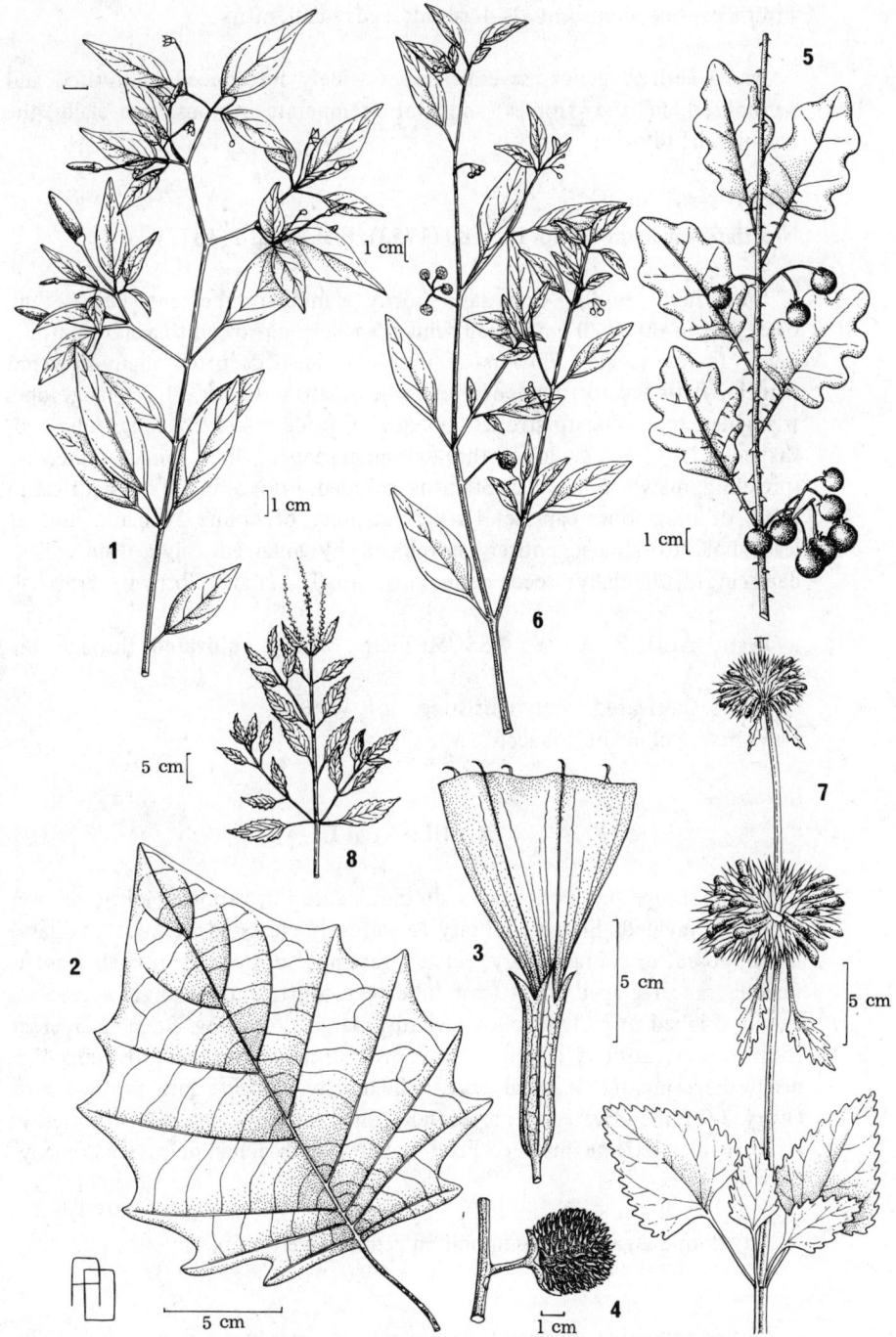

Fig. 33 1, *Capsicum frutescens*, habit. 2, *Datura metel*, leaf; 3, flower;
4, young fruit. 5, *Solanum indicum* var. *aldabrense*, habit.
6, *Solanum nigrum* var. *americanum*, habit. 7, *Leonotis
nepetifolia*, habit. 8, *Ocimum gratissimum*, habit.

5. Stamens 5, included or slightly exserted. Ovary 2–4-celled, ovules many. Fruit a capsule, dehiscent, 2–4-valved; seeds small, many.

An American genus, several species widely introduced, cultivated and naturalised in the tropics and warm-temperate regions, especially the commercial tobacco.

Nicotiana tabacum L., Sp. Pl.: 180 (1753); F.M.S.: 218 (1877).

Erect herbs to 0.7–1 m tall; shortly glandular-pubescent. Leaves thin, ovate, up to 60 × 20 cm, usually much smaller, narrowed to a short petiole, lower leaves sessile. Flowers in a much-branched, often many-flowered panicle, flattened to rounded at top when mature. Calyx c 1 cm long, lobes triangular, somewhat or strongly unequal. Corolla c 4–5 cm long, tube bell-shaped c 2–2.5 cm long, throat funnel-shaped; limb pink, somewhat spreading, margin 5-toothed. Stamens included. Fruit a broadly cylindrical to ovoid or subglobose capsule, 12–15 mm long, becoming 2-lobed at top, at least half to almost entirely enveloped by enlarged calyx, thin-walled, dehiscing septicidally; seeds numerous, small, rounded, brown, wrinkled.

Distr. ALDABRA: W; ASSUMPTION; widely cultivated through the tropics.

Notes. Cultivated, rarely persisting.

Vernac. 'tobac' or 'tobacco'.

4. SOLANUM L.

Herbs, shrubs or vines. Leaves alternate, entire or variously lobed, or even pinnately divided. Flowers solitary to variously clustered, terminal, axillary, leaf-opposed, or supra-axillary, clusters racemose or cymose; usually regular. Corolla saucer-shaped or at least tube very short, limb strongly spreading, deeply 5-lobed or-parted, rolled inwards, margins touching. Stamens 5, erect from throat; anthers closely converging, forming an apparently cone-like beak, dehiscing by 2 apical pores which may elongate into introrse slits. Ovary 2-(or in cultivated forms several-) celled, ovules many, placentas fleshy; style slender, stigma minute. Fruit a berry with juicy pulp; seeds many, flattish.

An enormous, almost cosmopolitan genus, largely tropical.

1. Leaves appearing pinnately compound; inflorescence a loose, raceme-like cyme   **2. S. lycopersicum**
1. Leaves entire or merely lobed, raceme umbelloid or condensed, if raceme-like, plant prickly
  2. Plant glabrous or almost so, stellate hairs absent; inflorescence long-pedunculate   **4. S. nigrum**

2. Plants with stellate hairs
  3. Plant not prickly; leaves more than 10 cm long; flowers much more than 1 cm across; fruit more than 5 cm long   **3. S. melongena**
  3. Plant prickly; leaves mostly less than 7 cm long; flowers less than 1 cm across; fruit c 1 cm long   **1. S. indicum**

1. Solanum indicum L., Sp. Pl.: 187 (1753).

var. **aldabrense** (C.H. Wright) Fosberg in K.B. 33: 141 (1978). Type: Aldabra, *Abbott* s.n. (K, holo.).

S. aldabrense C.H. Wright in B.M.I.K. 1894: 149 (1894); Schinz in A.S.N.G. 21: 89 (1897); Voeltzkow in A.S.N.G. 26: 551 (1902); Dupont, Report: 38 (1907); Hemsley in B.M.I.K. 1919: 126 (1919); Bitter in Fedde, Repert. 16: 110 (1923); Fosberg in P.T.R.S. B, 260: 218, 225 (1971); Renvoize in P.T.R.S. B, 260: 231 (1971).
S. indicum sensu Hemsley in B.M.I.K. 1919: 126 (1919), not L. (1753).

Prickly shrub to 1(–2) m tall, branching intricate, spreading; stems with sparse, minute, stellate hairs; prickles broad-based, somewhat recurved, short, varying in number but apparently always some present. Leaves elliptic to very broadly ovate or suborbicular, 3–4.5(–5) × 2–3.5(–4) cm, apex usually obtuse, rarely bluntly subacute, base truncate to obtuse, sometimes somewhat decurrent on petiole, margins sinuate to broadly 2–3-lobed, much more deeply lobed on young plants or fast-growing shoots, abundant stellate hairs especially beneath, variably sparsely prickly on both surfaces, prickles narrower based and straighter than those on stems, bases blackish-purple, distally yellow; petiole 1–2.5 cm long, usually with 1–5 recurved prickles. Inflorescence a slender, 1-sided cyme or perhaps more properly a cymiform raceme, slender, few-flowered, sometimes subumbellate, somewhat drooping; peduncle distinct but short, up to 1 cm long, occasionally with 1–2 prickles; pedicels to 1 cm long, or in fruit 1.5 cm, with or usually without 1–2(–3) prickles, the whole inflorescence sparsely to moderately covered with stellate hairs. Calyx 3–4 mm long, lobed about half-way, teeth triangular, moderately to densely covered with stellate hairs. Corolla white to pale lavender, 6–9 mm long, lobed nearly to base, tube 1–2 mm long, glabrous, lobes ovate-lanceolate, densely covered with stellate hairs outside, almost glabrous inside. Anthers linear-oblong, about 3 mm long, glabrous. Ovary globose, glabrous; style glabrous, exceeding anthers by about 2 mm. Fruit bright red, c 1 cm in diameter, juicy, glabrous; seeds white, discoid, somewhat eccentric, 2 mm across, with a rounded notch on one side. Fig. 33/5.

Generally regarded as a distinct species, but when compared with Asiatic material, especially from Ceylon, the type locality of *S. indicum*, the differences seem rather trivial.

Distr. ALDABRA: W, P, M, S, Esprit, Michel; endemic. Recorded from Assumption, Cosmoledo, Astove and St. Pierre by Dupont, l.c. but no material seen.

Notes. On Aldabra especially common on the coastal ridge of South Island in *Guettarda* woodland, less common on platin. Flowers throughout the year but mainly during the dry season. Fruit eaten and seed locally dispersed by tortoise, blue pigeon, dove and bulbul.

Vernac. 'anguive', 'bois zil mowa' or 'bringelle marronne'.

2. Solanum lycopersicum L., Sp. Pl.: 185 (1753).

Diffusely branched, glandular-pubescent herb, stems tending to be zigzag, nodes somewhat prominent. Leaves apparently pinnately compound, 'leaflets' 3–10 pairs with an odd terminal 'leaflet', 'leaflets' usually alternate, sometimes opposite, ovate or ovate-oblong, extremely variable in size, terminal leaflet larger, margins coarsely toothed. Cymes raceme-like, usually borne on nodes well above leaf-axils, hairy and shortly pubescent, glandular; pedicels prominently jointed not far below the calyx. Calyx deeply 5-lobed, persistent and much enlarged in fruit, lobes lanceolate, somewhat acuminate. Corolla yellow, lobes ovate-lanceolate, acuminate, somewhat exceeding calyx lobes. Stamens almost as long as corolla, anther-cells dehiscing by terminal pores which become slits. Fruit a shiny, bright red (or yellow) berry, 1.5–10 cm in diameter, 2–(or more-)celled, juicy; seeds in a jelly-like pulp, flat, disc-like, 3–5 × 2–4 mm, winged, prominently pubescent.

The garden tomato is considered by many botanists to belong to a separate genus, *Lycopersicon* Mill., but this seems to be no more distinct than many other groups in the vast genus *Solanum*.

Distr. ALDABRA: W; ASSUMPTION; cultivated through the tropical and temperate regions, sometimes naturalised.

Notes. Planted in gardens.

Vernac. 'tomate' or 'tomato'.

3. Solanum melongena L., Sp. Pl.: 785 (1953).

Robust herb, 1 m or more tall, with dense, stellate pubescence. Leaves large, broadly ovate to broadly oblong in outline, up to 25 × 15 cm, apex obtuse, base rounded to somewhat cordate, bilaterally very unequal, margins with several, broad, rounded lobes or entire, dense pale stellate hairs beneath, green with less dense stellate hairs above; petiole 1–3 cm long. Inflorescence 1–several-flowered, sessile to shortly pedunculate, pedicels 1–2 cm long; flowers partly bisexual, partly male. Calyx deeply lobed, covered with dense stellate hairs. Corolla purple or white, to 5 cm across, 5(–6)-lobed, lobes triangular-acute. Fruit a large, yellowish to almost black, obovoid or cylindrical berry, up to 15–20 cm long, borne on a much enlarged calyx, fruiting pedicel up to 7 cm long, thickened, deflexed; seeds light brown, discoid, 3 mm across.

Distr. ALDABRA: W; COSMOLEDO: M; cultivated throughout the tropical and temperate regions.

Notes. Cultivated.

Vernac. 'bringelle' or 'aubergine'.

4. Solanum nigrum L., Sp. Pl.: 186 (1753); F.M.S.: 214 (1877).

var. **americanum** (Mill.) O.E. Schulz in Urban, Symb. Antill. 6: 160 (1909).

Solanum americanum Mill., Gard. Dict. ed. 8, SOL (1768).
S. nodiflorum Jacq., Icon. Pl. Rar. 2; fig. 325 (1789); F.M.S.: 214 (1877);
   Schinz in A.S.N.G. 21: 90 (1897); Voeltzkow in A.S.N.G. 26: 551 (1902);
   Hemsley in B.M.I.K. 1919: 127 (1919).
S. nigrum sensu F.M.S.: 214 (1877); Dupont, Report: 38 (1907); Hemsley in
   B.M.I.K. 1919: 127 (1919); Fosberg in P.T.R.S. B, 260: 222, 225 (1971);
   Renvoize in P.T.R.S. B, 260: 229 (1971).

Erect to tangled, glabrous to slightly puberulent herb, up to 1 m tall. Leaves ovate, up to 5(−6) × 2.5 cm, apex acute to slightly acuminate, base decurrent well down the short petiole, margin entire to wavy or coarsely and bluntly toothed, blade thin. Flowers in few-flowered umbels; peduncles slender, appearing about half to two-thirds along the internodes; pedicels slender, c 4−6(−7) mm long; flowers and fruit nodding. Calyx deeply lobed, lobes ovate-oblong, not more than 1 mm long, apex obtuse, somewhat reflexed in fruit. Corolla white, up to 6 mm across, deeply lobed, lobes narrowly ovate. Fruit black, globose, 4−6 mm diameter; seeds white, sub-orbicular, c 1 mm across, netted. Fig. 33/6.

This is the small-flowered form with a very small calyx, reflexed in fruit, with drooping pedicels and slender inflorescence, that has variously been called *Solanum americanum* Mill. and *S. nodiflorum* Jacq. It differs somewhat from the European *S. nigrum* but certainly not specifically. The whole group is a subject of much disagreement.

Distr. ALDABRA: W, S (east); ASSUMPTION; COSMOLEDO: W; ASTOVE; widely distributed throughout the world.

Notes. On Aldabra only known from Settlement on West Island and Cinq Cases on South Island, in gardens or crevices on the platin. Whether or not the plant is indigenous is uncertain. It is most commonly found as a weed around settlements, but is also found in remote places where is appears native. If it had been brought by man, the fruit pigeons and doves might well have distributed the seeds rather effectively. Generally considered poisonous, but Ridgway says "edible leaves", and Veevers-Carter says "leaves eaten as spinach". Fosberg tasted the fruit and found it not very palatable. We would not recommend eating it. Observations few but flowers throughout the year if sufficient moisture available.

Vernac. 'brède martin'.

49. SCROPHULARIACEAE

Herbs or shrubs, rarely small trees, sometimes parasitic or saprophytic. Leaves alternate, or usually opposite or whorled; stipules absent. Flowers solitary, axillary, or in terminal or axillary racemes, spikes or heads; bisexual, usually irregular. Calyx usually (3—4—)5— lobed, lobes overlapping or margins touching. Corolla usually 4—5-lobed, usually more or less irregular, often strongly so. Stamens 4 or 2, rarely 5(1 often sterile), inserted on the corolla and alternate with the lobes; anthers 1—2-celled, opening lengthwise. Ovary superior, usually 2-celled, ovules numerous, placentation axile; style 1. Fruit a capsule or berry; seeds many.

A cosmopolitan family.

1. Mat-forming, fleshy herbs; flowers white, pink or mauve
  2. Corolla 5—6-lobed; stamens 4   **1. Bacopa**
  2. Corolla 4—5-lobed; stamens 2   **2. Bryodes**
1. Erect, parasitic herbs; flowers scarlet   **3. Striga**

1. BACOPA Aublet

Prostrate, glabrous, terrestrial or aquatic, perennial herbs. Leaves opposite, entire to divided. Flowers in terminal spikes or axillary and solitary. Calyx deeply 5-lobed, lobes unequal in width. Corolla bell-shaped, more or less equally 5-lobed. Stamens 4, 2 long, 2 short, inserted on the corolla tube. Ovary 2-celled. Capsule 4-valved; seeds numerous.

50 species in the tropics, sub-tropics and warm-temperate parts of the world.

Bacopa monnieri (L.) Pennell in Proc. Acad. Nat. Sci. Philad. 98: 94 (1946).

Lysimachia monnieri L., Cent. Pl. 2: 9 (1755).
Herpestis monnieri (L.) Kunth, Nov. Gen. et Sp. 2: 366 (1817); F.M.S.: 237 (1877); Schinz in A.S.N.G. 21: 90 (1897); Voeltzkow in A.S.N.G. 26: 551 (1902); Dupont, Report: 39 (1907); Hemsley in B.M.I.K. 1919: 127 (1919).

Small terrestial or acquatic herbs, stems up to 60 cm long, but usually much less in our area, often rooting at the nodes. Leaves, sessile, narrowly obovate, 2—10 (—20) mm long, margins usually entire, fleshy, noticably pitted when dry. Flowers axillary, solitary; pedicels up to 10 mm long; bracteoles 2, narrow, 2—4 mm long, just below each flower. Calyx 3—6 mm long, lobes very unequal. Corolla white, pale pink or pale blue, bell-shaped,

6—10 mm long. Capsule ovoid, 4—8 mm long; seeds pale brown, spindle-shaped or oblong, 0.5—0.6 × 0.3 mm, angular. Fig. 21/8—9.

Distr. ALDABRA: S (east); pantropical.

Notes. Found in the dried mud of shallow seasonal pools at the eastern end of South Island. Although sometimes minute, the plant is conspicuous when flowering.

2. BRYODES Benth.

Glabrous herb. Leaves opposite. Calyx bell-shaped, divided to the base, lobes 5, lanceolate. Corolla bell-shaped, lobes 4, short, rounded. Stamens 2, inserted low down in the corolla tube. Style curved, slender; stigma capitate. Capsule globose; seeds numerous.

1 species, in Mauritius, Madagascar and Aldabra.

Bryodes micrantha Benth. in DC., Prodr. 10: 433 (1846): F.M.S.: 241 (1877).

Tiny, prostrate, glabrous, moss-like herbs, moderately branched, often rooting from the nodes. Leaves opposite, sessile, linear — oblong, 2—3.5 × 0.4—0.8 mm, flat, apex acute or rounded. Flowers axillary, solitary, on slender pedicels up to 1.5 mm long. Calyx bell-shaped, up to 1 mm long, lobes acute to acuminate. Corolla white or pale mauve, bell-shaped, enclosed by the calyx. Capsule globose, 1.2 mm in diameter; seeds numerous, obovoid or kidney-shaped, 0.3 × 0.2 mm, 5 — grooved, wrinkled between the grooves. Fig. 21/10—11.

Distr. ALDABRA: S (east); Madagascar, Mauritius.

Notes. Known only from the tortoise turf at Cinq Cases, *Hnatiuk* 731763 (K). Plant moss-like and inconspicuous except when flowering.

3. STRIGA Lour.

Parasitic herbs, turning black when dried. Leaves alternate, entire, narrow. Flowers solitary, sessile, axillary or in terminal spikes. Calyx cylindrical, tube strongly ribbed, teeth 5, long. Corolla 2-lipped, curved, lower lip 3-lobed, upper lip entire or 2-lobed. Stamens 4, very short, in 2 pairs inserted in the corolla tube. Ovary oblong, ovoid; seeds numerous, minute.

40 species in tropical and southern Africa, Asia and Australia.

Striga asiatica (L.) O. Kuntze, Rev. Gen. 2: 466 (1891).

Buchnera asiatica L., Sp. Pl.: 630 (1753).

Striga hirsuta Benth. in DC., Prodr. 10: 502 (1846); F.M.S.: 242 (1877).

Slender, erect herbs, up to 40 cm high, rough-pubescent to almost glabrous. Leaves linear, 1–3.5 × 1–3 mm, apex acute or rounded. Calyx up to 10 mm long, strongly 10-ribbed, teeth narrow, quarter the length of calyx tube. Corolla scarlet in our area (elsewhere also yellow or white), up to 1 cm long, upper lip 2-lobed. Capsule ovoid, up to 6 × 2.5 mm, enclosed by the persistent calyx, splitting into 2 valves; seeds numerous, ellipsoid, 0.3 × 0.2 mm. Fig. 24/5.

Distr. ALDABRA: W; COSMOLEDO; widely distributed through the tropics of the Old World, introduced into the New World.
Notes. Recorded as parasitizing the roots of *Eragrostis subaequiglumis*
Vernac. 'herbe de feu' or 'herbe rouge'.

50. BIGNONIACEAE

Trees, shrubs or climbers, rarely herbs. Leaves opposite or whorled, rarely alternate, usually digitate or pinnate, the terminal leaflet sometimes tendril-like; stipules absent. Inflorescence cymose, rarely much reduced and then borne on trunk or larger branches; flowers often showy, bisexual, irregular. Calyx bell-shaped, truncate or 5- toothed. Corolla irregular, usually 5-lobed, lobes overlapping in bud. Stamens usually 4 with 1 staminode, inserted on corolla, rarely 2 with or without 3 staminodes, anthers 2-celled, the cells often one above the other. Ovary superior, (1–) 2-celled, placentation axile or in 1-celled ovaries, parietal; style 1, stigma 2-lobed. Fruit a 2-valved capsule or rarely indehiscent, then large and fleshy or hard with fleshy pulp; seeds often winged.

A large, mainly tropical family with many ornamental, cultivated genera.

TABEBUIA Gomes ex DC.

Trees or shrubs. Leaves opposite, simple or digitately compound. Flowers large and showy, in terminal clusters. Calyx tubular, toothed or lobed. Corolla funnel- or bell-shaped, 5-lobed. Stamens 4, in 2 pairs. Fruit a linear-oblong, tubular capsule; seeds numerous, winged.

A large tropical American genus of c 80 species, some species introduced throughout the tropics as ornamentals.

Tabebuia pallida (Lindl.) Miers in Proc. Roy. Hort. Soc. 3: 199 (1863).

Bignonia pallida Lindl., Bot. Reg. fig. 965 (1826).

Trees, up to 20 m high. Leaves 3–5-foliolate, occasionally reduced to 1-foliolate; petioles up to 10 cm long. leaflets stalked or sessile, elliptic to oblong, 7–15 cm long, apex acute or obtuse, slightly leathery, shiny above. Flowers on pedicels 2 cm long. Calyx 1 cm long. Corolla white, pink or purple, 5–7 cm long. Capsule cylindrical, 10–20 × 0.6–1 cm, longitudinally dehiscent; seeds oval-oblong, 7× 15 mm, with membranous wings at either end.

Distr. ASTOVE; native of Central America, introduced elsewhere.

Notes. Cultivated; only known from 1 collection, *Stoddart & Poore* 1280 (K).

51. ACANTHACEAE

Herbs, shrubs and climbers; nodes often prominent. Leaves opposite, simple; stipules absent. Flowers variously disposed, bisexual, irregular, rarely almost regular; bracts usually present, often conspicuous. Calyx 4–5-lobed or parted, usually appearing polysepalous. Corolla usually 4–5-lobed, usually strongly 2-lipped. Stamens 2–5, usually 4, often paired, inserted on corolla, disk present. Ovary superior, 2-celled, ovules 2-several per cell, axile placentation; style 1, slender, stigmas 2. Fruit usually a loculicidal capsule, rarely a drupe, elastically dehiscent, usually equipped with strong hook-like "retinacula" supporting the large, frequently disk-like seeds.

A large, mainly tropical family; includes many important and widely distributed ornamentals.

1. Shrubs; calyx 4-lobed (lobes not to be confused with the 4 fused
    bracteoles in *Hypoestes*); stamens 4–5, 2 fertile, 2 or 3 sterile, anthers
    2-celled; capsule 2– or 4-seeded   **2. Barleria**
1. Herbs; calyx 5-lobed; stamens 2 or 4, all fertile
  2. Stamens 2; capsule 4-seeded
    3. Anthers with 1 cell below the other, tailed; capsule enclosed by 4 fused
        bracteoles   **3. Hypoestes**
    3. Anthers 2-celled; capsule not enclosed by bracteoles   **4. Justicia**
  2. Stamens 4; capsule 4–8-seeded
    4. Flowers in terminal spikes; capsule 4-seeded   **1. Asystasia**
    4. Flowers axillary, solitary or clustered; capsule (4–)6–(–8) seeded
                                                                  5. Ruellia

1. ASYSTASIA Blume

Shrubs or herbs. Leaves entire. Flowers in terminal racemes, spikes or panicles. Calyx 5-lobed, divided almost to the base. Corolla funnel-shaped, lobes 5, rounded, subequal. Stamens 4, all fertile, anthers 2-celled. Ovary 2-celled, 2 ovules per cell. Capsule club-shaped, stalked; seeds 4, compressed.

A genus of 40 species throughout the Old World tropics.

Asystasia gangetica (L.)T.Anders. in Thwaites, Enum. Pl. Zeyl.: 235 (1864); F.M.S.: 247 (1877); Hemsley in B.M.I.K. 1919: 127 (1919); Renvoize in P.T.R.S. B, 260: 229 (1971), sens. lat.

Justicia gangetica L., Amoen. Acad. 4: 299 (1759).
Asystasia coromandeliana Nees in Wall., Pl. As. Rar. 3: 89 (1832).
Asystasia bojeriana Nees in DC., Prodr. 11: 166 (1847); Fosberg in P.T.R.S. B, 260: 224 (1971).

Perennial herbs, often much-branched; stems slender, delicate, up to 40 cm high, sparsely hairy. Leaves widely spaced, ovate-lanceolate to broadly ovate, $1-4 \times 0.8-2$ cm, sparsely hairy to subglabrous; subsessile or petiole short, up to 5 mm long. Raceme 1-sided, interrupted, up to 10 cm long, flowers few to many; pedicels short, 1 mm long, well spaced; bracts linear, 1 mm long. Calyx 5–6 mm long, shortly hairy, lobes 5, equal, linear, acuminate. Corolla white or cream, rarely blue or pinkish, lower lobe flecked with purple, trumpet-shaped, 15–18 mm long, sparsely hairy. Capsule club-shaped, 1.5–2 cm long, sparsely pubescent, longitudinally dehiscent; seeds 4, quadrangular or triangular, beaked, 3–3.5 mm long, roughly covered with wart-like protuberances. Fig. 34/1–3.

The above description applies only to plants in our area. African mainland plants are generally larger in all parts and more hairy. This species, together with its close allies, *A. bojeriana* Nees and *A. multiflora* Klotzsch are currently being revised. The status of the Aldabra flora material is at present uncertain; the name *A. gangetica* is here being used in the wide sense to embrace the 2 allied species.

Distr. ALDABRA: W, M (west), S; ASSUMPTION; ASTOVE; widespread through the tropics of the Old World, introduced into the New World.
Notes. Not common, usually found in pits and crevices of the coastal champignon. Flowers during the wet season but flower throughout the year if sufficient moisture available.
Vernac. 'herbe mange tout' or 'mange tout'.

2. BARLERIA L.

Shrubs or herbs. Leaves entire. Flowers in 1-sided cymes, or solitary and axillary or in terminal spikes. Calyx 4-lobed, inner pair of lobes usually smaller than the outer pair. Corolla tubular or funnel-shaped, 2-lipped, 5-lobed. Stamens 4–5, 2 fertile, 2 shorter and sterile and, when present, a fifth one reduced to a short staminode; anthers 2-celled. Ovary 2-celled, ovules 1-2 per cell; stigma 2-lobed. Capsule ovoid or oblong; seeds 2–4, hairy, hygroscopic.

A genus of c 230 tropical and subtropical species, mainly in the Old World.

Barleria decaisniana Nees in DC., Prodr. 11: 230 (1847); F.M.C. 182: 160 fig. 27/1–4 (1967).

Barleria sp. Hemsley in B.M.I.K. 1919: 127 (1919).

Shrub up to 60 cm high; branches smooth, glaucous. Leaves ovate-lanceolate, 5–10 × 2–4.5 cm, attenuate at the base, sparsely hairy to sub-glabrous; petiole up to 8 mm long. Flowers in small, dense or large, spreading, terminal, 1-sided cyme; bracts linear, often curved, 5–8 mm long, sparsely hairy. Calyx lobes entire, outer lobes broadly ovate-elliptic, obtuse, 10–13 × 5.7 mm, prominently veined, glabrous or sparsely hairy, inner lobes linear, obtuse, 7–10 × 1–2 mm, hairy. Corolla blue, trumpet-shaped, 2.5–3.5 cm long, lobes ovate. Fertile stamens 2, as long as the corolla; sterile stamens 2 and 1 staminode very short and inserted deep down in the corolla tube. Style up to 4 cm long. Capsule blackish, oblong, 10–15 × 6 mm, usually enclosed at the base by the persistent calyx, longitudinally dehiscent; seeds 4, reddish black, flattened-ovoid, 3–4 mm in diameter. Fig. 34/4–6.

Distr. ASTOVE; Madagascar and Comoros Is.
Notes. Recorded as growing under coconuts in the collection from Astove by *Stoddart & Poore* 1264 (K). A second collection, *Dupont* 118 states that it is a common weed but does not specify the island, which is presumed to be Astove.

3. HYPOESTES R. Br.

Shrubs or herbs. Leaves entire, rarely toothed. Flowers in sparse or dense, globose or elongate, terminal or axillary spikes or cymes, each flower enclosed by 4, conspicuous bracteoles fused for up to half of their length. Calyx shorter than the corolla, 5-lobed, lobes acute, free to the base or united for one third to two thirds of their length. Corolla tubular, 2-lipped, upper lip entire, indented or notched, lower lip 3-lobed. Stamens 2, both fertile, anthers 1-celled. Ovary 2-celled, ovules 2 per cell. Capsule stalked, oblong-ovoid or club-shaped; seeds 4.

A genus of c 150 species in the Old World tropics.

Hypoestes aldabrensis Baker in B.M.I.K. 1894: 150 (1894); Schinz in A.S.N.G. 21: 98 (1897); Voeltzkow in A.S.N.G. 26: 552 (1902); Dupont, Report: 39 (1907); Hemsley in B.M.I.K. 1919: 127 (1919). Type: Aldabra, *Abbott* s.n. (K. holo., US, iso.).

H. serpens sensu Hemsley in B.M.I.K. 1919: 127 (1919) not R.Br.

Much-branched, tufted perennial herb, 5–18(–25) cm high, or with decumbent stems up to 40 cm long. Leaves ovate, oblong or obovate, rarely orbicular, 5–30 × 5–20 mm, apex obtuse or acute, margins entire, somewhat fleshy, glabrous to sparsely hairy; subsessile or petiole short. Flowers in dense, globose, axillary or terminal whorls, rarely solitary, whorls usually interspersed with reduced leaves, each flower surrounded by a tubular, papery involucre 5–14 mm long, of 4 bracteoles fused for three-quarters of their length, 2 outer bracteoles broader and longer than the 2 inner bracteoles, glabrous to sparsely hairy outside, free apical part rounded, slightly inflated, 2 inner bracteoles linear-lanceolate, hairy at the apex. Calyx tubular, 3–5 mm long, lobes lanceolate, membranous, fused for half their length. Corolla blue, mauve or lilac, as long as or longer than the bracteoles, 13–16 mm long, upper lip entire, lower lip 3-lobed, central lobe speckled purple, covered with glandular and non-glandular hairs outside. Stamens slightly exserted from the corolla tube. Style slightly longer than the stamens. Capsule club-shaped, 6–8 × 3 mm, hairy towards the apex, longitudinally dehiscent; seeds 4, flattened, broadly obovate or oblong, 2 × 1.2 mm, beaked. Fig. 34/7–9.

This species is very closely related to *H. serpens* R.Br. from Madagascar and Mauritius, the chief differences being in the shape of the free apical part of the bracteole lobes, which in *H. serpens* are acute to acuminate and in *H. aldabrensis* are rounded. In addition the seeds in *H. serpens* are 1 mm long and in *H. aldabrensis* they are 2 mm long.

Distr. ALDABRA: W, P, M, S; ASSUMPTION; COSMOLEDO: W, M; ASTOVE; endemic.
Notes. Usually found growing near the coast in pits and crevices in coral limestone. Flowers primarily during the wet season, but throughout the year if sufficient moisture available.

4. JUSTICIA L.

Herbs or shrubs. Leaves simple, entire or wavy. Flowers solitary, in spikes, spicate-whorls or cymes. Calyx (4–)5-toothed, divided nearly to the base. Corolla 2-lipped, lower lip 3-lobed, upper lip entire or shortly 2-lobed. Stamens 2, 1 anther cell below the other and distinctly tailed. Ovary 2-celled, ovules 2 per cell. Capsule oblong-ovoid; seeds 4.

A genus of 300 species throughout the tropics and subtropics of the world.

Justicia procumbens L., Sp. Pl.: 15 (1753); Hemsley in B.M.I.K. 1919: 140 (1919).

Trailing annual herb, up to 15 cm high; stems wiry, often rooting at the lower nodes. Leaves elliptic-oblong, 1–2.5 × 5–15 mm, wide, appressed-hairy

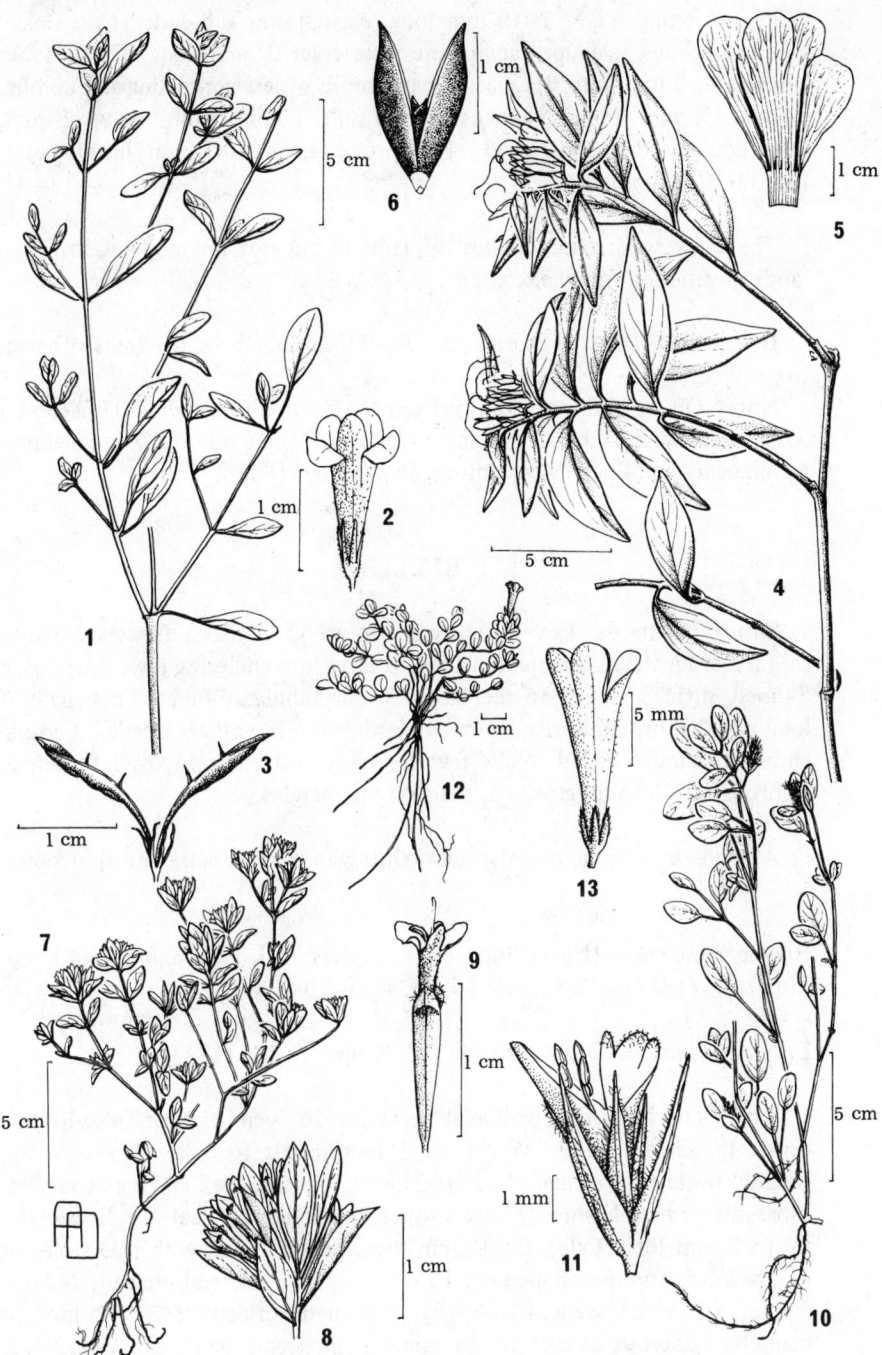

Fig. 34   1, *Asystasia gangetica*, habit; 2, flower; 3, fruit. 4, *Barleria decaisniana*, habit; 5, flower dissection; 6, fruit. 7, *Hypoestes aldabrensis*, habit; 8, inflorescence; 9, flower. 10, *Justicia procumbens*, habit; 11, flower. 12, *Ruellia monanthos*, habit; 13, flower.

above and beneath; petioles slender, 2–24 mm long. Inflorescence a terminal, ovate or oblong spike, 5–15 mm long; each flower subtended by a single, membranaceous, slender, acuminate bracteole, 3 mm long. Calyx lobes lanceolate, 5 mm long, apex acuminate, more or less concealing the corolla. Corolla 5 mm long. Stamens 2. Capsule oblong-ovoid, 3 × 2 mm, longitudinally dehiscent; seeds 4, depressed-ovoid, 0.7 mm in diameter. Fig. 34/10–11.

The above description applies to plants in our area and may not cover the range of variation found elsewhere.

Distr. ALDABRA: S (east); ASSUMPTION; islands of the Indian Ocean, tropical Asia.
Notes. On Aldabra only known from 1 collection, *Hnatiuk* 730702 (K) on coastal champignon between Cinq Cases and Flamingo Pool. One collection from guano deposits on Assumption, *Dupont* 103 (K).

5. RUELLIA L.

Shrubs or herbs. Leaves subentire or wavy, crenate. Flowers axillary, solitary or clustered; bracteoles small or large, not enclosing the calyx. Calyx 5-lobed, divided almost to the base. Corolla tubular, 5-lobed. Stamens 4, 2 long and 2 short, all fertile, shorter than the corolla, anthers 2-celled. Capsule club- or spindle-shaped; seeds few to many, orbicular to oval, flattened, glabrous except for hygroscopic hairs on the margins.

A genus of 200 or more species, throughout the tropics and subtropics.

Ruellia monanthos (Nees) Bojer ex T. Anders. in J. Proc. Linn. Soc. Lond. Bot. 7: 24 (1863); F.M.C. 182: 70, fig. 8/15–19 (1967).

Dipteracanthus monanthos Nees in DC. Prodr. 11: 125 (1847).

Small, lossely tufted, perennial herbs, up to 5 cm high or spreading by prostrate runners up to 25 cm long, subglabrous to hairy. Leaves small, broadly ovate to orbicular, 1–1.5(–2) × 0.5–1 cm, apex and base rounded; subsessile or petiole short. Flowers solitary; bracteoles 2, leaf-like, lanceolate, up to 8 mm long. Calyx (3–)4 mm long. Corolla white with pale violet or mauve lobes, trumpet-shaped, 10-15 mm long. Capsule club-shaped, 7–10 × 3 mm, glabrous; seeds (4–)6(–8), flattened, orbicular, 2–2.2 mm in diameter, glabrous except for marginal hygroscopic hairs. Fig. 34/12–13.

Distr. ALDABRA: M, S; ASSUMPTION; Madagascar.
Notes. Occurs in pits and crevices of the coastal champignon. Few observations but apparently flowers throughout the year if sufficient moisture available.

52. VERBENACEAE

Herbs, lianes, shrubs or trees; branchlets often 4-angled, often armed with thorns. Leaves opposite, or rarely whorled, simple or rarely compound; stipules absent. Inflorescence racemose or cymose. Calyx usually 4—5-lobed or -toothed, persistent. Corolla usually, salver-shaped 4—5-lobed, frequently irregular. Stamens same number as corolla lobes, or party reduced to staminodes. Ovary superior 2—8-celled, 1 ovule per cell, placentation axile. Fruit a capsule, drupe or berry.

A large, mostly tropical family; sometimes with *Avicennia* recognised as distinct.

1. Individual cymes within panicle subtended by 4 conspicuous violet bracts
3. Congea
1. Inflorescence not as above
  2. Flowers in heads or spikes
    3. Leaves entire, woolly beneath   **1. Avicennia**
    3. Leaves dentate or crenate-dentate, not woolly
      4. Flowers in heads   **4. Lantana**
      4. Flowers in elongate spikes   **6. Stachytarpheta**
  2. Flowers in cymes
    5. Stamens and styles strongly exserted   **2. Clerodendrum**
    5. Stamens and styles included or only slightly exserted   **5. Premna**

1. AVICENNIA L.

Trees and shrubs, with slender, erect, pencil-like pneumatophores arising in lines from the roots. Leaves opposite and decussate, entire, leathery, gland-dotted above, finely and densely pubescent beneath, exuding minute salt crystals beneath. Flowers opposite, in terminal and axillary pedunculate heads or very short spikes. Calyx small, 5-lobed. Corolla short, salver-shaped, 4-lobed. Stamens 4, inserted in throat of corolla, included. Ovary with a central, 4-winged placenta, ovules 1 per compartment; style 1, 2-lobed. Fruit oblique, fleshy, somewhat compressed; seeds 1, large, consisting of a naked embryo with 2, somewhat unequal cotyledons, at maturity freeing itself from the fruit wall and floating.

A pan-tropical genus of saline or brackish muddy shores and mangrove swamps.

Avicennia marina (Forssk.) Vierh. in Denkschr. Akad. Wien. Math. Nat. 71: 435 (1903); Hemsley in B.M.I.K. 1919: 128 (1919); Vesey-Fitzgerald in J. Ecol. 30: 11, 14 (1942); F.M.C. 174 bis.: 2 (1956); Stoddart & Wright in Atoll Res. Bull. 118: 28 (1967); Fosberg in P.T.R.S. B, 260: 219, 224 (1971); Renvoize in P.T.R.S. B, 260: 230 (1971); Macnae in P.T.R.S. B, 260: 239, fig. 5 (1971).

Seceura marina Forssk., Fl. Aegypt. Arab.: 37 (1775).

Avicennia officinalis sensu F.M.S.: 257 (1877); Baker in B.M.I.K. 1894; 150 (1894); Schinz in A.S.N.G. 21: 89 (1897); Voeltzkow in A.S.N.G. 26: 55 (1902); Dupont, Report: 39 (1907), not L. (1753).

Shrub or small tree, to 6 m tall; twigs slender, nodes prominent. Leaves ovate to elliptic, 5–6(–8) × 2–2.5(–4) cm, apex and base acute, stiff, sub-leathery, densely grey- or yellowish- to white-pubescent beneath; petiole 4–10 mm long. Inflorescence of 1 terminal cyme and 2 axillary cymes in the axils of the terminal pair, sometimes 1 in each axil at penultimate node; terminal peduncle up to 2 cm long, lateral peduncles shorter; flowers few, c 6, in opposite and decussate pairs, the lowest pair opening first, subtended by overlapping woolly bracts and bracteoles. Calyx deeply 5-lobed, lobes resembling bracteoles but larger. Corolla yellow, tube 1.5–2 mm long, glabrous, lobes 4, oval c 3 mm long, obtusely pointed, hairy except along the margins. Anthers suborbicular, filaments very short, in sinuses. Ovary ovoid, densely silvery-pubescent; style thick, erect, very short with 2, erect to some-what recurved, ovate lobes, becoming erect after anthesis. Fruit greyish-green, broadly oval to orbicular, thick but somewhat flattened, c 2.5 × 2.5 cm, beaked by persistent style and stigmas, woolly, falling from tree into water and opening, releasing the naked embryo, cotyledons suborbicular. Fig. 35/1–2.

Distr. ALDABRA: W, P, M, S, Esprit, Michel, Moustique; COSMOLEDO: W, M; ASTOVE; a widespread Indo-Pacific mangrove plant.

Notes. A common component of mangrove swamps and saline tidal flats throughout the Aldabra group, except Assumption, which has no suitable habitat. Flowers during the dry season.

Vernac. 'manglier blanc'.

2. CLERODENDRUM L.

Shrubs, rarely scandent. Leaves opposite, entire or toothed. Flowers in axillary cymes or in terminal panicles, often showy. Calyx of 5 sepals variously united, margins lobed, toothed or truncate. Corolla slightly to conspicuously irregular, tube narrow, straight or curved, lobes 5, often large. Stamens 4, at maturity long-exserted. Ovary incompletely 4-celled, 1 ovule per cell; style long-exserted, shortly 2-fid. Fruit a globose or obovoid drupe, often 4-lobed, pyrenes 1–4, hard, fleshy, juicy, or scanty and dryish.

A rather large pan-tropical genus, containing a number of widely cultivated ornamentals.

Clerodendrum glabrum E. Meyer, Comment. Pl. Afr. Austr. 1: 273 (1838).

var. **minutiflorum** (Baker) Fosberg in K.B. 33: 193 (1978). Type: Aldabra, *Abbott* s.n. (K, holo., US, iso.).

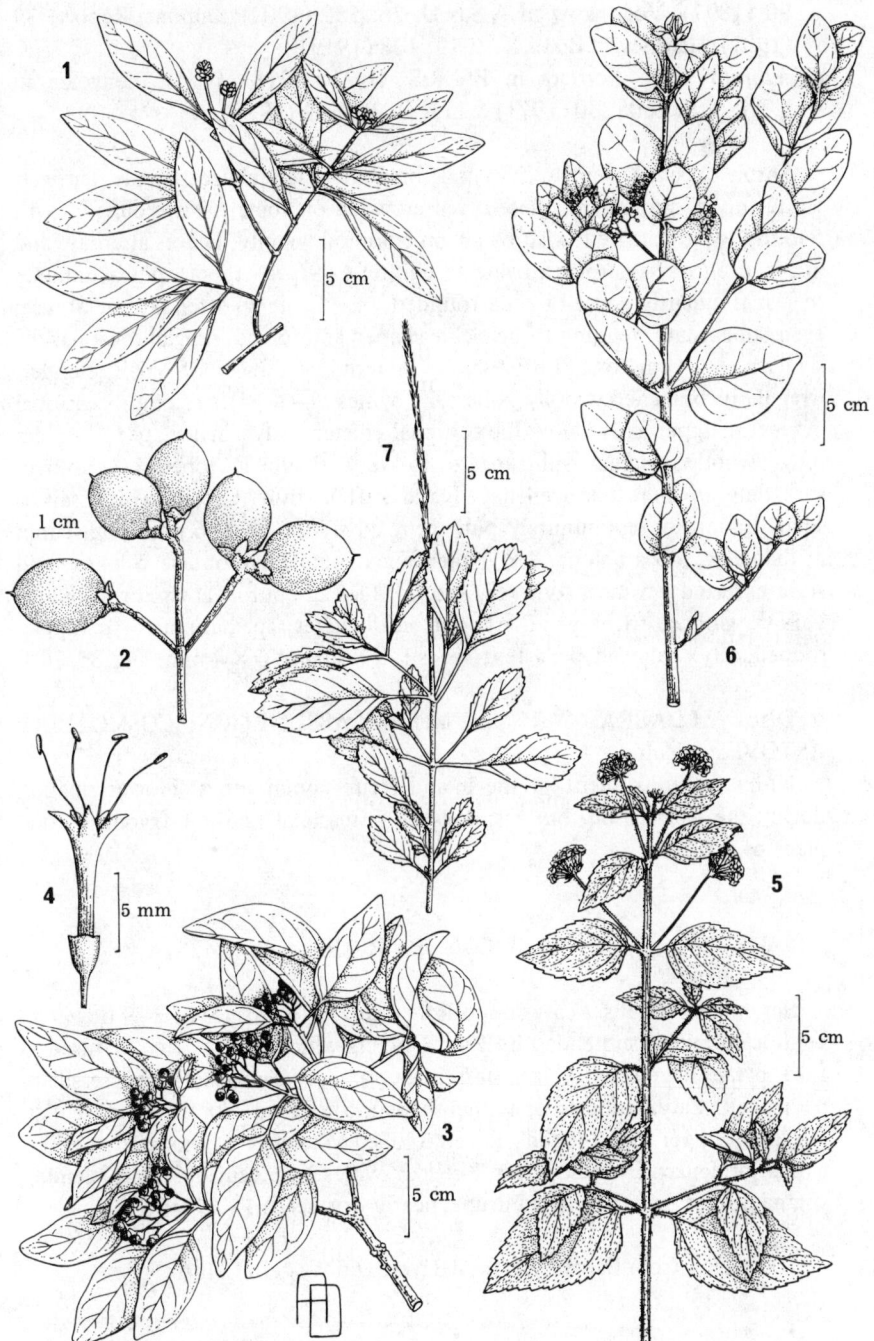

Fig. 35   1, *Avicennia marina*, habit; 2, fruit. 3, *Clerodendrum glabrum*,
         habit; 4, flower. 5, *Lantana camara*, habit; 6, *Premna obtusifolia*,
         habit. 7, *Stachytarpheta jamaicensis*, habit.

C. minutiflorum Baker in B.M.I.K. 1894: 150 (1899); Schinz in A.S.N.G. 21: 90 (1897); Voeltzkow in A.S.N.G. 26: 552 (1902); Dupont, Report: 39 (1907); Hemsley in B.M.I.K. 1919: 128 (1919).

C. glabrum sensu Fosberg in P.T.R.S. B, 260: 218 (1971); Renvoize in P.T.R.S. B, 260: 230 (1971) & in K.B. 30: 151 (1975).

Shrub or small tree, to 5 m tall; branchlets tending to be pale, lenticels white, nodes prominent, these often crowded on some twigs, young growth tending to be pale yellowish or cream coloured, woolly. Leaves alternate and in 3's, even on same plant, ovate to elliptic 6–9 × 3–6 cm, apex acuminate to less frequently acute or even rounded, base acute to attenuate or obtuse, even subcordate, tending to be sickle-shaped and folded, subglabrous; petiole slender, 1–2 cm long. Inflorescence a terminal cluster of small, slender, irregularly branched, woolly-pubescent cymes, 2–3(–4) cm long, occasional cymes in upper axils below the terminal cluster. Calyx bell-shaped, 2–3 cm long, woolly, margin subtruncate to very shallowly lobed or toothed, spreading to almost saucer-shaped and stiff in fruit. Corolla white, salver-shaped, more or less minutely pubescent outside, tube 5–6 mm long, 1 mm in diameter, lobes oblong, 2 × 1 mm, apex rounded to obtuse. Stamens and style exserted c 5 mm; style very shortly 2-fid and purple at tip. Fruit orange yellow, globose to obovoid, c 6 mm in diameter, apex rounded to nipple-shaped, calyx enlarged, persistent; seeds 1, top-shaped, 7× 4 mm. Fig. 35/3–4.

Distr. ALDABRA: W, P, M, S, Mentor; ASSUMPTION; COSMOLEDO; ASTOVE; endemic.

Notes. A constituent of the inland scrub communities. Flowers mainly during the wet season, but responds to unseasonal rains. A favourite food plant of tortoise.

3. CONGEA Roxb.

Scrambling shrubs. Leaves opposite, entire. Inflorescence of 3–9-flowered, peduncled cymes subtended by 3 or 4 bracts which may be free or fused at base, cymes aggregated in large terminal or axillary panicles; flowers sessile or pedicelled. Calyx tubular or funnel-shaped, 5-toothed. Corolla tube slender, 2-lipped, upper lip 2-lobed, lower 3-lobed. Stamens 4, exserted; anthers almost orbicular. Ovary superior, 2-celled, ovules 2 per cell; style long, slender, stigma 2-fid. Fruit an obovoid drupe, nearly dry; seeds 1.

10 species, from Bangladesh to Malaysia and Western China.

Congea griffithiana Munir in Gard. Bull. Singapore 21: 285, fig. 3 (1966).

Scrambling climber; branchlets tawny-haired. Leaves opposite, elliptic to broadly ovate, 3–12 × 3–6 cm, apex acute to acuminate, base wedge-shaped, leathery to papery, glabrous above, pubescent beneath; petiole 5–10 mm long. Panicle up to 45 cm long, branches opposite, spreading, tawny-haired;

cymes 5-flowered, bracts 4, violet, oblanceolate, up to 3 × 1.7 cm, free and narrowed to the base, pale purple-hairy above, grey-brown pubescent beneath; flowers sessile. Calyx bell-shaped, 5-lobed, 4—5 mm long, densely hairy. Corolla tube barely exserted, throat hairy, otherwise glabrous. Stamens 4, exserted, filaments up to 1 cm long. Ovary obovoid, glabrous, apex glandular; style longer than the filaments, stigma 2-lobed. Drupe obovoid, 4 × 3 mm, enveloped by persistent calyx.

Distr. ASSUMPTION; native in Burma, Thailand and Malaysia, introduced elsewhere as an ornamental.

Notes. Known only from 1 collection, *Dupont* 203 (K), without data, presumably planted as an ornamental.

4. LANTANA L.

Shrubs, aromatic sometimes scandent. Leaves opposite, crenate or serrate, glandular. Flowers bracteate, in axillary, pedunculate heads or short spikes. Calyx 2-lipped. Corolla irregularly salver-shaped, tube narrow below, dilated upward, pubescent without, limb irregular, 4—5-lobed, changing colour as flower develops. Stamens 4, in middle of corolla tube, in pairs. Ovary 2-celled, 1 ovule per cell. Fruit drupaceous, pyrenes united into a 2-seeded putamen.

A tropical American genus with 1 ubiquitous, ornamental species which frequently naturalizes itself as a most agressive and persistent weed.

Lantana camara L., Sp. Pl.: 627 (1753); F.M.S.: 253 (1877).

Shrub, prickly or not. Leaves opposite, triangular-ovate, 5—7 × 3—4.5 cm, apex acute or slightly acuminate, base truncate or subcordate, margins crenate-serrate; rough on both surfaces; petiole c 1 cm long. Umbellate heads axillary, in the upper axils, peduncles 3 cm long, heads of flowers enveloped by a whorl of bracts, each flower subtended by a green, leaf-like, narrowly lanceolate bract, outer flowers opening first. Calyx very short. Corolla variously coloured, cream to bluish-pink, noticeably changing colour with age, salver-shaped. Drupes bluish-lead-colour, globose, densely clustered on swollen club-shaped receptacle, individual drupes 3—4 mm across, receptacle after fruits are shed, thick, fleshy, cylindrical; seed 1, subglobose 3 mm diameter, somewhat angular. Fig. 35/5.

The Aldabra plant has prickly stems, so would fit the var. *aculeata* (L.) Moldenke, and has white flowers with orange-yellow centres, the whole flower turning bright pink in age.

Distr. ALDABRA: W; tropical America, now cultivated and naturalised throughout the tropics.

Notes. Planted at Settlement, rather chlorotic, but flowering and fruiting freely. Observations few but appear to flower throughout the year but mainly during the wet season. Fruit eaten by bulbuls.

Conditions appear to be unsuitable for the establishment of this species which can become a serious obnoxious weed, nevertheless it should be carefully watched or preferably eliminated.

Vernac. 'lantana' or 'vieille fille'.

5. PREMNA L.

Shrubs or small trees, usually aromatic. Leaves opposite, simple, entire or somewhat serrate. Flowers in terminal or axillary, corymbiform, much-branched cymes, jointed to cyme branchlets. Calyx subtruncate, 2-lipped or 4—5-toothed. Corolla 2-lipped, upper lip entire, lower lip 3-parted. Stamens 4, inserted in corolla throat. Ovary 4-celled, 1 ovule per cell; style somewhat exserted. Fruit a drupe with 1 pyrene; seeds 1—4.

A fairly large Old World genus.

Premna obtusifolia R.Br., Prodr.: 512 (1810); Renvoize in P.T.R.S. B, 260: 230 (1971).

P. integrifolia L., Mant. 2: 252 (1771); Hemsley in B.M.I.K. 1919: 128 (1919).

P. serratifolia L., Mant. 2: 253 (1771); F.M.S.: 254 (1877); Hemsley in B.M.I.K. 1919: 128 (1919).

Shrub or small tree, up to 5 m tall. Leaves opposite, oblong to somewhat obovate or ovate, 4.5—11 × 2.5—6.5 cm, apex bluntly subacute to obtuse or rounded, margins entire (elsewhere may be toothed); petiole 0.5—1 mm long. Cymes broad, much branched, terminal but sometimes becoming axillary by development of a branch from node at base of peduncle, shortly pubescent, 2, tiny, triangular bractlets at each branch; flowers fragrant. Calyx bell-shaped, c 2 mm long, very shallowly and somewhat irregularly 4-toothed, teeth obtuse. Corolla whitish to pale green, tube short, lobes 4, c 2 mm across, 1 lobe larger, apex rounded, throat with a slight or prominent white beard. Stamens short, well-exserted; anthers kidney-shaped. Style very slightly exceeding stamens. Fruit blue-black, globose c 3 mm in diameter, fleshy; seed kidney-shaped, 1.5 × 1 mm. Fig. 35/6.

Distr. ALDABRA: W,P,M,S,Esprit,Michel;ASSUMPTION;COSMOLEDO: M; ASTOVE; a widespread Indo-Pacific species that reaches the islands off the coast of East Africa but not on the mainland.

Notes. It is generally distributed in mixed scrub and *Guettarda* scrub, and in dwarfed form on coastal champignon. Flowers throughout the year. Fruit eaten by pigeons.

Vernac. 'bois sureau'.

6. STACHYTARPHETA Vahl

Herbs or shrubs. Leaves opposite. Spikes terminal, subtended by 2 branches, elongate, rhachis more or less thickened, bearing the flowers in grooves, only a few flowers opening at a time, each with an external bract. Calyx tubular. Corolla tube narrow below, curved outward from the groove at anthesis, lobes 5. Stamens 2. Ovary 2-celled, 1 ovule per cell; style persistent, slender. Fruit enclosed in calyx tube, longer than wide, splitting into 2 nutlets, enclosed by the persistent bract.

A widespread tropical weedy genus.

1. Rhachis of spike 3–4 mm thick; leaf margins with low, rather blunt teeth; flowers pale, bluish   **1. S. jamaicensis**
1. Rhachis of spike c 2 mm thick; leaf margins with prominent, acute teeth; flowers deep indigo or purplish-blue   **2. S. urticaefolia**

1. Stachytarpheta jamaicensis (L.) Vahl, Enum. Pl. 1: 206 (1804); F.M.C. 174: 22, fig. 3 (1956).

Verbena jamaicensis L., Sp. Pl.: 19 (1753).
Stachytarpheta indica sensu F.M.S.: 251 (1877); Hemsley in B.M.I.K. 1919: 128 (1919), not (L.)Vahl (1804).

Semi-decumbent, herbaceous, perennial herb, sometimes semi-woody near base, glabrous or almost so. Leaves oblong to ovate-oblong, up to 5 × 3.5 cm, apex obtuse or rounded, base obtuse or rounded, tending to be decurrent on the rather short petiole, margins crenate-serrate, dull almost greyish green (but not at all pubescent); petiole up to 4 cm long. Spike becoming 3–4 mm thick, wiry, ascending, up to 30 cm long; bracts ovate-lanceolate, sharply and strongly acuminate, base much narrower than rhachis. Corolla (in Aldabra material) light blue or somewhat purplish, trumpet-shaped, tube c 7 mm long, lobes slightly irregular, 8–10 mm across. Capsule elliptic-oblong, 3.5× 1.5 mm, enclosed by the dry calyx, finally splitting; nutlets 2, elliptic-oblong, 3 × 0.6 mm. Fig. 35/7.

Distr. ALDABRA: W, M. (east), S (west and south); ASSUMPTION; ASTOVE; pantropical weed, native of tropical America, introduced elsewhere.
Notes. Occurs mainly on coastal champignon. Flowers throughout the year but mainly during the wet season. Seeds locally dispersed by tortoises.
Vernac. 'epi-bleu queue de rat' or 'verveine bleue'.

2. Stachytarpheta urticaefolia Sims, Bot. Mag. 3: 1848 (1816).

Erect perennial herb, similar to *S. jamaicensis*, tending to be woody, branching angles rather acute, glabrous except for scattered hairs on under sides of leaf midribs. Leaves ovate, apex acute, base obtuse, decurrent on the petiole, margins prominently and sharply serrate, dark green, petioles short. Spikes slender when in flower, 2 mm thick when older, bracts ovate-lanceolate, usually half or even more the width of the spike, acute acuminate. Corolla deep blue or purple-blue.

Distr. ASTOVE; pantropical weed.
Notes. A rare weed, *Ridgway* 38b (US); *Fosberg* 49700 (K, US).

53. DICRASTYLIDACEAE

Herbs or usually shrubs. Leaves opposite or whorled, rarely alternate, entire; stipules absent. Inflorescence terminal, spicate, capitate or compound or flowers axillary. Calyx 4–8-lobed. Corolla 4–8-lobed, regular or irregular. Stamens 3–8, attached to the corolla. Ovary superior, 2-celled, ovules 2 per cell, axile placentation; style slender, shortly 2-fid. Fruit dry, rarely drupaceous, indehiscent; seeds 2, with endosperm.

A small family of 14 genera in the southern tropics of the Old World.

Close to *Verbenaceae* from which it chiefly differs in the endospermic seeds.

NESOGENES A.DC.

Herbs or dwarf shrubs; glandular. Leaves opposite, decussate. Pedicels 1–5, axillary, 1-flowered, or flowers subsessile. Calyx 5-fid. Corolla irregular, bell- or funnel-shaped, 5-lobed. Stamens 4, in 2 pairs, anthers with 2 basal lobes. Ovary 2-celled, ovules 1 per cell, erect; style 1, stigma capitate. Fruit drupaceous, 2-celled; seeds 2.

A small genus of 6 species from East Africa to the Pacific.

This genus, at present placed in *Dicrastylidaceae*, has close affinities with certain genera in *Scrophulariaceae* rather than *Verbenaceae*, where it is sometimes placed.

Nesogenes dupontii Hemsley in J. Linn. Soc. Bot. 41: 314, fig. 14 (1913) & in B.M.I.K. 1919: 127 (1919); Renvoize in P.T.R.S. B, 260: 231 (1971). Types: Aldabra, *Dupont* 97; *Thomasset* 261; Assumption, *Dupont* 106 (all K, syn.).

Nesogenes sp. nov., Hemsley in J. Linn. Soc. Bot. 41: 315 (1913) & in B.M.I.K. 1919: 127 (1919).

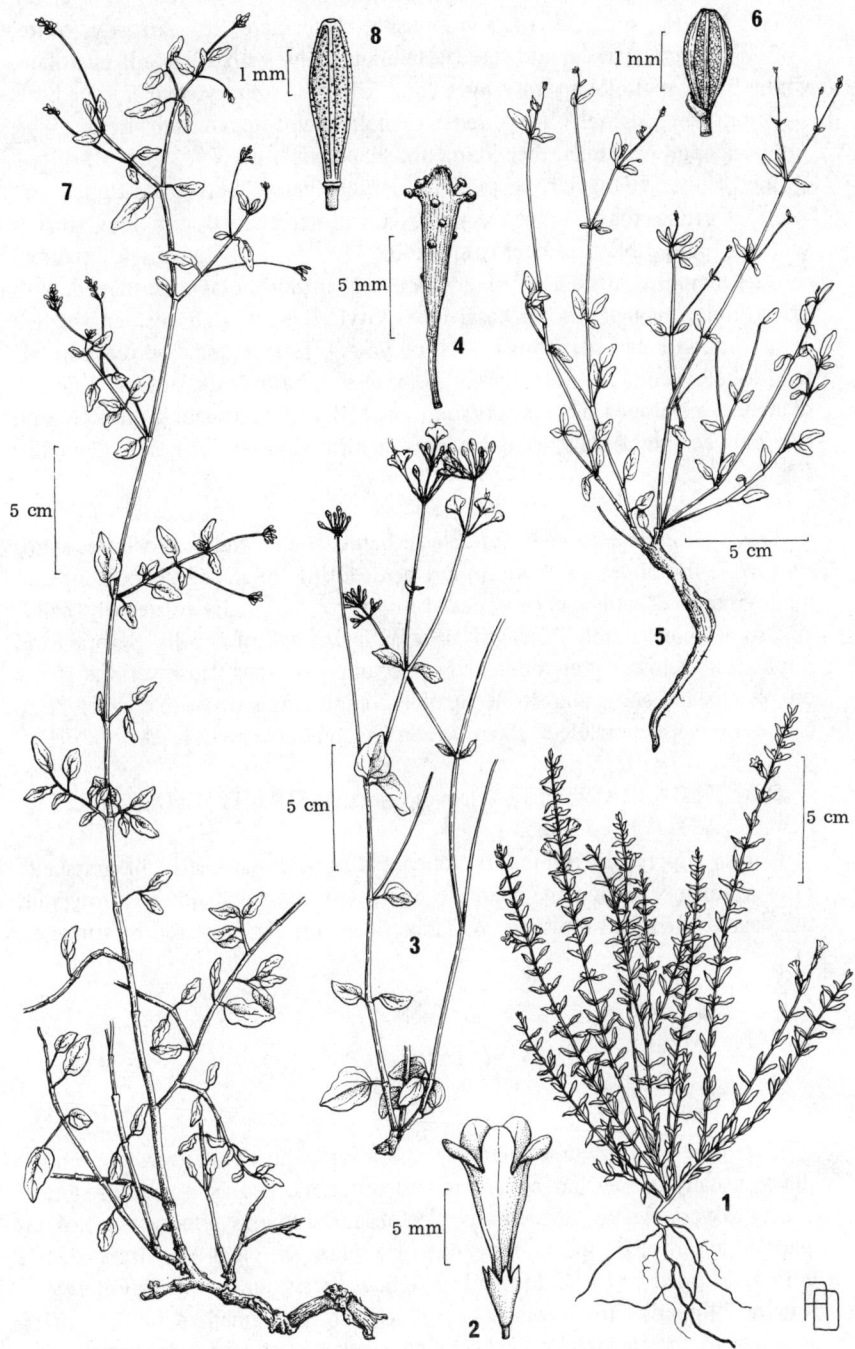

Fig. 36 1, *Nesogenes dupontii*, habit; 2, flower. 3, *Boerhavia africana*, habit; 4, fruit. 5, *Boerhavia crispifolia*, habit; 6, fruit. 7, *Boerhavia repens* var. *maris-indici*, habit; 8, fruit.

Prostrate, decumbent, or ascending herbs; glandular, densely to sparsely hairy. Leaves opposite or subopposite, lanceolate to narrowly ovate, 1–2 × 0.4 cm, apex acute, base attenuate, hairs with enlarged, pustulate, white bases; petiole up to 4 mm long. Flowers solitary, axillary, pedicels 1–2 mm long, densely hairy, most of hairs gland-tipped and viscid. Calyx funnel-shaped or tubular, becoming urn-shaped in fruit, 3 × 2 mm, 10-ribbed, 5-lobed, lobes triangular, very acute, densely hairy, hairs gland-tipped and viscid. Corolla tube mauve, slender, funnel-shaped, 10 mm long, throat swollen, lobes orbicular, erect to spreading, overlapping. Stamens 4, 2 pairs of different lengths, filaments strongly incurved, anthers horse-shoe-shaped, with or without small points on basal lobes. Style slender, glabrous, lengthening somewhat with age, persisting at least on young fruit; stigma exserted, capitate to 2-lobed. Fruit an oval, slightly compressed, hard drupe or nut, 2 mm in diameter, enveloped by the persistent calyx, 2-celled, a slight groove on each side between the 2 cells; seeds 1, yellow-brown, globose, 1.5 mm in diameter. Fig. 36/1–2.

This species is extremely variable in habit, from slender, prostrate, stems rooting at the nodes to prostrate but not rooting, to strongly ascending and much-branched; pubescence varies from sparse stiff hairs to densely hairy, also somewhat variable in length of calyx and shape of corolla. Several local populations show a preponderance of certain variations. However, the plants on West Island show almost all possible combinations of these characters, so it does not seem possible to separate convincingly at varietal level.

Distr. ALDABRA: W, P, M, S, lagoon islets; ASSUMPTION; COSMOLEDO: N.W. Is.; ASTOVE; endemic.

Notes. Widely distributed on open platin and pavé, also in grassland. Flowers mainly during the mid-late wet season, but will flower throughout the year if sufficient moisture available. Seeds locally dispersed by tortoises.

54. LABIATAE

Mostly herbs, sometimes creeping, more rarely shrubby, usually aromatic; stems usually somewhat square in cross-section. Leaves opposite, simple, entire to very deeply divided; stipules absent. Flowers usually bisexual, in whorled racemes or spikes, more rarely axillary, in pairs or solitary. Sepals united, 2-lipped or (4–)5-toothed or -lobed. Petals united, strongly irregular, usually 2-lipped, with the lower lip much larger. Stamens 4 or 2, in pairs, attached to corolla tube. Ovary 4-celled, ovules 1 per cell; style short, stigma minute. Fruit usually 4 nutlets.

A large family, well represented in dry regions, temperate or tropical; many species are of economic importance for flavouring or medicinal use.

1. Robust, annual herbs; flowers brilliant orange, borne in dense clusters at intervals along the flowering stems **1. Leonotis**
1. Stout, perennial, shrubby herbs; flowers white, borne in numerous whorls evenly distributed along the flowering stems **2. Ocimum**

1. LEONOTIS (Pers.) R.Br.

Robust herbs. Leaves opposite and decussate. Flowers in dense whorls clustered at intervals on a long axis. Calyx 8–10-toothed. Corolla tube longer than the calyx, 2-lipped, the lower lip 3-lobed and deflexed. Stamens 4.

A genus of c 40 species, mostly tropical and southern Africa.

Leonotis nepetifolia (L.) Ait. f. in Ait., Hort. Kew ed. 2, 3: 409 (1811); F.M.S.: 261 (1877).

Phlomis nepetaefolia L., Sp. Pl.: 586 (1753).

Robust, erect annuals, 1–3 m tall, stems pubescent, often deeply grooved on each side. Leaves ovate, 4–11 × 2–8 cm, apex acute, base tapering, margins strongly toothed, sparsely to densely pubescent; petiole up to 10 cm long. Flowering spikes terminal, up to 1 m or more long, bearing dense globose clusters of flowers, up to 6 cm in diameter at intervals of up to 30 cm; flowers subtended by stiff, bristle-like, spiny bracts up to 2 cm long; bracts occasionally leaf-like, lanceolate, up to 8 cm long, with petiole of equal length. Calyx 2-lipped, upper lip with 1 spine, lower lip with 5 spines, persistent, sparsely to densely hairy. Corolla tube brilliant orange, up to 2.5 cm long, pubescent. Calyx persistent in fruit, up to 2 cm long, light brown; nutlets 4, light brown with purple-brown mottles, trigonal, obovate, 4 × 2 mm. Fig. 33/7.

Distr. ALDABRA: W; ASSUMPTION; pantropical.
Notes. Associated with areas of habitation, possibly introduced as an ornamental.
Vernac. 'dacca' or 'monte au ciel'.

2. OCIMUM L.

Herbs or shrubs. Leaves opposite. Flowers in numerous whorls on a long axis. Calyx 5-lobed. Corolla tube scarcely longer than the calyx, 2-lipped, lower lip 4-lobed, seldom larger than the upper. Stamens 4.

A genus of c 150 species, in the tropical and warm temperate parts of the world.

1. Leaves coarsely toothed; fruiting calyx pouched at the base
                                                    3. O. gratissimum
1. Leaves entire or shallowly toothed; fruiting calyx scarcely pouched at the base
   2. Stems glabrous or sparsely pubescent
      3. Corolla at least twice as long as the calyx   **1. O. basilicum**
      3. Corolla scarcely exceeding the calyx   **2. O. canum**
   2. Stems with stiff hairs   **4. O. sanctum**

All 4 species have been introduced as condiment or medicants. They are rather similar in appearance apart from the distinguishing characters indicated in the key, consequently only 1 species is described.

1. Ocimum basilicum L., Sp. Pl.: 597 (1753).

Distr. ALDABRA: W; pantropical.
Notes. Introduced as a condiment at Settlement; 1 record, *Fosberg* 49527 (US).
Vernac. 'basilic de France' or 'toc maria'.

2. Ocimum canum Sims, Bot. Mag. 51, fig. 2452 (1824); F.M.S.: 258 (1877).

Distr. ASTOVE; pantropical.
Notes. Introduced as a condiment; 1 record, *Ridgway* 43 (US).
Vernac. 'basilic'.

3. Ocimum gratissimum L., Sp. Pl: 1197 (1753); F.M.S.: 258 (1877); Hemsley in B.M.I.K. 1919: 128 (1919).

Aromatic perennial herbs or shrubs, up to 2 m high. Leaves ovate or elliptic, (4–)8–14 × 3–6 cm, apex acute, base tapering, margins toothed; sparsely pubescent; petiole 2–5 cm long. Infloresence of terminal and axillary spikes, 6–12(–20) cm long. Flowers opposite. Calyx pouched, persistent, up to 4 mm long, upper lobe broad, rounded, reflexed, lateral lobes reduced to small, acuminate teeth, 2 lower lobes fused except for their acuminate tips. Corolla white, as long as the calyx, lips subequal, Stamens 4. Calyx persistent and expanded in fruit; nutlets 4, spherical, 1.5 mm in diameter. Fig. 33/8.

Distr. ALDABRA: W; pantropical.
Notes. Introduced as a condiment, 1 record, *Dupont* 219 (K).
Vernac. 'basilic grande feuille'.

4. Ocimum sanctum L. Mant. 1: 85 (1767).

Distr. ALDABRA: W; pantropical.

Notes. Introduced as a condiment, 1 record, *Fosberg* 49465 (US).
Vernac. 'basilic petite feuille'.

55. NYCTAGINACEAE

Herbs, shrubs, woody scramblers, or trees, nodes often enlarged. Leaves
usually opposite, simple ; stipules absent. Flowers usually regular or slightly
irregular, bisexual or in some genera unisexual, often enveloped by a whorl of
bracts. Perianth of 1 series, petaloid, united. Stamens 1-many, hypogynous,
free or united, often unequal. Ovary superior, 1-celled, ovules basal; style 1,
stigma 1. Fruit an achene enclosed in an anthocarp (persistent perianth base).

A small tropical and subtropical family, predominantly New World, with
1 important ornamental genus, *Bougainvillea*, which has not yet been
recorded from the Aldabra group of islands.

1. Trees without spines or climbing shrubs with spines   **3. Pisonia**
1. Plants herbaceous
  2. Stout, erect herbs; flowers showy, subtended by a whorl of united bracts
           2. Mirabilis
  2. Slender, prostrate or ascending herbs; flowers small, subtended by single
    bracts   **1. Boerhavia**

1. BOERHAVIA L.

Herbs, rarely somewhat woody or somewhat scandent; nodes prominent.
Leaves opposite. Flowers in axillary cymes or terminal, cymose panicles.
Perianth (perigone) single, constricted above the hypanthium, upper part
funnel- to bell-shaped. Stamens 2-several. Ovary superior, closely invested by
the hypanthium; style unbranched. Fruit (anthocarp) club-shaped or ellipsoid,
usually (3–), 5– or 10-furrowed, furrows between ridges and even the ridges
frequently glandular-viscous.

A pantropical genus of many species, often distinguished with difficulty.

1. Elongate, semi-scandent; flowers with white funnel-shaped perianth limb;
  anthocarp with large, gummy glands, ribs 10   **1. B. africana**
1. Stems ascending or prostrate, not at all vine-like; flowers with pink, bell-
  shaped, perianth limb; anthocarp with tiny, sessile or stalked glands or
  none, ribs 5
  2. Peduncles axillary, distal ones not noticeably paniculate   **3. B. repens**
  2. Flowers in terminal, leafless, diffuse panicles   **2. B. crispifolia**

1. Boerhavia africana Lour., Fl. Cochinch. 1: 16 (1790).

Boerhavia plumbaginea Cav., Ic. Pl. 2: 7, fig. 112 (1793).
Commicarpus plumbagineus (Cav.) Standl. in Contr. U.S. Nat. Herb. 18: 101 (1916).
Commicarpus africanus (Lour.) Cuf. in Bull Jard. Bot. Brux. 23, Suppl.: 79 (1953).

Elongate, procumbent, even scandent herb; stems somewhat glaucous, finely striated, tending to woodiness at base, internodes rather elongated, nodes not conspicuously enlarged. Leaves subcordate to cordate, apex obtuse to acute, with a short projecting tip, margins somewhat irregular; petioles much shorter than blades. Peduncles terminal and in upper axils, bearing 1 or 2 whorls of flowers on short pedicels. Flowers c 6–7 mm long, lower third of perianth club-shaped, abruptly contracted above, upper two-thirds, white, very slender below, dilated upward, then flaring, somewhat irregular. Stamens 3, exserted. Stigma conspicuously exserted. Anthocarps angled or deflexed at top of pedicels, subcylindrical, 3.5 × 0.7 mm, tending to be somewhat irregular, weakly 10-ribbed, with large, sticky glands around the uppermost part. Fig. 36/3–4.

Very distinct from the other species in our flora; with its several relatives segregated by some as the genus *Commicarpus* Standl.

Distr. ASSUMPTION; tropical and southern Africa and Asia.
Notes. Occurs on the southern coastal sand plateau; 1 collection, *Frazier* 801 (K, US). Possibly introduced by man, or more likely by sea-brids.

2. Boerhavia crispifolia Fosberg in Smithsonian Contr. Bot. 39: 15 (1978). Type: Aldabra, South Island, near Cinq Cases Camp, *Fosberg* 48873 (US, holo., K. L, Mo, iso.).

B. elegans sensu Hemsley in B.M.I.K. 1919: 128 (1919), not Choisy (1849).

Tap-root thick, to 8(–12) mm, branched, rarely nearly simple; stems several, from slightly thickened root crown, prostrate to somewhat ascending, branched, especially near base, minutely pubescent, greyish to white below, purplish distally. Leaves ovate to oblong, mostly less than 10(–20) × 5(–10) mm, apex obtuse to acute, margins usually strongly undulate or crisped, upper surface usually purplish- to brownish-green, beneath pale, blades containing abundant cystoliths or crystals and a few rhaphide-bundles; petiole 2–6 mm long, rarely more. Axillary cymes absent; panicle slender, 10–30 cm long, minutely pubescent, main axis with branches on alternate sides at alternate nodes in lower part, 2 leaf-like bracts at each lower node, distal nodes tending more and more to develop 2 branches, bracts on distal nodes reduced to ovate, ciliate scales; flowers in glomerules of 3–6, sessile, subtended by ovate-lanceolate, glandular-ciliate, scale-like bractlets, central flower opening first. Flowers pink, c 2.5 mm long, very strongly constricted, part above constriction bell-shaped to cylindrical, 1.5 mm long, soon falling, minutely pubescent, slightly 4—5-lobed, lobes indented. Stamens 2—3(—4), filaments strongly curved, anthers broader than long, each half almost orbicular. Style subequal with stamens, curved in upper part, 2.5—3 mm long, stigma strongly depressed-globose. Anthocarp narrowly ellipsoid to subclub-shaped, c 2 × 0.7 mm, ends rounded to subtruncate, ribs 5, broad, smooth, intervals glandular to almost or quite glabrous. Fig. 36/5—6.

Most parts of the plant are marked by white, short-linear or elliptic rhaphide-bundles; especially conspicuous in the flowering perianth and in the intervals between the ribs of the anthocarp.

This species is notably variable in habit, especially in the degree to which the inflorescence is developed, and in leaf size and shape. Fairly uniform in floral and fruit morphology, varying only in distribution and amount of pubescence on the flowering perianth, slightly in fruit shape, and considerably from glabrous to densely but shortly minutely glandular pubescent in the intervals between the ribs of the anthocarp.

An abnormal form is occasionally found with the whole plant reduced, often to a small, prostrate, closely branched mat, with sterile flowers not strongly constricted in the middle as in the normal plant, rather minutely 'vase-shaped' in appearance. No fruits are produced. Intermediate plants are found with both normal and abnormal flowers. The cause of the abnormality is not at all known; it is possible that a fungus may be involved. It has been found on Aldabra near Cinq Cases *Fosberg* 48929 (US), *Renvoize* 769 & 981 (both K, US) and 'Takamaka turn-off', *Renvoize* 869 (K, US).

Distr. ALDABRA: W, M, S; endemic.

3. Boerhavia repens L., Sp. Pl.: 3 (1753).

Herbaceous plant with thickened root and several prostrate stems radiating from the root crown, sparingly to much branched. Leaves mostly ovate to oblong, apex obtuse to acute, base rounded to truncate to subcordate, usually green above, pale or even white beneath; petiole short. Flowers borne in axillary pedunculate cymes or pseudo-umbels, usually from alternate axils, leaves often reduced distally, creating a paniculate appearance, but with a clear central axis; peduncle branched only near tips, branches umbellately or subumbellately branched, or reduced to a glomerate cluster of flowers. Perianth strongly constricted below the middle, lower part glandular, ribs 5, upper part petaloid, pink, bell-shaped, shortly 5-lobed. Stamens 2—4, subequal with perianth. Style also subequal. Anthocarp club-shaped to ellipsoid, rounded at apex, 5-ribbed, very glandular.

A widespread and rather variable Indo-Pacific and African species, whose varieties have not been clearly distinguished.

var. **maris-indici** Fosberg ex Fosberg in K.B. 33: (1978). Type: Astove, Grande Anse, *Fosberg* 49677 (US, holo., K, iso.).

Boerhavia diffusa sensu Baker in B.M.I.K. 1894: 150 (1894); Heimerl in
    A.S.N.G. 21: 83 (1897); Voeltzkow in A.S.N.G. 26: 550 (1902); Dupont,
    Report: 39 (1907); Hemsley in B.M.I.K. 1919: 128 (1919), not L (1753)
B. diffusa L. forma *psammophila* Heimerl in A.S.N.G. 21: 83 (1897);
    Voeltzkow in A.S.N.G. 26: 550 (1902). Type: Aldabra, *Voeltzkow* 34
    (FR, holo.)
B. repens L. var. *maris-indici* Fosberg in Smithsonian Contr. Bot. 39: 9
    (1978), nom. inval., typus non cit.

Root slender, spindle-shaped; stems very prostrate, minutely pubescent,
internodes mostly 1–2 cm long except toward base where they may be up to
4–5 cm, striate. Leaves small, 1–1.5(–2) × 0.7–1.0(–1.5) cm, becoming
smaller distally, apex obtuse, base truncate to subcordate, margins somewhat
wavy, green, minutely glandular-pubescent above, pale beneath, petiole
slender, up to 1 cm long. Peduncles very slender, 2–3(–4) cm long, branched
above, minutely pubescent, bracts lanceolate. Perianth with 5 ribs, smooth
or somewhat glandular, intervals glandular, constriction extreme, limb whitish
to purple, bell-shaped, 1.5–2 × as long as lower part, minutely pubescent.
Stamens 4(–5?). Style subexserted; stigma discoid, peltate. Anthocarp
ellipsoid to somewhat club-shaped, 2.5–3 × 0.7 mm, shortly stalked.
Fig. 36/7–8.

Distr. ALDABRA: Mentor Is., islet in Gionnet Channel; COSMOLEDO;
ASTOVE, and other western Indian Ocean islands.

2. MIRABILIS L.

Herbs (rarely subshrubs), usually perennial. Leaves opposite, simple, entire
or nearly so. Inflorescence cymose, axillary or usually terminal, open or
congested, ultimate branches bearing whorls of 4–5 united bracts; flowers
1–10. Perianth corolla-like, usually somewhat constricted above the ovary,
limb funnel- or bell-shaped to saucer-shaped. Stamens 3–5, filaments free
or united at base, anthers in pairs of different length, usually exserted. Ovary
ellipsoid or globose, closely invested by lower part of the perianth; style
slender, stigma capitate, exserted. Anthocarp globose or obovate, adherent
to testa of the single seed, surface smooth to variously rugose, round to
slightly 5-angled.

A genus of 30 or more species from tropical to warm-temperate America;
1 species widely introduced and now pantropical.

Mirabilis jalapa L., Sp. Pl.: 177 (1753); F.M.S.: 262 (1877).

Erect herb to 50 cm; root thick; stems forked. Leaves opposite, triangular-ovate to broadly triangular-subcordate, up to 10 × 6 cm, apex acute to acuminate, glabrous to minutely pubescent; petiole up to 4.5 cm. Peduncles terminal, short, cymes congested or glomerate, with many leafy bracts; involucres bell-shaped, deeply 5-lobed; flowers 1. Perianth 3–5 cm long, tube gradually dilated upward to a broad limb 2–3.5 cm wide, shallowly 5-lobed, variously coloured, red, yellow, white or variegated, in naturalised forms often deep purplish-red, collapsing in the afternoon. Stamens 5. Anthocarp dark brown to black, obovoid to subglobose, 10 × 5 mm, prominently wrinkled.

Distr. ALDABRA: W; native of tropical America, now widely naturalised and pantropical.

Notes. Planted as a medicinal and ornamental plant at Settlement; 1 collection, *Hnatiuk* 731517 (US).

Vernac. 'belle de nuit' or 'four o'clock'.

3. PISONIA L.

Trees, shrubs or scrambling, spiny lianas. Leaves opposite, simple. Flowers in alternately branched cymes, whitish, not all showy, without involucres, 1 minute bractlet at base of perianth, dioecious. Male flowers: perianth obconical-bell-shaped; stamens 6–8, rarely more, exserted, anthers in unequal pairs. Female flowers: perianth tubular, not constricted; ovary ovoid to oblong, sessile, style short, stigma slightly exserted, minutely pubescent. Anthocarp spindle- to club-shaped, 10 × 3 mm, smooth, 5-angled or with 5, spiny or glandular keels, frequently sticky.

Pantropical genus of c 50 species.

1. Climbing shrub, with hooked spines   **1. P. aculeata**
1. Tree, without spines   **2. P. grandis**

1. Pisonia aculeata L., Sp. Pl.: 1026 (1753); F.M.S.: 263 (1877).

Climbing shrub, up to 10 m or more high; stems slender, armed with hooked, axillary spines. Leaves opposite, oblong, 4–12 × 3–6 cm, apex acute, leathery, yellow-green, glabrous or pubescent; petiole up to 2 cm long. Male flowers: sessile or pedicels very short, arranged in dense axillary corymbs; perianth green, pubescent, 2.5–4 mm long; stamens 6–8, exserted. Female flowers: panicles lax, pedicels long; perianth green, pubescent, 2–3 mm long. Anthocarp narrowly oblong, 10–12 × 5 mm, ribs 5, prominently glandular. Fig. 37/1–3.

Sterile specimens can be confused with *Scutia myrtina* (Family 21.3) which has smaller leaves which are usually rounded or indented at the apex, dark green and shiny above, and shorter petioles.

Distr. ALDABRA: S (east); widespread throughout the tropics.

Notes. Only known from 2 sterile collections; *Gibson* 5 & 9 (both K) in mixed scrub at Basin Frigate.

2. Pisonia grandis R.Br., Prodr.: 422 (1810); Vesey-Fitzgerald in J. Ecol. 30: 12, 15 (1942); Fosberg in P.T.R.S. B, 260: 216, 225 (1971); Renvoize in P.T.R.S. B, 260: 230, 235 (1971).

P. macrophylla (Boj.) Choisy in DC., Prodr. 13 (2): 446 (1849); F.M.S.: 263 (1877); Dupont, Report: 39 (1907).

P. costata sensu Hemsley in B.M.I.K. 1919: 128 (1919), not (Boj.) Choisy (1849).

Tree or shrub to 8 m high; trunk thick, pale, very soft; bark pale grey or brown. Leaves opposite, sometimes subopposite or even alternate on same tree, elliptic to oblong or broadly ovate, 9–30 × 6–18 cm, apex usually acuminate, base acute to rounded, truncate, or even subcordate, glabrous or with some hairs beneath when young, especially along midrib and in domatia in vein axils; petiole 0.5–4 cm long. Cymes up to 8 cm long; pedicels short; flowers frequent. Perianth whitish, funnel-shaped, c 5 mm long, furrows 5, lobes 5, short ovate. Male flowers: perianth margin somewhat flaring. Female flowers: perianth contracted around style, ridges strongly glandular with stalked glands. Anthocarp club-shaped, 10 × 3 mm, stalked, strongly 5-ridged, ridges with short, spine-like, stalked very sticky glands, fruiting cymes expanding, open, to 10–15 cm in diameter, falling after fruit Fig. 37/4–7.

Dist. ALDABRA: W, P, M, S; ASSUMPTION; COSMOLEDO: W, M; ASTOVE; widely distributed on the islands of the Indian and Pacific Oceans.

Notes. Found on small rocky lagoon islets and on coastal dunes, uncommon, rather scattered; a small grove of large trees on Aldabra at Point Hodoul may possibly have been planted, similarly at Dune Patatas. Observations few, but appears to flower during the dry season. Fruit adhere to plumage of birds.

Vernac. 'mapou' or 'bois mapou'.

56. AMARANTHACEAE

Annual or perennial herbs or shrubs, rarely trees or climbers. Leaves alternate or opposite, simple, usually entire; stipules usually absent. Flowers in heads, spikes, racemes or panicles, rarely solitary, subtended by membranaceous or spinose bracts and bracteoles. Perianth of 3–5 sepals,

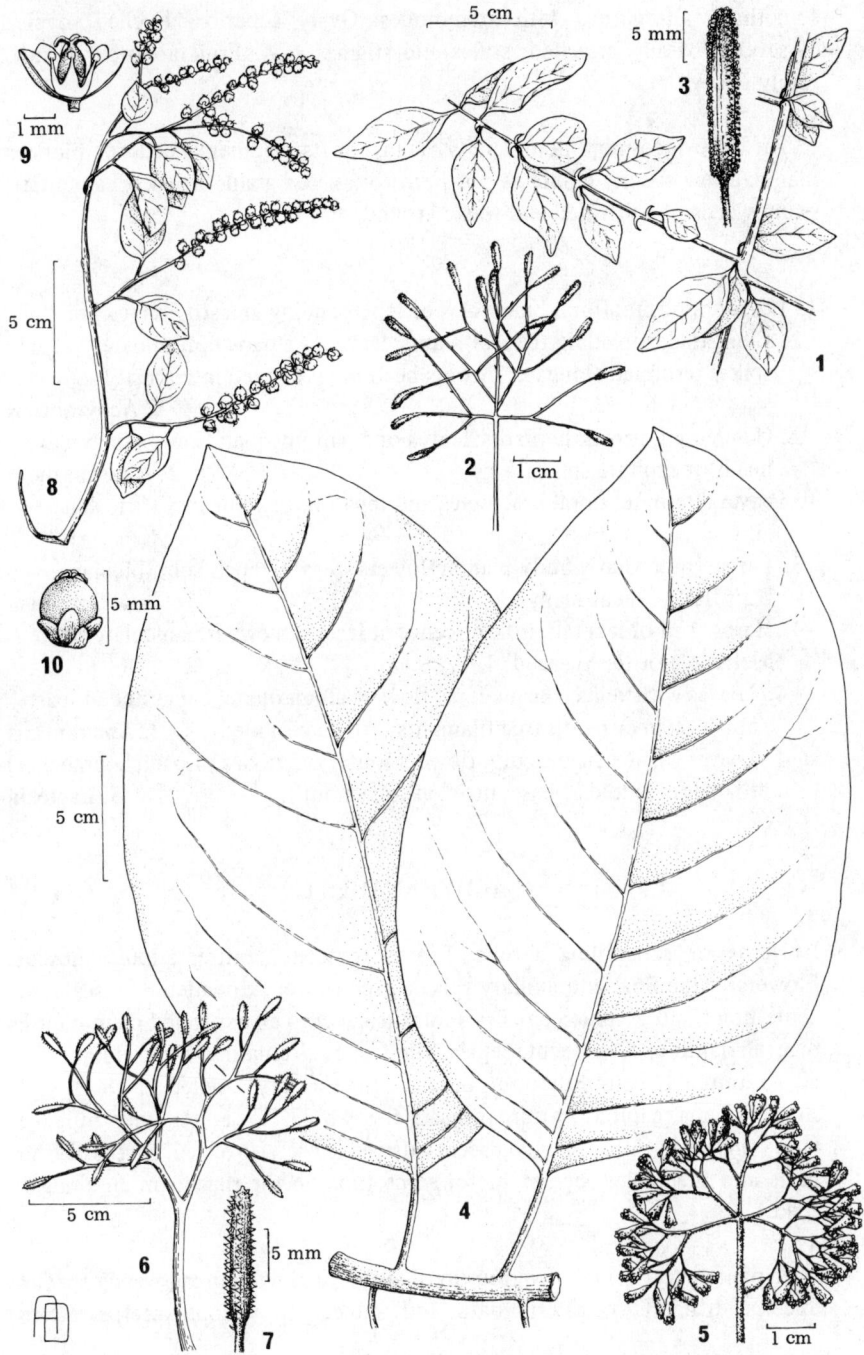

Fig. 37   1, *Pisonia aculeata*, habit; 2, fruiting inflorescence; 3, fruit.
4, *Pisonia grandis*, leaves; 5, flowering inflorescence; 6, fruiting
inflorescence; 7, fruit. 8, *Deeringia polysperma*, habit; 9, flower;
10, fruit.

usually membranaceous. Stamens opposite sepals, usually variously united, sometimes alternating with staminodes. Ovary superior, 1-celled, ovules 1—several, basally attached; styles and stigmas 1—4. Fruit a utricle or nut, rarely fleshy.

An almost cosmopolitan family, many of its members weedy or pioneer plants; a few species edible as pot herbs or as seed grains, a few ornamentals, many weeds of cultivated and waste ground.

1. Leaves opposite; floral bracteoles or tepals ending in a stiff, sharp point
    2. Elongated, spreading to erect plants; leaves of a pair approximately equal; spikes terminal, elongate; flowers becoming reflexed in maturity
                                                                1. Achyranthes
    2. Condensed, prostrate plants; leaves of a pair unequal; flowers in axillary heads or capitate spikes   **2. Alternanthera**
1. Leaves alternate; floral bracteoles and tepals not ending in a stiff, sharp point
    3. Large, somewhat woody plants; flowers spicate; tepals suborbicular; fruit fleshy, seeds many   **4. Deeringia**
    3. Herbs, 1 m or less tall; inflorescence at least somewhat paniculate; fruit a utricle, not fleshy, seeds 1
        4. Leaves with veins conspicuous; flowers glomerulate, congested at least above; stamens with free filaments; utricle wrinkled   **3. Amaranthus**
        4. Leaves with veins obscure; flowers solitary at nodes of panicle branches; filaments united at base; utricle thin, smooth   **5. Lagrezia**

1. ACHYRANTHES L.

Herbs or scrambling shrubs. Leaves opposite, entire; stipules absent. Flowers in terminal and axillary spikes, each flower subtended by 1 scale-like acute bract and 2 spinescent bracteoles, each with an expanded membranous base. Perianth of 5 very acute sepals. Stamens 5, alternating with 5 staminodes, all 10 united at base into a cup or short tube. Ovary 1-celled, ovule 1; style short, stigma capitate; after anthesis the whole flower strongly reflexed. Ovary maturing into an indehiscent utricle, closely invested by the flower-parts and bracteoles, persistent for some time on the rhachis in an elongate spike.

A pantropical genus of various local and several widespread weedy species. Species difficult to discriminate and subject to various interpretations.

Achyranthes aspera L., Sp. Pl.: 204 (1753); F.M.S.: 268 (1877).

Stiff, sparsely branching herbs, sometimes becoming somewhat woody below (var. *fruticosa*), stems 4-angled. Leaves appressed-pubescent above and beneath. Spikes terminal, bearing buds, flowers, and fruits at the same time, fruiting portion of spike stiffish; bract acuminate, persistent, bracteoles 2, at least half as long as perianth, persistent. Perianth purple, after anthesis closing tightly around fruit, fruiting perianth segments with stiff, sharp point. Fruit, plus perianth and bracteoles strongly reflexed, falling or readily detached when mature, clinging to clothing, hair or feathers.

The species contains several widespread varieties, of which 2 appear to be native in the Aldabra group of islands; neither var. *aspera* nor var. *pubescens* (Moq.) C.C. Townsend, both pantropical weeds, are known from the Aldabra group.

1. Stems reddish, woody at the base, semiclimbing   **a. var. fruticosa**
1. Stems grey, not woody, erect   **b. var. velutina**

a. var. fruticosa (Lam.) Boerlage in Ned. Kruidk. Arch. Ser. II, 5: 423, fig. 6/2 (1889).*

A. fruticosa Lam., Encycl. Méth. Bot. 1: 545 (1783).
A. aspera sensu Baker in B.M.I.K. 1894: 150 (1894); Schinz in A.S.N.G. 21: 83 (1897); Voeltzkow in A.S.N.G. 26: 549 (1902); Dupont, Report: 40 (1907); Hemsley in B.M.I.K. 1919: 129 (1919).
Achyranthes sp., Renvoize in P.T.R.S. B, 260: 231 (1971).

Herbs or shrubs, stems usually reddish or reddish brown, up to 5 m long, erect when young, later subscandent or reclining, usually woody below, densely pubescent. Leaves narrowly elliptic or elliptic-obovate, rarely broadly obovate, 7 × (1.5) 2 × 3 cm, apex acute to acuminate, base wedge-shaped, appressed-pubescent above and beneath, more densely beneath; petiole short, 0.5–1.5 cm long. Spikes mostly solitary, terminal, occasionally somewhat paniculate, elongating to 30–35(–45) cm, rachis green, thinly to notably appressed-pubescent, flowering from the base upwards. Mature fruit reflexed, with associated perianth and bracteoles, 4–5 mm long. Fig. 38/1–2.

Distr. ALDABRA: W, P, M, S (west), Michel, Mentor; ASSUMPTION; COSMOLEDO: W, NW; ASTOVE; pantropical.
Notes. Usually local and scattered, rarely abundant in most of the plant communities. Flowers mainly during the wet season, but throughout the year if sufficient moisture is available.
Vernac. 'herbe sergent'.

b. var. velutina (Hook. & Arn.) C.C. Townsend in K.B. 29: 473 (1974).

A. velutina Hook. & Arn., Bot. Beechey Voy.: 68 (1832).

---

* Some botanists, including C.C. Townsend, consider that var. *fruticosa* cannot be satisfactorily separated from var. *pubescens*.

An erect or spreading sparsely-branched herb, not known to clamber, or to reach more than 1 m or so in length, stems and leaves densely grey pubescent; rachis of the spike densely woolly. Fig. 38/3.

Distr. ALDABRA; COSMOLEDO: W, M; ASTOVE; islands of the Indo-Pacific.

Notes. The Aldabra citation is based on Townsend loc. cit. but no specimens seen. Few collections, Cosmoledo: *Vesey-Fitzgerald* 5990 (K) and *Thomasset* s.n. (K); Astove: *Veevers-Carter* 41 (EA) and *Fosberg* 49707 (K, US).

2. ALTERNANTHERA Forssk.

Herbs of various habit, sometimes woody at base. Leaves opposite, entire, pairs equal or often unequal; stipules absent. Flowers in axillary or terminal heads, sessile or pedunculate; flowers bisexual, each subtended by 1 bract and 2 bracteoles. Perianth parts 5, unequal. Stamens 3 or 5, alternating with (pseudo-) staminodes, all united to form a short cup; anthers 1-celled. Ovary 1-celled, ovules 1; style 1, short or absent, stigma capitate. Fruit a flattened utricle; seeds 1, smooth, somewhat compressed.

A genus of more than 150 species, mainly in tropical America.

Alternanthera pungens Kunth, Nov. Gen. et Sp. 2: 206 (1818).

Prostrate herb; stems slender, radiating from a strong root crown, more or less hairy (Assumption plant prominently white-hoary). Leaves opposite, 1 of a pair larger, oval to obovate or suborbicular, up to $1-1.5 \times 0.5-1$ cm, apex obtuse to rounded, base narrowed; petiole short, appressed hairy. Flowers in sessile, axillary, short-cylindrical, pale straw-coloured heads, prickly to touch; bract a thin, membranaceous, nerveless, ovate scale, c 2 mm long; bracteoles 2, lanceolate, 3 mm long, apex sharply spine-tipped, translucent, opposite each other in 1 plane on adaxial side. Adaxial tepal oblong, translucent, c 3 mm long, apex rounded, jagged with a short spine-tip, abaxial tepals tough, ovate-lanceolate, 5 mm long, apex acuminate, prolonged into a rigid spine, lateral tepals ovate, c 4 mm long, weakly spine-tipped or scarcely so. Staminal cup very short, anthers minute. Ovary globose; stigma sessile. Seeds 1, retained in the corky utricle which falls with the bracteoles attached. Fig. 38/4.

Distr. ASSUMPTION; a weed of tropical American origin, now pantropical.
Notes. Found at main settlement, sea level, *Frazier* 813 (US) and *Blackmore* 77418 (K).
Vernac. 'brède emballage'.

3. AMARANTHUS L.

Annual or perennial herbs. Leaves alternate; stipules present or absent,

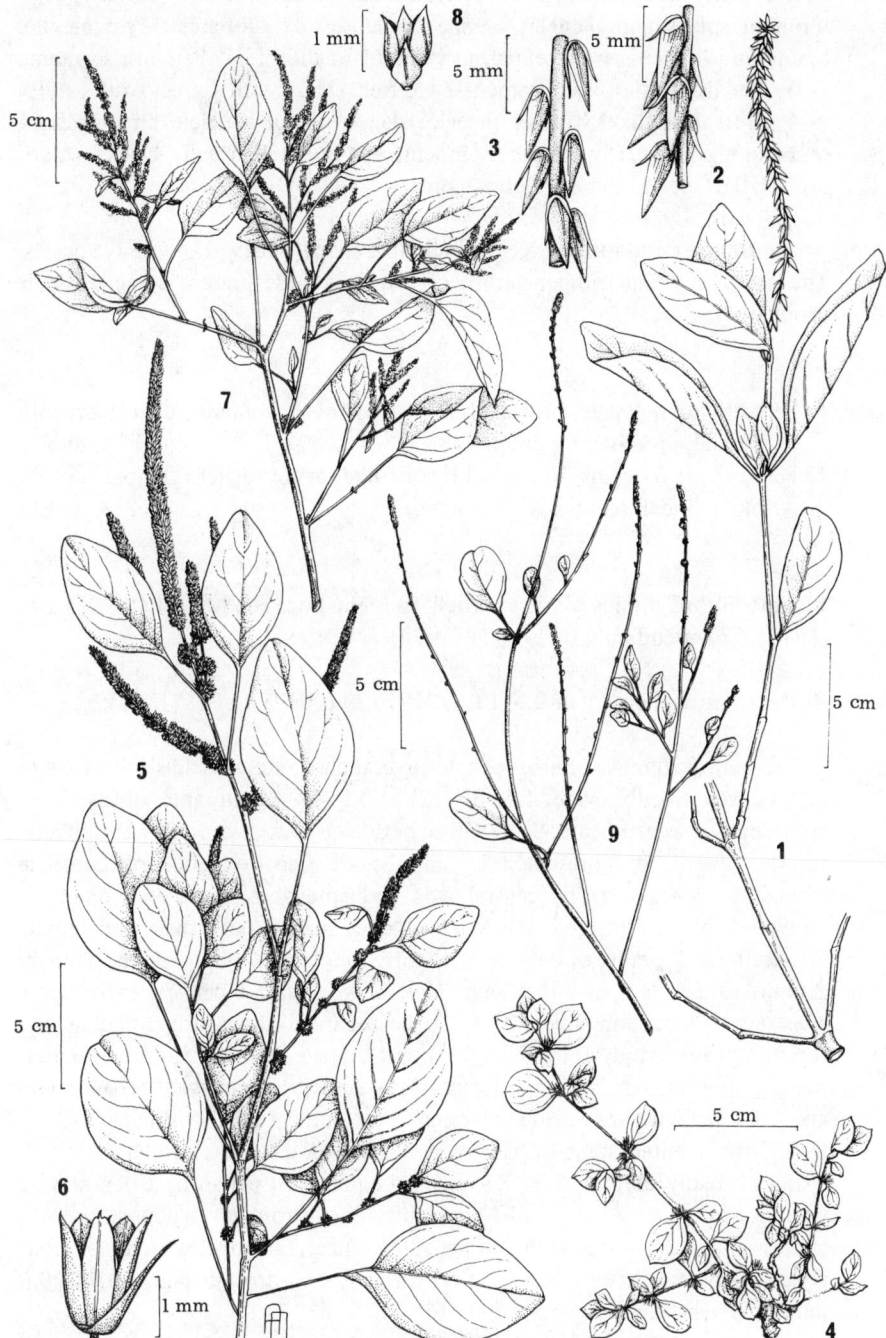

Fig. 38 1, *Achyranthes aspera* var. *fruticosa*, habit; 2, detail of flowers.
3, *Achyranthes aspera* var. *velutina*, detail of flowers.
4, *Alternanthera pungens*, habit. 5, *Amaranthus dubius*, habit;
6, female flower with bract. 7, *Amaranthus viridis*, habit; 8, flower.
9, *Lagrezia oligomeroides*, habit.

ovate, bristly or spine-tipped. Flowers usually in small, axillary clusters, or in terminal spikes and panicles, sessile or subsessile, subtended by scale-like bracts and bracteoles; unisexual, monocious or dioecious. Perianth segments 1–5, greenish to purplish. Stamens 1–5, free. Ovary with 1 erect ovule; styles 2–4, short, beak-like, stigmas linear or slender. Fruit a utricle, circumscissile or breaking irregularly, or rarely indehiscent; seeds 1, usually black or dark brown, flattened to almost globose, erect.

An almost cosmopolitan genus including many widespread weedy species. The taxonomy and nomenclature of some of these present some difficult problems.

1. Perianth of 5 tepals; bracts prominent, firm, acuminate; utricle broadly ellipsoidal, prominently circumscissile   **1. A. dubius**
1. Perianth of 3 tepals; bracts thin, scale-like, acute; utricle globose, deeply wrinkled, indehiscent   **2. A. viridis**

1. Amaranthus dubius Mart. ex Thell. in Mém. Soc. Sci. Nat. Cherb. 38: 203 (1912); Townsend in K.B. 29: 471 (1974).

A. tristis sensu Hemsley in B.M.I.K. 1919: 129 (1919), not L. (1753).

Erect to decumbent herb, simple or branched, stems reddish, up to 1 m tall. Leaves broadly ovate, 1.5–8(–12) × 0.5–5(–8) cm, apex obtuse, base contracted to subtruncate, veins white; petiole as long as or longer than blade; stipules absent or represented by an obscure flap of tissue. Inflorescence paniculate; with a strong central axis and spreading branches, glomerules crowded above, scattered below, small ones in leaf-axils, male and female flowers mixed, bracts as long as or slightly longer than the perianth, strongly acuminate. Tepals 5, 2–3 mm long. Male flowers: all 5 tepals narrowly ovate, apex acute to acuminate, tending to be incurved at maturity, making the flower appear ovoid, 1 tepal larger, boart-shaped; stamens 5, filaments flat. Female flowers: tepals oblong to spatulate, subequal, somewhat broader near apex, 1 tepal with apex acute-acuminate, remaining 4 tepals with apex obtuse to slightly indented, spine-tipped; styles 3, beak-like; stigmas slender, exserted. Utricle broadly elliptic, 1 × 0.5 mm, subequal with perianth, differentiated into an upper and a lower half, these differently wrinkled to almost smooth, circumscissile at middle, each half broadly conical, the upper with 3 persistent styles; seeds glossy dark brown, very broadly tear-drop-shaped, c 0.7 × 0.5 mm, only very slightly compressed. Fig. 38/5–6.

Distr. ALDABRA: W; ASSUMPTION; COSMOLEDO: M; ASTOVE; a weed of American origin, now pantropical.
Notes. Occurs around areas of settlement.
Vernac. 'brède malabar'.

2. **Amaranthus viridis** L., Sp. Pl. ed. 2: 1405 (1763).

A. gangeticus sensu Dupont, Report: 40 (1907), not L. (1759).

Erect or decumbent herbs, up to 50 cm high, glabrous or almost so. Leaves ovate to triangular- or rhombic-ovate, up to 7 × 5 cm, apex obtuse to acute drying with prominent white veins beneath, veins rather obscure above; petioles up to 6 cm long, slender, as long as or longer than blades; stipules minute, ovate-lanceolate or vestigial. Inflorescence a terminal, branched panicle; flowers in sessile or subsessile glomerules, a few flowers in the leaf axils; male flowers few, female flowers numerous; bracts thin, scale-like, ovate, apex acute, not spine-tipped, margins broad, transparent. Tepals 3, light green, 2 broadly ovate-elliptic, 1 narrower, oblong. Male flowers: stamens 3, slightly exceeding perianth, anthers oblong. Female flowers: ovary enlarging, exserting the 3, beak-like styles with their recurved papillate-stigmas. Utricle globose c 1 mm in diameter, prominently and closely wrinkled; seed 1, glossy dark brown, discoid-ovate, c 0.8 mm in diameter. Fig 38/7–8.

Distr. ALDABRA: W, S (east); ASSUMPTION; COSMOLEDO: M; a weed of tropical American origin, now pantropical, extending into the warm temperate regions.

Notes. Established in gardens and around settlements in the islands of the Aldabra group, with one clump on the south-east coast of Aldabra away from settlement.

Vernac. 'brêde malabar'.

4. DEERINGIA R.Br.

Herbs to large shrubs, scandent or straggling. Leaves alternate; stipules absent. Inflorescence of terminal or axillary spikes or racemes; each flower subtended by 1 small membranous bract and bracteoles; flowers bisexual or if unisexual, dioecious. Tepals (4–)5. Stamens (4–)5, opposite the tepals, very slightly united at base, staminodes absent. Ovary with few to many ovules on slender stalks; style short, 2–3(–4) branched, stigmas linear or slightly thickened. Fruit baccate; seeds several–many, small.

A small genus extending from East Africa to Australia, China and the Philippines to Taiwan and the Ryuku Islands.

Deeringia polysperma (Roxb.) Moq. in DC., Prodr. 13 (2): 236 (1849); Renvoize in P.T.R.S. B, 260: 230 (1971).

Celosia polysperma Roxb., Fl. Ind. 2: 511 (1824).
Deeringia celosioides sensu Hemsley in B.M.I.K. 1919: 129 (1919), not R.Br. (1810).
D. indica Zoll. ex Moq. in DC. Prodr. 13 (2): 236 (1849); Hemsley in B.M.I.K. 1919: 129 (1919) not *D. indica* Retz. ex Bl. (1825).

Straggling or semi-scandant shrub, up to 3 m tall; stem thick, soft-wooded. Leaves ovate to oval or sub-orbicular, 2–3(–7.5) × 1.5–2(–4.5) cm, apex obtuse to shortly acuminate, base contracted; petiole slender, 1–2 cm long. Flowers in loose, axillary spikes, elongating up to 10 cm; bracts oblong to ovate, 1–1.5 mm long, apex obtuse; bracteoles smaller, broadly ovate. Tepals pale yellowish-green, very broadly ovate or suborbicular, c 2 mm long, apex rounded, midrib red. Stamens 5, anthers falling before exsertion of stigmas. Ovary sub-globose, style branches 3(–4), linear-spatulate, inner surfaces stigmatic, well exserted from perianth. Fruit white, fleshy, translucent, depressed-globose, c 5 mm across or smaller; seeds many, glossy black, kidney shaped, c 1 × 0.7 mm, compressed, concentrically chequered, (fruit and seed description from other than Aldabra material). Fig. 37/8–10.

Specimens from Aldabra have usually much smaller leaves than those from other areas.

Distr. ALDABRA: W, P, M, S, Esprit; across to India, Malesia, north to the Philippines and Formosa.

Notes. A rather uncommon understory shrub in scrub communities or on platin. Flowers throughout the year if sufficient moisture is available.

5. LAGREZIA Moq.

Herbs, erect to decumbent, branches alternate. Leaves alternate; stipules absent. Flowers in a narrow panicle with a strong rhachis, panicle branches spike-like; flowers bisexual, each subtended by 1 scale-like bract and 2 bracteoles. Tepals 5, equal, glabrous. Stamens 5, opposite the tepals, staminodes absent; bases of filaments united to form a cup, free part awl-shaped; anthers 2-celled. Ovary 1-celled, ovule 1; style short, 2–3-branched, stigmas spreading. Utricle smooth, circumscissile or breaking irregularly or indehiscent; seeds 1, vertical, with a minute notch on one edge.

A small genus confined to the Western Indian Ocean, especially Madagascar.

Lagrezia oligomeroides (C.H. Wright) Fosberg in K.B. 29: 262 (1974). Type: Aldabra, *Dupont* 56 (lectotype, selected by Fosberg, loc. cit., K).

Apteranthera oligomeroides C.H. Wright in B.M.I.K. 1918: 203 (1918); Hemsley in B.M.I.K. 1919: 129 (1919).
Celosia madagascariensis sensu Hemsley in B.M.I.K. 1919: 129 (1919), not Poir. (1804).
Lagrezia madagascariensis sensu Fosberg in P.T.R.S. B, 260: 225 (1974); Renvoize in P.T.R.S. B, 260: 231 (1914); Grubb in P.T.R.S. B, 260: 355 (1974), not (Poir.) Moq. (1849).

Erect or spreading herb, up to 15 cm tall; branches alternate. Leaves alternate, ovate to linear, up to 2 × 1.4 cm, apex rounded, base attenuated sub-fleshy; petiole up to 2.5 cm long. Inflorescence a terminal, narrow panicle, rachis stout, with alternate, conspicuously jointed, short branches, a flower between the branches, each branch composed of up to 6–8, swollen, globose or pear-shaped segments, giving the branch the appearance of a string of beads, each segment subtended by an ovate, scale-like bract; flowers 1 per node, or at least at the upper nodes, and slightly congested at tips. Tepals 5, pinkish, bluntly-acute, more or less hooded, membraneous. Stamens 5, opposite and somewhat shorter than the tepals; anthers very broadly ovate, connective bright rose pink, filaments orange-yellow, expanded and united at base into a shallow cup. Ovary yellow, depressed-globose; style short, 2-fid. Utricle red, ovoid, c 1 mm long, partly exserted when developed, indehiscent, somewhat adherent to seed; seed 1, shiny black, erect, discoid, slightly compressed, c 0.8 mm across, minutely chequered. Fig. 38/9.

In the discussion of this species (Fosberg, l.c. 1974) one character was not mentioned that strengthens its distinction from *L. madagascariensis* (Poir.) Moq. This is the lack of a circumscissile dehiscence of the utricle. The utricle wall is very thin and delicate, and the whole is retained within the closely enveloping, mature tepals, which are shed with the fruit.

Forms with succulent leaves and inflorescences found along the coasts exposed to the trade winds where salt spray is abundant; non-succulent forms occur elsewhere and can be induced to succulence by exposure to salt spray.

Distr. ALDABRA: W, P, M, S, Esprit, lagoon islets; ASSUMPTION; COSMOLEDO: NW. Is.; also Cousin Island in the Seychelles.

Notes. Flowers during the late wet season; fruits mature during the mid-dry season.

57. CHENOPODIACEAE

Herbs or shrubs, often succulent. Leaves opposite or alternate, simple; stipules absent. Flowers in spikes, racemes or panicles, rarely cymose. Perianth of 1 series of 2–5, usually scale-like or fleshy, but not membranous, segments, or these vestigial. Stamens opposite perianth segments. Ovary superior; styles 1–3. Fruit a 1-seeded utricle or with adherent, variously united flower-parts.

An almost cosmopolitan family, predominantly halophytic or xerophytic, with many ruderal or nitrophilous species; includes several edible plants such as the beet, spinach, and quinoa, as well as several ubiquitous weeds.

ARTHROCNEMUM Moq.

Leafless herbs or sub-shrubs, stems succulent, jointed, joints cylindrical. Flowers in terminal spikes, in groups of 3, hidden by broad scales, spikes somewhat thicker than sterile parts of stems, flowers all hermaphrodite or lateral flowers of a triad male. Perianth angular, with a 1-sided swelling, opening irregularly or with 3–5 lobes at top, not winged. Stamens 1–2. Fruit free from perianth, but included in it; pericarp thin, free from the erect compressed seed, endosperm abundant.

A small genus, rather ill-distinguished from *Salicornia*, with a few species in warm saline regions.

Athrocnemum pachystachyum (Bunge ex Ung.-Sternb.) A. Chev. in Rev. Int. Bot. Appl. 2: 748 (1922)*

Salicornia pachystachya Bunge ex Ung.-Sternb., Vers. Syst. Salicorn. 5: (1866); Hemsley in B.M.I.K. 1919: 129 (1919); F.T.E.A. Chenopod.: 21 (1954). *Arthrocnemum glaucum* sensu Renvoize in P.T.R.S. B, 260: 230 (1974), not (Del.) Ung.-Sternb. (1866).

Much-branched, annual shrublet, glaucous, with many dense spikes thicker than stems. Central flower in a group hermaphrodite, lateral flowers male, perianth truncate. Stamens 1. Seed compressed, vertical.

This specimen resembles very much what has usually been called *A. glaucum* (Del.) Ung.-Sternb. but is not very adequate; annotated by Moss as cf. *Salicornia perrieri* A. Chev. (a synonym of *S. pachystachya* according to Tölken l.c.). Its identity is very doubtful and it should be looked for and recollected if possible.

Distr. ALDABRA; coast of Tanzania, Mozambique, Natal and Madagascar.
Notes. Probably occurs in the saline flats fringing the lagoon. Sole record, without data, *Fryer* 49 (K).

58. LAURACEAE

Evergreen or deciduous trees and shrubs, rarely parasitic climbers, usually aromatic. Leaves alternate, simple, rarely reduced; stipules absent. Inflorescences various; flowers bisexual, rarely unisexual, rarely dioecious, regular. Perianth of 2 series of usually 3 segments each, segments similar, sepaloid. Stamens up to 4 whorls of 3 each, 1 or more whorls often sterile, anthers dehiscing by flap-like valves; disk present or absent. Ovary superior; ovule 1, pendulous, placentation more or less parietal. Fruit fleshy, baccate or drupaceous, usually basally enveloped in a cupule formed by an enlarged, toughened perianth base, rarely completely enveloped.

---

* Both Brenan in F.T.E.A. Chenopod. (1954) and Tölken in his critical revision of *Arthrocnemum* and *Salicornia* in Bothalia 9: 255–307 (1967) treat this as *Salicornia pachystachya*.

A large tropical family with a few temperate representatives, including a number of economic species, such as the avocado, and cinnamon, as well as the classical laurel or bay, *Laurus nobilis*.

CASSYTHA Osb. ex L.

Essentially leafless, parasitic, cord-like, twining climbers attaching to their hosts by numerous lateral haustoria. Leaves reduced to minute scales. Flowers in axillary spikes. Fruit a drupe.

A tropical genus of few species.

Cassytha filiformis L., Sp. Pl.: 35 (1753); F.M.S.: 292 (1877); Dupont, Report: 40 (1907); F.M.C. 81: 85 (1950).

Plants slender, tough, tangled, yellow to green, string-like stems bearing haustoria, young stems pubescent. Spikes 1 to 4 cm long, rachis pubescent; flowers scattered on rachis, globular, c 1 mm across, bases invested by a number of overlapping, scale-like, minutely ciliate bracts. Perianth persistant, white, in 2 whorls of 3, outer whorl small, scale-like, minutely ciliate; inner whorl larger, broadly oblong or suborbicular, concave, obtuse. Stamens 9; staminodes similar but smaller and thinner. Ovary becoming enclosed by the enlarged cup-like receptacle crowned by the persistent perianth. Fruit enclosed by the enlarged white and fleshy receptacle to form a white, globose, drupe-like structure, 5 mm in diameter. Fig. 39/1.

Distr. ALDABRA: W, S (east); ASSUMPTION; COSMOLEDO: W, M; ASTOVE; pantropical.

Notes. On Aldabra frequent at the east end of South Island, one record only for West Island. Parasitic on *Flacourtia, Thespesia, Guettarda*, and other plants, including itself. Haustoria seem to be produced wherever a *Cassytha* stem comes in contact with a living plant surface, including its own stems. Few observations but apparently flowers during the dry season. Fruit eaten and seeds locally dispersed by birds.

Vernac. 'liane sans fin'.

59. HERNANDIACEAE

Mostly trees and shrubs. Leaves alternate, simple or rarely palmately compound, palmately veined; stipules absent. Inflorescence a corymbiform or paniculate cyme or thyrse; flowers unisexual or bisexual, if unisexual usually monoecious. Perianth in 2 series of 4–10 sepaloid segments. Stamens 3–5, opposite outer sepals, anthers 2-celled; 1–2 whorls of staminodes often present. Ovary inferior, 1-celled, ovules 1, pendent; style 1, stigma 1. Fruit a nut, enveloped by the expanded inflated receptacle, or a samara.

A genus with a number of species scattered through the tropics.

Hernandia nymphiifolia (Presl) Kubitzki in Engl., Bot. Jahrb. 90: 272 (1970).

Biasolettia nymphiifolia Presl, Rel. Haenk. 2: 142 (1835).
Hernandia peltata Meissn. in DC., Prodr. 15, 1: 263 (1864); F.M.S.: 293 (1877).

Tree, trunk massive; bark pale. Leaves alternate, peltate, ovate-lanceolate, 7–30 × 6–22 cm, apex shortly acuminate or obtuse, base rounded to subcordate, rather leathery, 8–9 nerved; petiole 5–17 cm long, attached 1–3 cm within margin of leaf. Cymes axillary, up to 25 cm long, branches subtended by small bracts; involucre of 4 bracts subtending each triad of flowers. Flowers tomentulose, subglobose, 3–3.5 mm. Male flowers: 3-merous, perianth segments in 2 whorls of 3, obovate, subequal; stamens 3. Female flowers: 4-merous, set in cup-shaped involucel, perianth segments 8; ovary flask-shaped, neck recurved. Enveloping involucel of fruit white to crimson, inflated; nut black, ellipsoid, 2.5–3 × 1.7–2.3 cm, faintly 8–ribbed, apex with an abrupt, low, rounded boss. Fig. 39/2.

This species has often been confused with *H. sonora* L. from the Antilles, which has 3–4-merous male flowers and conspicuously ribbed fruit (see Kubitzki in Engl., Bot. Jahrb. 89: 122–157 (1969) for a revision of the genus),

Distr. Astove; Indo-Pacific strand tree.
Notes. Only 1 tree known, at Settlement on Astove, probably planted; *Fosberg* 49686 (K, US), *Veevers-Carter* 82 (EA).
Vernac. 'bois blanc'.

60. LORANTHACEAE

Usually partially parasitic shrubs or rarely small trees; branches not jointed; attachment to host external. Leaves opposite or rarely alternate; stipules absent. Inflorescences axillary, of various forms; flowers usually bisexual. Perianth whorls 2. Calyx much reduced. Stamens opposite petals. Ovary inferior, ovules 1. Fruit a 1-seeded berry with viscid pulp.

An abundant pantropical family, with few to many genera, depending on one's taxonomic viewpoint.

BAKERELLA van Tiegh.

Glabrous shrubs; with branches lying close to those of host tree, adhering by 1 principal and numerous secondary haustoria. Leaves mostly opposite or

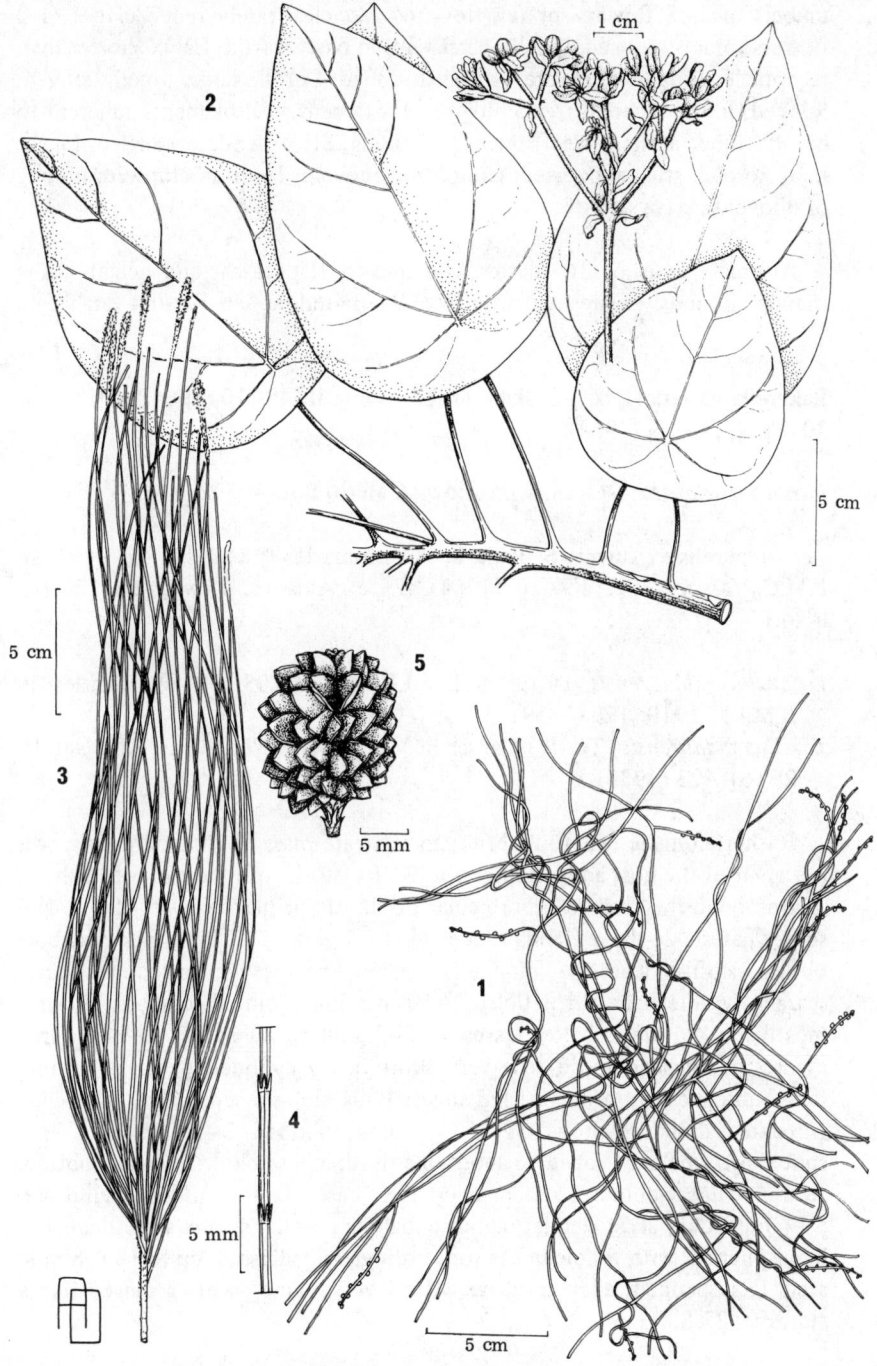

Fig. 39 1, *Cassytha filiformis*, habit. 2, *Hernandia nymphaeifolia*, habit.
3, *Casuarina equisetifolia*, habit; 4, detail of branchlet; 5, fruit.

subopposite, rarely alternate, leathery; stipules absent. Flowers in short umbels of 1–5 flowers, or few-flowered fascicles, rarely reduced to 1 or 2 flowers; bracts reduced, usually persistent on base of fruit. Calyx shorter than receptacle, usually subentire or slightly lobed. Corolla large, usually curved, 5-lobed, lobes coherent, remaining erect. Stamens with filaments adherent to corolla tube, anthers erect, linear or oblong. Style slender, 5-angled, longer than corolla, stigma exserted, globose or obovoid. Fruit an ellipsoid, oblong or obovoid berry; seeds 1.

A small genus of 16 species, endemic to Madagascar and neighbouring islands; dubiously segregated from the Asiatic-Indonesian *Taxillus* van Tiegh.

Bakerella clavata (Desr.) S. Balle in Adansonia II, 4: 110 (1964); F.M.C. 60: 20, fig. 1/11–13 (1964).

Loranthus clavatus Desr. in Lam., Encycl. Méth. Bot. 3: 598 (1789).

var. **aldabrensis** (Turrill) S. Balle in Adansonia II, 4: 110, fig. 4/13 (1964); F.M.C. 60: 22. fig. 2/9–10 (1964). Type: Aldabra, *Thomassett* 229 (K, holo.).

Loranthus aldabrensis Turrill in B.M.I.K. 1918: 203 (1918); Hemsley in B.M.I.K. 1919: 129 (1919).
Taxillus aldabrensis (Turrill) Danser in Verh. Akad. Wet. Amst. Afd. Nat. II, 29 (6): 123 (1933).

Rather elongate, glabrous shrub; an intricate mass of roots spreading over a branch of the host and sending root-like haustoria into the bark of the host. Leaves opposite, oval to suborbicular or elliptic or narrow-ovate-elliptic and somewhat sickle-shaped and folded, up to 5–9 × 2–5 cm, apex and base obtuse, softly leathery, pinnately nerved, nerves inconspicuous above, scarcely visible beneath; petioles 3–10 mm long, jointed to stems. Flowers mostly on old wood below leaves, in 1–2 axillary, subsessile, 2–3-flowered cymes; peduncle reduced to a very short fleshy cylinder; pedicel 4–5 mm long subtended by a cup-shaped bract. Buds club-shaped. Calyx collar-like, truncate, ciliolate. Corolla c 2.5–3 cm long, tube red, 2–2.5 cm long, limb split down one side, lobed to about one third, lobes yellowish, ovate, obtuse. Anthers linear-oblong, attached by the base. Ovary inferior, ellipsoid, 3–4 mm long; style slender, about equalling corolla, stigma same diameter, hemispherical with a hole in the top. Fruit green, ellipsoid, up to 18 × 8 mm; seed 1, suspended, apex truncate with 5 very blunt points arranged like a star. Fig. 27/8.

Distr. ALDABRA: M (east), S, Michel; Madagascar.
Notes. Parasitic on *Sideroxylon inerme* L., *Premna obtusifolia* R.Br. and also, according to *Thomasset* 229 on *Apodytes dimidiata* E. Mey. ('bois none') and *Dupont* 107 on *Macphersonia hildebrandtii* O. Hoffm. ('tamarin bâtard'); locally common.
Vernac. 'mangoula'.

61. VISCACEAE

Partially parasitic shrubs and herbs; stems appearing jointed because of constrictions at nodes, branches usually decussate; attachment to host internal. Leaves opposite, rarely whorled, or lacking; stipules absent. Inflorescence a catkin-like spike or reduced to a fascicle or solitary flower, axillary; flowers dioecious. Perianth of 1 whorl. Stamens opposite perianth parts or rudimentary. Ovary inferior, ovules 1. Fruit a berry with extremely viscid pulp; seeds 1.

The justification for the separation of *Viscaceae* from *Loranthaceae* is given by Kuijt in Brittonia 20: 136–147 (1968).

A small, mainly tropical family.

VISCUM L.

Shrubs, green. Leaves present or reduced to scales. Flowers in few-flowered cymes or solitary or fascicled, axillary or terminal, monoecious or usually dioecious. Calyx rudimentary or reduced to a mere rim. Petals 3—4-lobed, small. Male flowers: anthers adherent to base of petals. Female flowers: petals persistent or falling, on rim of cup-like receptacle; ovary inferior, style short or absent. Fruit variously coloured.

Genus of c 100 species of mostly warmer parts of Old World.

Viscum triflorum DC., Prodr. 4: 279 (1830); F.M.S.: 134 (1877); Hemsley in B.M.I.K. 1919: 130 (1919); F.M.C. 60: 71 (1964); Renvoize in P.T.R.S. B, 260: 231 (1971).

Shrubs; nodes enlarged, internodes cylindrical, branching to form globose clusters. Leaves opposite, narrowly to broadly elliptic, to 6 × 2 cm, apex obtuse to rounded, base gradually narrowed into petiole, leathery, 3 main veins from base, forking, becoming obscure; petiole short, 2–3 mm long. Flowers in pairs; peduncles 2–3 mm long, with 2 scale-like bracts at summit subtending 2 very short (1 mm or less) pedicels. Flowers not seen, believed to be yellowish-green. Fruit pale green to white, subglobose to broadly elliptic, extremely sticky within, c 4 × 3 mm, crowned by a ring of rudimentary sepals and subpersistent, lanceolate petals and a short stylar beak; seeds 1, black, ovoid or ellipsoid, c 2–2.5 × 1.5–2 mm, tipped with a very shortly stalked disk-like appendage. Fig. 27/9.

Distr. ALDABRA: S, Michel; E. Africa, Comoros Is., Seychelles and Mascarene Is.

Notes. Parasitic on *Pemphis acidula* Forst. Few observations but apparently flowers during the dry season.

Vernac. 'bois mamai' or 'gui du pays'.

62. EUPHORBIACEAE

Habit various; plants often with milky sap. Leaves simple, 1- or 3-foliolate, rarely palmately compound; stipules usually present. Inflorescence basically cymose but may be strongly reduced to a cup-like involucre (cyathium), enclosing reduced flowers, these variously clustered; flowers functionally unisexual, regular. Male flowers: sepals usually 5, free; petals 5 or reduced in number or absent; stamens same number or twice the number of the petals, or reduced even to 1, often with intrastaminal disk and/or rudimentary ovary. Female flowers: perianth 2- or 1-whorled or absent; ovary superior, sessile or appearing shortly stalked, (2-)3-(4- or rarely more-)celled, ovules 1 or 2, per cell, pendulous, styles usually 3, simple or branched, stigmas (2–)3 or more. Fruit a capsule, often elastically dehiscent, or a drupe; seeds with endosperm, often with an appendage (caruncle).

An enormous almost cosmopolitan family, preponderantly tropical, very diverse in habit and also in technical botanical characters.

Wielandia elegans Baill., a Seychelles endemic, has been reported from Aldabra by Hemsley in B.M.I.K. 1919: 130 (1919) on the basis of a specimen preserved at Kew (*Fryer* 84). This specimen is unquestionably correctly identified. However, the species has been looked for on Aldabra but not found. Judging by its habitat on Silhouette Island, Seychelles (*Fosberg* 52202 (K, US)), it seems very unlikely that it occurs on Aldabra. In all probability the Fryer specimen was mislabelled and came from the Seychelles. It is not described here. Recent specimens from Aldabra labelled *Wielandia* are *Margaritaria*.

1. Leaves large, peltate, conspicuously and sharply palmately lobed
                                                                      6. Ricinus
1. Leaves smaller, usually not more than 10 cm long, neither peltate nor palmately lobed
  2. Leaves opposite; stipules interpetiolar; milky sap present   **2. Euphorbia**
  2. Leaves alternate or closely crowded, spirally arranged
    3. Succulent shrub; branchlets zigzag; flowers borne in slipper-shaped cyathia; milky sap present   **Pedilanthus**
    3. Not as above; milky sap present or absent
      4. Branchlets thick, smooth; leaves spirally crowded, in rosettes at tips of branches; flowers dioecious, borne in round cyathia in short clusters during leafless period; milky sap present   **2. Euphorbia**
      4. Not as above; milky sap absent
        5. Leaves conspicuously 2-ranked   **5. Phyllanthus**
        5. Leaves not 2-ranked
          6. Leaves ovate; flowers in catkin-like spikes, female flowers subtended by conspicuous fan-like veined bracts   **1. Acalypha**
          6. Leaves obovate; flowers in axillary fascicles or solitary in axils
                                                                   3. Margaritaria

1. ACALYPHA L.

Shrubs and herbs; milky sap absent. Leaves alternate, simple, tending to be 3(−5)-nerved from base, margins usually serrate or crenate; stipules present. Flowers unisexual, monoecious or dioecious; inflorescence spicate or rarely capitate or paniculate, spikes usually axillary, either of 1 sex only or both with male flowers distal. Male flowers: sepals 4(−5), margins touching; stamens usually 8. Female flowers: usually bracteate; sepals 3−5, small; ovary (2−)3 cells, 1 ovule per cell, styles (2−)3, branches few to many, very fine. Fruit splitting into 3, 2-valved cocci.

An enormous pantropical and warm-temperature genus.

1. Plants woody; leaves ovate, petiole usually shorter than blade; male and female flowers in separate spikes, often on separate plants
A. claoxyloides
1. Plants herbaceous; leaves broadly rhombic, petioles exceeding blades; both sexes of flowers in same spike **2. A. indica**

1. Acalypha claoxyloides Hutch. in B.M.I.K. 1918: 205 (1918); Hemsley in B.M.I.K. 1919: 130 (1919); Fosberg in P.T.R.S. B, 260: 220, 225 (1971); Renvoize in P.T.R.S. B, 260: 231 (1971); Fosberg in K.B. 29: 263 (1974). Type: Aldabra, *Abbott* (K, lecto., US, isolecto., selected by Fosberg in K.B., loc. cit. (1974)).

Claoxylon sp., Baker in B.M.I.K. 1894: 150 (1894); Schinz in A.S.N.G. 21: 86 (1897); Voeltzkow in A.S.N.G. 26: 550 (1902); Dupont, Report: 40 (1907).
Acalypha fryeri Hutch. in B.M.I.K. 1918: 205 (1918); Hemsley in B.M.I.K. 1919: 130 (1919). Type: Aldabra, *Fryer* 92 (K, holo.).
A. aldabrica Pax & Hoffm. in Engler, Pflanzenr. 4, 147 (16): 136 (1924). Type: Aldabra, *Abbott* (B, holo. †).

Shrubs (0.5−)1−1.5(−3) m tall, irregularly branched, glabrous except for youngest growth, whole plant beset by scattered, large, resinous glands; buds covered by conspicuous, dense, white wool. Leaves very bright green, extremely variable in size and shape, usually ovate or oblong or elliptic, 15 × 5(−7) cm, usually much smaller, apex acute to acuminate, base obtuse, rounded to slightly subcordate, margins crenate, almost glabrous, venation prominent; petiole 1−4.5 cm long. Male and female spikes noticeably different, borne on different plants but male plants may have a few reduced female spikes and vice versa. Male spikes: sessile or peduncle short; spikes 1−several cm long, woolly; flowers on pedicels, not crowded, clusters subtended by convex, scale-like bracts, woolly without, glands in clusters, conspicuous, stalked; sepals 4, scale-like, convex; stamen-filaments short, anthers broadly 'V'-shaped. Female spikes: peduncles very slender, bearing 1−5(−7) large, leaf-like bracts; bracts widely separated, cordate or kidney-shaped-cordate, up to 15 × 20 mm, folded, gland-dotted, margins obscurely crenate, each bract subtending 2–3 flowers; calyx 3-lobed, lobes small, triangular; ovary densely hairy; style few-branched, included in bracts (in all specimens seen). Fruit globose, c 3 mm in diameter, not or scarcely hairy, densely yellow-glandular, valves 2-fid; seeds grey, ovoid 2 × 1 mm, smooth, with a linear scar with ridge as long as seed. Fig. 40/1–4.

Distr. ALDABRA: W, P, M, S, Esprit, Michel, lagoon islets; ASSUMPTION; COSMOLEDO: W, M; ASTOVE; endemic.

Notes. A widespread constituent of coastal scrub and *Pemphis* scrub on champignon. Buds form at end of the wet season and may open in response to unseasonal rains but main flowering during the early wet season. Seeds locally dispersed by fody.

2. Acalypha indica L., Sp. Pl.: 1003 (1753); F.M.S.: 314 (1877).

Herb, rarely somewhat woody at base, a few to 50 cm tall, erect. Leaves broadly rhombic or rhombic-oval, up to 4 × 3–3.5 cm, apex obtuse but with short, sharp tip, base obtuse to truncate, margins shallowly serrate, thinly and minutely pubescent, especially on the nerves; petiole exceeding blade, to 6 cm, slender; stipules minute. Flowers monoecious, in spikes 2–3(–7) cm long, mostly female, but distal 1 cm or less male. Male flowers crowded. Female flowers: bracts obovate-funnel-shaped but lower margins free, 5–8 × 5.8 mm, sessile on rachis, thinly leaf-like, upper margins remotely and shallowly toothed, 1(–2) flowers per bract; stigmas white, exserted from bracts. Fruit globose, strongly 3-lobed, 1.5 mm in diameter, appressed-hairy; seed, grey when mature, obovoid, 1 × 0.6 mm, smooth. Fig. 40/5–6.

Distr. ALDABRA: W; ASSUMPTION; COSMOLEDO: M; ASTOVE; widespread through the tropics of the Old World.

Notes. Common around settlements and disturbed places. Few observations but apparently flowers throughout the year if sufficient moisture available.

Vernac. 'herbe chatte'.

2. EUPHORBIA L.

Herbs, shrubs or small trees of various habit; milky sap present. Leaves alternate, opposite, or much-reduced or even absent; stipules present or absent. Flowers unisexual, monoecious or rarely dioecious; male flowers reduced to 1 naked stamen on a pedicel; female flowers reduced to 1 naked ovary or with a reduced, 3-lobed perianth; flowers subtended by fused bracts forming a cup-like structure (cyathium) bearing at or near margin 1 or more glands, glands with or without appendages; cyathia solitary and axillary or usually in cymosely branched, terminal or axillary clusters. Fruit a 3-celled capsule, partially or wholly exserted from the cyathium on the somewhat enlarged pedicel, capsule dehiscing septicidally and loculicidally and elastically; seeds 1 per cell.

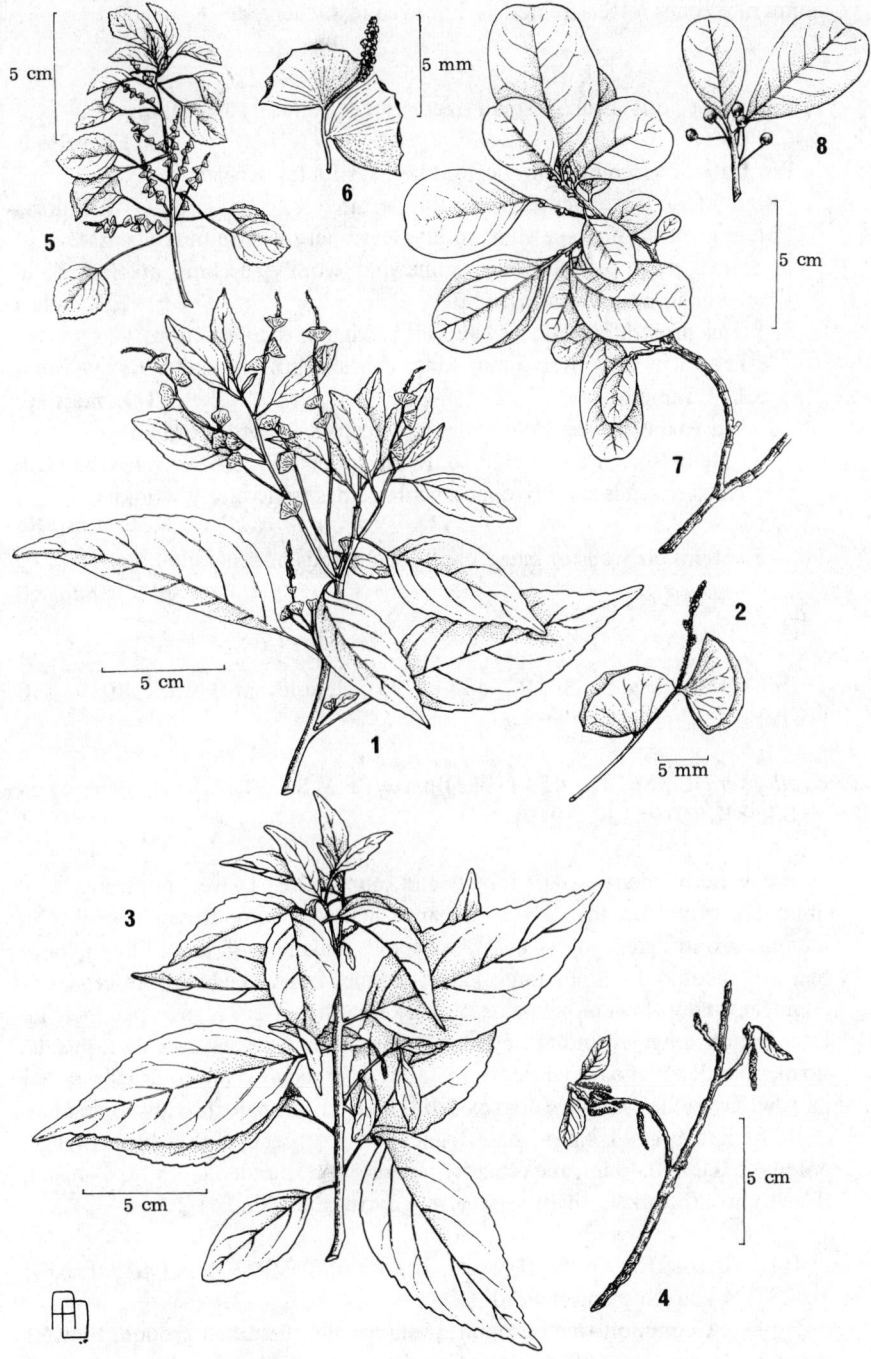

Fig. 40 1, *Acalpha claoxyloides*, habit, bisexual plant; 2, detail of flowering
branch; 3, habit, male plant; 4, detail of flowering branch.
5, *Acalypha indica*, habit; 6, detail of flowering branch.
7, *Margaritaria anomala* var. *cheloniphorbe*, habit; 8, fruits.

An enormous genus, widely distributed throughout the tropics and temperate zones; often treated as 2 or even many genera.

1. Erect dioecious shrub or small tree; stipules minute, soon falling
                                                          5. E. pyrifolia
1. Prostrate or ascending monoecious herbs; stipules persistent
    2. Stem hairy; leaves acute, noticeably serrate   **1. E. hirta**
    2. Stem glabrous or somewhat woolly; leaves entire or minutely serrate
        3. Stems erect or ascending, somewhat woolly, arching at tips; gland
           appendages conspicuous, white   **2. E. indica**
        3. Stems prostrate; gland appendages lacking or inconspicuous
            4. Leaves mostly over 5 mm long; capsule entirely glabrous; cyathium
               1–1.5 mm high   **3. E. mertonii**
            4. Leaves mostly less than 5 mm long; capsule glabrous or hairy
                5. Stems somewhat woolly only on upper side; capsule hairy only on
                   angles; seeds sharply quadrangular, sides transversely wrinkled
                                                          4. E. prostrata
                5. Stems glabrous or hairy; capsules glabrous or generally hairy
                                                          6. E. stoddartii

1. Euphorbia hirta L., Sp. Pl.: 454 (1793); Hemsley in B.M.I.K. 1919: 130 (1919).

E. pilulifera L., Sp. Pl.: 454 (1753) partly; F.M.S.: 303 (1877); Hemsley in B.M.I.K. 1919: 130 (1919).

Erect herb; hairy, hairs erect and appressed. Leaves opposite, very obliquely ovate, up to c 2 × 1 cm, apex acute or sub-obtuse, base obtuse, margins serrate, glabrous or nearly so above, usually with a red blotch, hairy beneath; petiole 1–2 mm long, hairy; stipules awl-shaped. Inflorescences in compact, many-flowered clusters, usually less than 1 cm across; peduncle up to 5–6 mm long, ascending; cyathia bell-shaped, hairy; glands dark purple, small, round; appendage white, from about as broad as gland to usually rather narrow. Ovary hairy; styles 3, apex 2-fid, hairy. Fruit pale, broadly triangular-ovoid, up to c 1 × 1.2 mm, base truncate, hairy; seed dull reddish orange, oblong, c 0.8 × 0.4 mm, apex slightly tapering and rounded apically, 4-angled, sides somwhat sunken, slightly traversely corrugated. Fig. 41/1.

Dist. ALDABRA: W, M, S (west), Esprit; ASSUMPTION; COSMOLEDO: M; ASTOVE; a pantropical weed.
Notes. A common weed around dwellings and disturbed ground. Flowers throughout the year if sufficient moisture available. Seedlings observed beneath birds' nests suggests possible local dispersal by birds.
Vernac. 'Jean Robert'.

2. Euphorbia indica Lam., Encycl. Méth.Bot. 2: 423 (1786).

The subglabrous form, var. *indica*, is widespread in the Indian Ocean area but is not known from the Aldabra group. A pubescent variety seems well established on Assumption.

var. **pubescens** Pax in Engler, Bot. Jahrb. 19: 117 (1895).

E. hypericifolia sensu F.M.S.: 303 (1877); Hemsley in B.M.I.K. 1919: 130 (1919), not L. (1753).

Erect to somewhat decumbent herb or slightly woody at base. Leaves oblong, up to 15 × 8 mm, apex rounded or obtuse, base unequally obtuse to wedge-shaped or subcordate, margins minutely serrate and minutely ciliate; petiole c 1 mm long; stipules interpetiolar, shallowly triangular. Cyathia in loose cluster subtended by 2 reduced leaves at summit of peduncle, cyathia bell-shaped, almost glabrous, glands yellowish-green to dark purple, small, with a white, kidney-shaped appendage. Ovary hairy; stigmas purple, minute, capitate. Fruit strongly 3-lobed, as wide as or wider than high, c 2 mm wide, hairy; seeds grey, quadrangular-oblong, 1 × 0.6 mm, obtusely pointed at one end, truncate at other, smooth. Fig. 41/2.

Our material is rather inadequate, but it seems to be the widespread Indian Ocean plant belonging to *E. hypericifolia* in the widest sense, but probably for now best kept separate.

Distr. ASSUMPTION; islands of the Indian Ocean.
Notes. Probably introduced, known only from 3 collections, *Dupont* 292 (K), *Stoddart* 1089 (K) and *Frazier* 746 (US).

3. **Euphorbia mertonii** Fosberg in K.B. 33: 181 (1978). Type: Aldabra, South Island, Point Hodoul, *Fosberg* 49045 (US, holo., G, K, MO, NY, iso.).

Probably perennial, entirely glabrous, prostrate herb; tap-root strong, stems greenish to purplish, several to many, forming a loose mat, wiry, slender, branched, up to 35 cm long, drying glaucous. Leaves maroon-purple, oval to broadly obovate, or somewhat oblong, up to 10(−12) × 5(−8) mm, apex rounded, base noticeably obliquely rounded to subcordate, margins entire to obscurely and minutely serrate; petiole 1−1.5 mm long; stipules awl-shaped, base broad, 2−4 on a side, or variously united into 2 or 1, with several awl-shaped processes on margin. Cyathia 1−6, terminal or on uppermost nodes, fertile branchlets usually with strongly reduced leaves; peduncles 1−3 mm long; cyathia bell-shaped, 1−1.5 × 1−1.5 mm, marginal lobes triangular to lanceolate, fringed or slightly so, glands 4−5, disk-like to kidney-shaped, margins white, very narrow. Ovary broadly oblong, strongly 3-lobed, reflexed. Fruit almost square in side view, or slightly narrowed upwards, c 1−1.2 × 1.2 mm; seeds dull salmon pink, ovoid-quadrangular, 0.8 mm long, apex rounded or obtuse, base sub-truncate, shallowly and irregularly wrinkled. Fig. 41/3−4.

Distr. ALDABRA: W, S; endemic.

Notes. Locally very common on pitted rough coral limestone, especially where a little coral sand has accumulated.

4. **Euphorbia prostrata** Ait., Hort. Kew. 2: 139 (1789); F.M.S.: 302 (1877); Hemsley in B.M.I.K. 1919: 130 (1919).

Annual or possibly short-lived perennial herbs; purple or purplish-green in colour, prostrate, several or many stems radiating from a tap root, much-branched, the upper side of stems prominently shortly woolly-haired. Leaves broadly obovate, oblong or suborbicular, up to 5 mm long, apex rounded, base somewhat obliquely subcordate, margins glandular-serrate, scattered hairs near margin; petiole less than 1 mm long; stipules interpetiolar, triangular, with a narrow, oblong, blunt, hairy tip, base broad, margins glandular. Cyathia very small, solitary, axillary, mostly on dwarf, congested, lateral branchlets; peduncles short; glands minute, transversely oblong, without appendages. Ovary with white hairs on angles, glabrous between. Fruit triangularly depressed-globose, c 1 × 1.2 mm; angles with white hairs; seeds light pinkish grey, half-elliptic or half-ovate, in cross-section, c 0.8 × 0.4 mm, deeply corrugated. Fig. 41/5–6.

Distr. ALDABRA: W; ASSUMPTION; a pantropical weed.

Notes. Common on bare coral and sandy soils around settlements. Plant eaten and seed locally dispersed by tortoises.

Vernac. 'traînasse'.

5. **Euphorbia pyrifolia** Lam., Encycl. Méth. Bot. 2: 419 (1786); F.M.S.: 303 (1877); Fosberg in K.B. 29: 264 (1974).

E. abbottii Baker in B.M.I.K. 1894: 150 (1894); Schinz in A.S.N.G. 21: 86 (1897); Voeltzkow in A.S.N.G. 26: 550 (1902); Dupont, Report: 40 (1907); Hemsley in B.M.I.K. 1919: 130 (1919); Fosberg in P.T.R.S. B, 260: 218, 225 (1971); Renvoize in P.T.R.S. B, 260: 231 (1971). Type: Aldabra, *Abbott* (K, holo.).

Deciduous, erect, glabrous shrub or small, flat-topped tree, to 3–4 m; bark smooth; branches forking to give a short upper shoot and long lower shoot, branchlets thick. Leaves spirally arranged, crowded at branch-tips, elliptic to elliptic-spathulate and apex acute, to 10 × 2.5 cm, or more rarely, broadly elliptic to obovate and apex rounded, 6–8 × 4 cm, base wedge-shaped to attenuate; petiole short; stipules triangular, 1–2 cm long, soon falling. Flowers dioecious, appearing when plant is bare of leaves, just before new leaves appear; cyathia in cymes, borne in a subterminal pseudo-whorl, cymes up to as many as 9, unequal in length, forking several times, 2 membranous, broadly oblong bracts at each ramification and closely subtending each terminal cyathium, bracts c 2–3 × 1.5–2.5 mm, apex rounded, base pouched and half enclosing node, cyathia sessile in forks and terminal on branchlets of

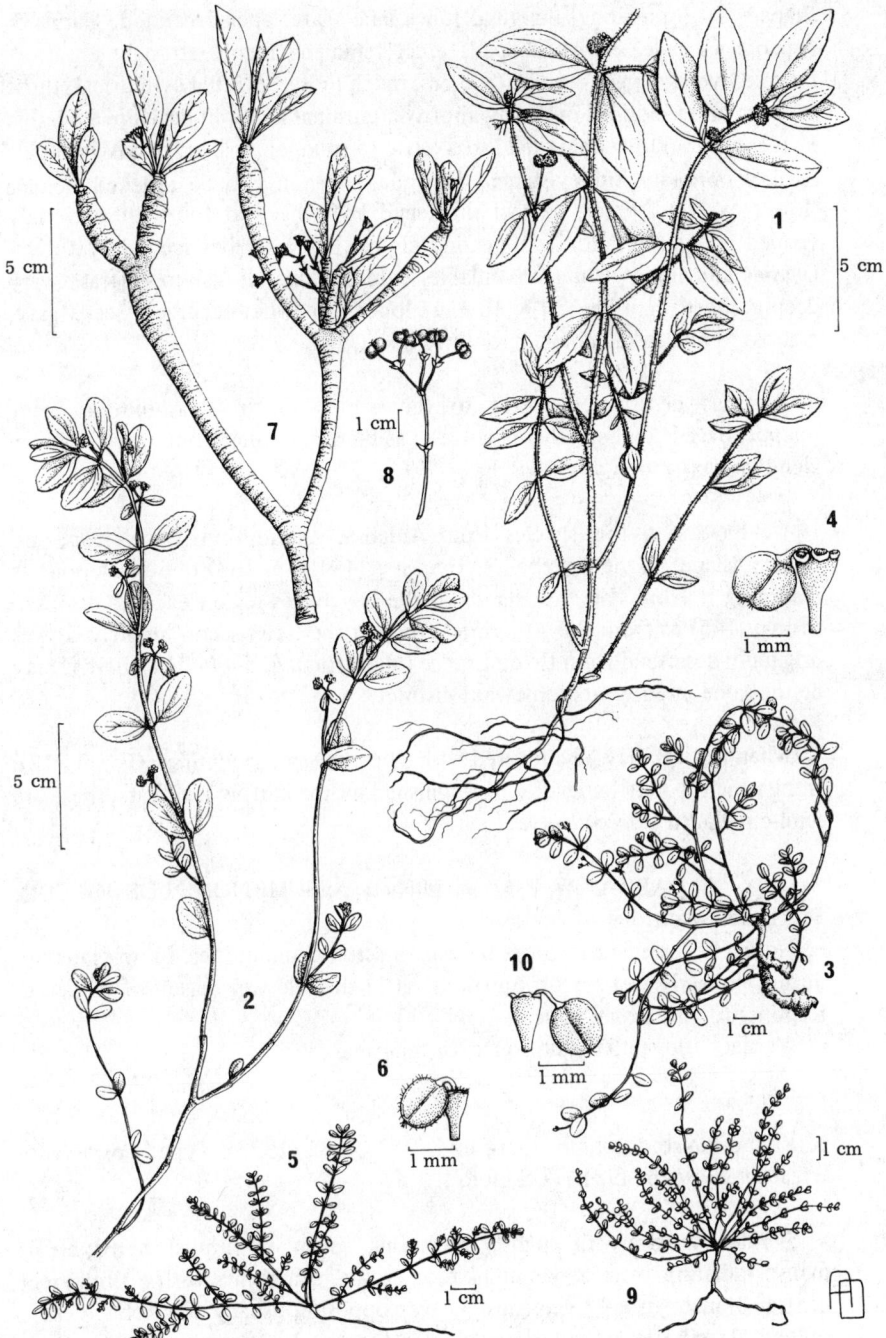

Fig. 41   1, *Euphorbia hirta*, habit. 2, *Euphorbia indica* var. *pubescens*, habit.
3, *Euphorbia mertonii*, habit; 4, female flower. 5, *Euphorbia
prostrata*, habit; 6, female flower. 7, *Euphorbia pyrifolia*, habit;
8, fruits. 9, *Euphorbia stoddartii*, habit; 10, female flower.

cyme. Male cymes short, c 2 × 2 cm, congested, male cyathia cylindrical-bell-shaped, 2–3 mm long, marginal lobes lanceolate, apex lacerated, glands 5, appendages pale, kidney-shaped, erect; stamens many, strongly exserted. Female cymes longer, up to 5 × 5 cm, much looser, cyathia cylindrical, those in forks often reduced or nearly abortive, terminal and well-developed cyathia in forks, 2 mm long, marginal lobes ovate to rounded, frilled; glands with pale or dark, broadly kidney-shaped appendages tending to be reflexed; female flower exserted, erect, pedicel thickened and ringed at top with a strongly crisped collar (perianth?), appearing almost like a smaller second cyathium. Ovaries and intact fruits unavailable, dehisced capsule valves suggest a very deeply lobed capsule, c 4 × 10 mm, lobes somewhat recurved; seeds grey, globose, 3 mm in diameter, smooth. Fig. 41/7–8.

Previous descriptions refer to the glands of the cyathium as being unappendaged. Perhaps this is correct, as there is no sharp boundary between gland and expanded portion.

We have seen this species from Aldabra, Assumption, Cosmoledo, and Cousin Island in the Seychelles. Hemsley (1919, p. 130) quotes Dupont as recording it from "all the islands of the Seychelles region except Gloriosa" and (p. 145) as from the Amirantes, though the latter seems unlikely. It was originally described from Ile de France (Mauritius). *E. daphnoides* Balf.f. may be the same but appears somewhat distinct.

When leafless may be confused with *Operculicarya gummifera* (Family 22), from which it may be readily distinguished by the narrow leaf scars; these are semi-circular in *Operculicarya*.

Distr. ALDABRA; W, P, M, S, Michel; ASSUMPTION; COSMOLEDO; Seychelles, Mauritius.
Notes. Widespread through the island scrub communities. Flowers mainly during the early wet season but buds set in the late wet season may open in response to unseasonal rains.
Vernac. 'tonga', 'tanghin rouge' or 'fangame'.

6. **Euphorbia stoddartii** Fosberg in K.B. 33: 182 (1978). Type: Cosmoledo, Wizard I., *Fosberg* 49816 (US, holo.).

Slender, prostrate or slightly ascending, green or purplish herbs; stems many, radiating from a perennial root crown, sometimes with a tiny, erect central branch, usually glabrous. Leaves opposite, usually somewhat fleshy, oblong to elliptic or suborbicular, mostly 3 × 1–2 mm or smaller, apex round, base unequally subcordate, margins entire, slightly cartilaginous; petiole very short; stipules several at a node, awl-shaped. Cyathia axillary, solitary; peduncles very short; cyathia bell-shaped, less than 1 mm long, glands dark, (4–)5, orbicular to somewhat elongated, appendages absent or very narrow, white. Ovary glabrous or slightly to strongly hairy. Capsule triangular but not sharply so, c 1.2 × 1.2 mm, apex rounded, crowned with short, spreading, thick style branches, base truncate, glabrous or hairy; seed white, grey or pink, oblong, weakly quadrangular in cross-section, c 1 × 0.5 mm, one end truncate, the other bluntly pointed, weakly wrinkled. Fig. 41/9—10.

Distr. ALDABRA: W, M, S, Esprit, Michel; ASSUMPTION; COSMOLEDO: W, M; ASTOVE; endemic.
Notes. Common, especially in sandy places. Plant eaten and seed locally dispersed by tortoises.

3. MARGARITARIA L.f.

Trees and large shrubs; milky sap absent; branching not "phyllanthoid" (phyllanthoid branching has the branches dimorphic, with smaller, often deciduous, flowering branches with 2-ranked leaves, arising from ordinary branches with reduced, spirally arranged leaves). Leaves alternate, spirally arranged, entire; stipules small, scale-like. Flowers dioecious, axillary, often on dwarfed, lateral branchlets, solitary or in fascicles, generally 4-merous. Ovary (3—)4—5-celled, ovules 2 per cell. Fruit a drupe or indehiscent capsule, stone bony to thin and almost parchment-like, usually 2—4(—5)-celled; seeds 2 per cell, aril fleshy, testa bony.

A small genus distributed in tropical America, tropical and subtropical Asia, Madagascar and Aldabra, Mauritius and Australia.

Margaritaria anomala (Baill.) Fosberg in K.B. 33: 185 (1978).

var. cheloniphorbe (Hutch.) Fosberg in K.B. 33:185 (1978). Type: Aldabra, *Abbott* s.n. (K, holo.).

Phyllanthus cheloniphorbe Hutch. in B.M.I.K. 1918: 204 (1918); Hemsley in
   B.M.I.K. 1919: 204 (1919); Fosberg in P.T.R.S. B, 260: 218, 285 (1971);
   Renvoize in P.T.R.S. B, 260: 231 (1971).
P. anomalus sensu Baker in B.M.I.K. 1894: 150 (1894); Schinz in A.S.N.G.
   21: 86 (1897); Voeltzkow in A.S.N.G. 26: 550 (1902); Dupont, Report:
   40 (1907), not strictly (Baill.) Muell. Arg. (1863).

Shrub or small tree, 1—6 m tall, crown broadly rounded; bark dark to light grey or pinkish-grey; branchlets tending to be slightly zigzag and gnarled, nodes enlarged under petiole scars. Leaves broadly obovate, c 5—6(—10) × 3—3.5(—6) cm, apex conspicuously rounded, base narrowed to or slightly decurrent on petiole, margins entire, dull, rather light green, pale beneath; petiole up to 1 cm long, flattened, with somewhat semicircular buds in axils; stipules falling very early, thin, scale-like, oblong or ovate to lanceolate (on young shoots), leaving short scars by leaf scar. Flowers dioecious. Male flowers: in "dense" to few-flowered axillary fascicles, subtended by lanceolate, translucent bractlets; pedicels 3—4 mm long; perianth segments yellow, 4, in 2 pairs, outer pair broadly oblong, inner orbicular, less than 1 mm long, margins translucent; stamens 4, opposite and about equalling perianth segments, from inside disk, anthers broadly oblong; disk glandular, fused to bases of perianth segments. Female flowers: several in an axil of a young leaf, subtended by a broadly oblong or suborbicular bractlet c 3 mm long; pedicels slender, 7–10 mm long, thickening near summit; perianth segments 4, in 2 pairs, ovate-orbicular, outer pair smaller, c 2 mm long, margins translucent; disk saucer-shaped. Ovary broadly ovoid, slightly shorter than perianth; stigmas 4, subsessile or style very short, fleshy, thickly awl-shaped, spreading to reflexed. Fruit subglobose to globose, 5–7 × 5 mm, hard-drupaceous, calyx sub-persistent, stone 2-celled; seeds 4 × 3 mm, 3-angled, dorsally convex. Fig. 40/7–8.

Vegetatively this species can be confused with *Ludia mauritiana* (Family 6/2) from which it may be distinguished by the distinct cartilaginous margin to the leaves and from *Sideroxylon inerme* (Family 41) by the distinctly netted appearance to the undersurface of the leaves, with the viens at c 80° to the midrib instead of 45° in *Sideroxylon*. During the dry season it may also be confused with *Erythroxylum acranthum* (Family 11) which has compressed instead of round branchlets.

Distr. ALDABRA: W, P, M, S, Esprit, Michel, lagoon islets; COSMOLEDO: M; ASTOVE; endemic.

Notes. A constituent of the inland scrub communities. Flowers during the early wet season or in response to unseasonal rains. Seeds locally dispersed by blue pigeon.

Vernac. 'si là'.

4. PEDILANTHUS Neck. ex Poit.

Shrubs, tending to be fleshy; sap milky. Leaves alternate; stipules gland-like. Flowers borne in narrow and asymmetrically prolonged, "shoe-shaped" cyathia, the elongate part (appendix) extending laterally from the base of the main part or tube; glands 4; male flowers numerous, pedicelled, reduced to single, naked stamens; 1 female flower present, reduced to a single ovary jointed to pedicel, with or without a reduced 3-lobed perianth. Ovary 3-celled, 1 ovule per cell; styles united into a column, stigmas 3, free. Fruit a 3-celled capsule, splitting into 3 cocci; seeds 1 per coccus.

A small tropical American genus.

Pedilanthus tithymaloides (L.) Poit. in Ann. Mus. Nat. Hist. Nat. 19: 389 (1812).

Euphorbia tithymaloides L., Sp. Pl.: 453 (1753).

Erect, dark green, slightly pubescent, succulent shrub; much branched, smaller branchlets noticeably zigzag. Leaves alternate, borne in 2 rows on outer angles of zigzag stem, broadly ovate, 4(–10) × 2.5–3(–7) cm, apex acuminate, base broadly acute, obtuse to subtruncate or subcordate, margins obscurely crenate to entire; subsessile or petiole very short; stipules dark, very small. Cyathia borne on condensed, terminal or axillary, dwarf branchlets; peduncles slender, subtending bracts soon falling, whole inflorescence somewhat pubescent. Cyathium red, lanceolate in outline, 10–15 mm long, with a hump on upper side; style and stamens exserted from tip of cyathium. Fruit subglobose, 7 × 7 mm, cocci ovoid, 4 mm in diameter; seeds grey, top-shaped, 2 × 1.2 mm with flared, reddish-brown base (fruit and seeds described from non-Aldabra material).

Distr. ALDABRA: W; ASSUMPTION; COSMOLEDO: M; ASTOVE; a native of tropical America, now widely cultivated throughout the tropics.

Notes. Planted around settlement sites. Few observations but apparently flowers in mid-late dry season.

Vernac. 'bois mal gaz'.

5. PHYLLANTHUS L.

Herbs, shrubs or rarely trees; milky sap absent; branching various, but branches often phyllanthoid, i.e. dimorphic, with spreading, slender, fertile branchlets with 2-ranked leaves originating from ordinary, usually ascending, branches with spirally arranged leaves which are often reduced to scale-like leaves (cataphylls). Leaves alternate, simple, entire, in some species closing together at night, resembling compound leaves; petiole usually short; stipules small. Flowers monoecious or rarely dioecious, strongly differentiated, axillary, solitary or fasciculate, disk present. Male flowers: perianth segments (4–)5–6; stamens 3–6, filaments free or united in a column. Female flowers: perianth segments 4–6; ovary usually 3-celled, ovules 2 per cell, styles free or united, entire or 2-fid, stigmas awl-shaped or enlarged. Fruit usually a capsule, more rarely baccate frequently splitting into cocci, or indehiscent; seeds 2 per cell, with 2 flat sides, the dorsal side vertically convex, not at all fleshy, without a caruncle.

An enormous, highly variable pantropical genus.

1. Shrubs 1–2 m or more tall   **2. P. casticum**
1. Herbs, much smaller
  2. Plants purple,very delicate; male flowers with pedicels exceeding those of female; stamen filaments free distally; glands or disk-lobes of female flowers 3, very thin, soon disappearing   **4. P. mckenzei**
  2. Plants green, not noticeably delicate; male flowers with pedicels subequal to or shorter than female; stamen filaments united to top; glands of female flowers 5 or 6
    3. Main stem leafless, leaves oblong, rounded at both ends; solitary female flowers distal; perianth segments broadly white-margined   **1. P. amarus**
    3. Main stem leafy, leaves elliptic, narrowed to base; solitary female flowers proximal; perianth segments narrowly or scarcely white-margined
                                 3. P. maderaspatensis

Phyllanthus aff. *urinaria* of Schinz in A.S.N.G. 21: 85 (1897) and Voeltzkow in A.S.N.G. 26: 550 (1902) could be of any of the 3 herbaceous species; *Voeltzkow* 40 referred to by Schinz and Voeltzkow is *Sarcostemma viminale* (Family 45/3), which is unlikely to have been confused with *Phyllanthus*. Dupont, Report: 40 (1907) also refers to *Phyllanthus urinaria* L. on Aldabra, which likewise cannot be correctly indentified.

1. Phyllanthus amarus Schum., Beskr. Guin. Pl.: 421 (1827).

P. niruri sensu F.M.S.: 309 (1877) partly? not L. (1753).

Erect herb, glabrous to obscurely roughened, light or slightly dull green, branching phyllanthoid, more or less horizontal. Leaves oblong, up to c 6 × 3 mm, apex and base rounded; periole short; stipules ovate-lanceolate, subentire. Flowers axillary, usually in fascicles of 2–3, 1–2 male, 1 female, or distal female solitary, pedicels present; perianth segments 6, oblong to oblong-ovate, alternate segments slightly wider, margins broad, white-margined. Male flowers: stamens 3, filaments united, anthers sessile on column. Female flowers: larger than male, styles 3, short, spreading, persisting in fruit. Fruit depressed-globose, 2–2.5 mm across, smooth; seeds 6, light reddish tan, 3-sided, dorsally convex, 1 × 0.5 mm, convex surface with 5–6 longitudinal ridges, and fine cross-lines, flat sides with 3–4, concentric, low ridges and fine cross-lines. Fig. 41/1.

Distr. ALDABRA: W; ASSUMPTION; COSMOLEDO; M: ASTOVE; a pantropical weed.
Notes. Established as a weed around settlement sites.
Vernac. 'curanellia'.

2. Phyllanthus casticum Willem. f. in Usteri, Neue Ann. der Bot. 18: 55 (1796); F.M.S.: 311 (1877); Hemsley in B.M.I.K. 1919: 130 (1919).

Spreading, mound-shaped shrub to 4 m high; branches slender, arching and drooping, leafless; flowering branchlets very slender, 15–25 cm long, leafy or occasionally leafless or with reduced leaves, leafless branchlets up to 10 cm long, both ultimately falling, leaves also falling, both forms found on same plant, on some only the leafy branchlets. Leaves ovate to elliptic, up to 5 × 2 cm, apex bluntly sub-acute to obtuse, base obtuse; petioles c 2–2.5 mm long; stipules firm, scale-like, minute, ovate. Flowers red, dark red or reddish, in fascicles of few −10(−12) at a node, in leaf axils or on naked dwarf branchlets, male and female flowers in same fascicle; pedicels 1–2(−3) mm long, those of female flowers thickening and elongating to 2–4 mm in fruit, thread-like, usually pubescent, sometimes glabrous; bracteoles scale-like, at base of pedicels. Male flowers: perianth segments 4 or usually 5, broadly oblong or suborbicular, strongly overlapping, glabrous or pubescent; stamens 4–5, filaments variously partially united. Female flowers: perianth segments similar, styles 3–5, variously 2-fid, tending to be recurved or twisted when dry, united or not at base, even on same plant. Fruit reddish, pea-like,

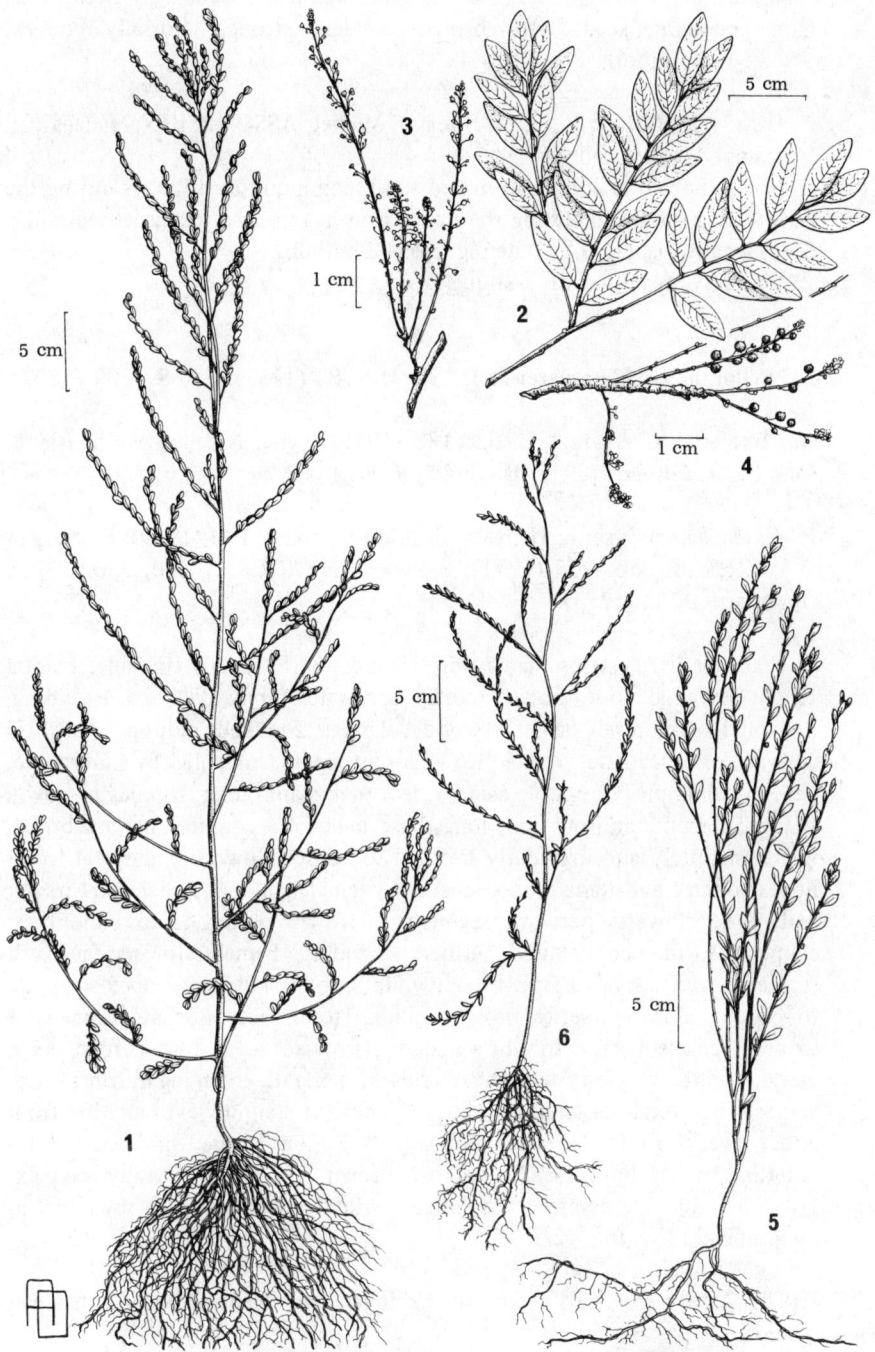

Fig. 42   1, *Phyllanthus amarus*, habit. 2, *Phyllanthus casticum*, leaves;
3, flowering branches; 4, fruiting branches. 5, *Phyllanthus*
maderaspatensis, habit. 6, *Phyllanthus mckenzei*, habit.

somewhat depressed-globose, c 4 × 2.5 mm, 3–5 celled, slightly fleshy with hard endocarp; seeds dark brown, 3-sided, strongly dorsally convex, 1 × 0.5 mm, smooth. Fig. 42/2–4.

Distr. ALDABRA: W, P, M, S (east), Michel; ASSUMPTION; Madagascar, Mascarene Is., Seychelles.

Notes. Locally common in mixed scrub communities. Flowers during the early wet season and during the dry season in response to unseasonal rains. Seeds locally dispersed by blue pigeons and bulbuls.

Vernac. 'kirganellie' or 'castique'.

3. Phyllanthus maderaspatensis L., Sp. Pl.: 982 (1753); F.M.S.: 309 (1877).

var. **frazieri** Fosberg in K.B. 33: 188 (1978). Type: Aldabra, South Island, Cinq Cases, *Fosberg* 48977 (US, holo., K, iso.).

P. maderaspatensis sensu Hemsley in B.M.I.K. 1919: 130 (1919); Fosberg in P.T.R.S. B, 260: 225 (1971); Renvoize in P.T.R.S. B, 260: 230 (1971), not strictly L. (1753).

Glabrous herb; stems many, very slender, 5–15 (–20) cm long, from a tough perennial root crown, mostly prostrate, rarely erect or ascending, branched, very rarely slightly woody. Leaves 2-ranked, elliptic to slightly obovate or lanceolate, 4–6 × 1.6–2.5 mm, apex acute, slightly mucronate, densely white-pitted; petiole slender, less than 1 mm long; stipules narrowly triangular-acuminate, c 1 mm long, base unequally cordate, margins broad, white, minutely and irregularly toothed to entire. Flowers axillary, at lower nodes solitary and female, more distally 1 female and 1 or 2 male flowers per axil. Male flowers: perianth segments narrowly oblong, apex subobtuse; stamens 3, filaments united, anthers spreading. Female flowers: perianth segments with 1 series sessile, broadly elliptic to broadly ovate, apex subacute to obtuse, alternate series broadly elliptic to broadly obovate, somewhat longer, apex subacute to obtuse, base narrowed to a short, broad claw, margins white, very narrow or nearly absent, perianth enlarging in fruit; styles very short, broad, ascending, 6 linear glands (or staminodes) radiating from under ovary. Fruit depressed-globose, to 3 mm across, smooth, with 6 radiating paler lines; seeds tan to brown, 3 sided, dorsally convex, 1 × 0.5 mm, transversely wrinkled, wrinkles arranged side by side in longitudinal rows. Fig. 42/5.

The Astove material, though in habit near var. *maderaspatensis* has the seeds of var. *frazieri*.

Distr. ALDABRA: W, M (east), S; ASTOVE; endemic.

Notes. It is abundant on South Island in the Cinq Cases area, on rough limestone, where it grows in pits and crevices and on platin, where it is heavily grazed by tortoises. It is much less common on Middle Island and West Island, and on Astove. On these areas it is not much affected by grazing.

In the Cinq Cases area and at Anse Mais, where tortoises are abundant, all plants seen were very depressed, with many prostrate stems. On Middle and West Islands and Astove where tortoises are fewer, or lacking, all plants seen were erect or ascending, with a developed main stem with branches. The difference in habit seems obviously the result of difference in grazing pressure. Flowers throughout the year. Plant eaten and seed locally dispersed by tortoises.

4. Phyllanthus mckenzei Fosberg in K.B. 33: 189 (1978). Type: Aldabra, South Is., NW Cinq Cases, *Fosberg* 49159 (US, holo., K, iso.).

Phyllanthus sp., Fosberg in P.T.R.S. B, 260: 223, 225 (1971).

Delicate, annual herb, purplish, glabrous; either main stem slender, to 20(−35) cm tall with scattered, very slender, fertile branches or suppressed, with several, prostrate, fertile branches radiating from slender root-crown. Leaves alternate, subsessile, oblong or elliptic, c 4 × 2 cm; apex and base rounded; stipules scale-like, white or brown, narrowly to broadly triangular, c 0.5 mm long, apex acuminate. Flowers axillary, 2 or rarely 3, with 1 female and/or 2 male flowers, flower clusters distal on branches, lower or towards base mostly solitary female flowers, or none; pedicels of male flowers long, thread-like, of female flowers short. Male flowers: perianth segments 6, c 1 mm long, alternate segments ovate or narrowly oblanceolate, subacute and broadly oblong or obovate, apex obtuse to rounded, broadly white-margined; stamens 3, filaments united in lower half, free parts radiating and bearing anthers. Female flowers: perianth segments as in male flowers but inflated with age; glands 3, membranous, broadly spathulate-subtruncate, opposite longer perianth segments; styles 3, short, flattened, radiating, stigmas expanded, flat, entire to very slightly 2-fid, appressed to developing or mature fruit. Fruit strongly depressed-globose, to 1.5 × 0.6−0.8 mm, with 6 radiating lines, septicidally dehiscent into 3 cocci which then separate into twisted valves; seeds dull orange to salmon or dull chocolate brown, 2 sides flat, inner angle sharp, strongly longitudinally convex, inner angle 0.4−0.6 mm long, Fig. 42/6.

Distr. ALDABRA: S; COSMOLEDO: W, NW; ASTOVE; endemic.
Notes. An ephemeral herb known only from South Island of Aldabra and from Wizard and North West Islets, Cosmoledo Atoll and from Astove. It is locally common on desiccating marl flats, and in pits and crevices of slightly elevated coral limestone. The depressed, stemless habit of most of the South Island plants may most likely be due to heavy grazing by tortoises.
Vernac. 'canelli rouge'.

6. RICINUS L.

Glabrous shrubs. Leaves alternate, palmately lobed and veined, peltate; petiole present; stipules united, falling. Flowers monoecious, in terminal racemes, male and female flowers in same racemes, male below female. Male flowers: petals absent; stamens many. Female flowers: petals absent; ovary 3-celled, ovules 1 per cell, styles 3, deeply 2-fid. Fruit splitting elastically into 2-valved cocci; seeds with caruncle.

A genus of 1 pantropical weedy and cultivated species.

Ricinus communis L., Sp. Pl.: 1007 (1753); F.M.S.: 316 (1877); Schinz in A.S.N.G. 21: 86 (1897); Voeltzkow in A.S.N.G. 26: 550 (1902); Renvoize in P.T.R.S. B, 260: 228 (1971).

Shrub or, when young, a coarse herb, to several m tall. Leaves 7–11-lobed, to 15 cm or more across, lobes sharp, toothed; petiole long; stipules forming a cap enclosing bud, membranous, falling, leaving a scar completely encircling node. Racemes to 13 cm long at maturity. Flowers green. Ovary covered by non-rigid, prickle-like processes; stigmas 6, red, awl-shaped. Fruit 3-lobed, 12–15 × 12–15 mm, covered by non-rigid, prickle-like processes; seeds mottled brown and grey, obovoid, somewhat compressed, 8 × 6 mm, smooth, with a conspicuous fleshy knob (caruncle) at the small end.

Distr. ALDABRA: W; ASSUMPTION; COSMOLEDO: M; ASTOVE; pantropical, cultivated and widely naturalized.

Notes. Occurs sparingly around settlement sites. Seeds poisonous but yielding an oil used in medicine and as a lubricant for machinery.

Vernca. 'palma Christi', 'iricin', 'tantan' or 'caster-oil plant'.

63. URTICACEAE

Habit various. Leaves simple, alternate or opposite; stipules usually present. Inflorescence usually cymose, capitate or rarely spicate, or flowers solitary; flowers usually unisexual, very reduced. Perianth parts 4–5, sepaloid, free or united. Male flowers: stamens opposite perianth parts, in bud bent inwards, spreading the pollen by elastically straightening and dehiscing. Female flowers: ovary superior or inferior, ovules 1; apparently basal. Fruit a nut or drupe.

A large mostly tropical and subtropical family.

1. Monoecious herbs	**1. Laportea**
1. Dioecious trees	**2. Obetia**

1. LAPORTEA Gaud.

Annual or perennial herbs, rarely shrubs; hairs irritant. Leaves simple, toothed, alternate; petiole present; stipules united at the base, free at the apex, not persistent. Male and female flowers on the same plant, rarely on separate plants; flowers small, in loose or dense clusters. Male flowers: perianth segments 4–5, equal; stamens 4–5. Female flowers: perianth segments 4, unequal; ovary ovoid. Achenes ovoid, compressed, sessile or reflexed on a short stalk.

A genus of 22 species, throughout the tropics.

Laportea aestuans (L.) Chew in Gard. Bull. Singapore 21: 200 (1965).

Urtica aestuans L., Sp. Pl. ed. 2,: 1397 (1763).

Small, delicate, annual herbs, up to 10 cm high. Leaves ovate, 2–3.5 × 1.5–2.5 cm, apex acute, base shallowly cordate, margins toothed, teeth ciliate, membranous, hairy; petioles 1–5 cm long, flexuous. Male and female flowers on the same plant; inflorescences up to 2 cm long, shortly stalked, pale; male and female flowers yellow-green, in dense or loose clusters. Male flowers: up to 1.5 mm long with a few glandular hairs from the apex. Female flowers: 0.5 mm long, lateral perianth segments enclosing the ovary, dorsal and ventral segments much smaller. Achenes 1–2 × 1–2 mm, readily disarticulating from the pedicels, more or less covered by the lateral perianth segments. Fig. 43/1–3.

This description applies only to Aldabra plants; elsewhere this species is often much taller with larger leaves and longer inflorescences.

Distr. ALDABRA: S (east); pantropical, but absent from Pacific islands.
Notes. Found growing in rock crevices near the south-east coast, *Renvoize* 978 (K, US) & 1137 (L). Flowers during the mid-late wet season.

2. OBETIA Gaud.

Hairy shrubs or small trees. Leaves alternate, margins lobed or coarsely toothed; petiole present; stipules large, free. Male and female flowers on separate trees, in stalked clusters in the axils of the leaves. Male flowers: perianth 5-lobed, lobes ovate, obtuse, pubescent; stamens 5; ovary rudimentary. Female flowers: perianth 4-lobed, lobes unequal, oval or round, glabrous; ovary globose, becoming oblique at maturity, stigma subsessile. Fruit a single ovoid, dry achene.

A genus of 5 species in tropical and southern Africa, Madagascar and the Mascarenes.

Obetia ficifolia Gaud., Bot. Voy. Bonite, fig. 82 (1844); F.M.S.: 273 (1877).

O. morifolia sensu Hemsley in B.M.I.K. 1919: 131 (1919), not Baker (1883).

Stout spreading trees, to 6 m high; bark grey, smooth, soft. Leaves and inflorescences usually confined to the branch tips, which are often swollen; old leaf and stipule scars prominent. Leaves light green, ovate to broadly ovate, 6–20 × 4–20 cm, 3-lobed, rarely 5-lobed, base cordate, margins toothed, membranous, usually densely pubescent below, sparsely hairy above, petioles up to 14 cm long, pubescent; stipules reddish-brown, ovate to narrowly-ovate or lanceolate, papery, up to 15 mm long, rarely to 20 cm long. Male flowers pink, 2–3 mm across. Female flowers yellowish-green, 1.5–2 mm across. Achene compressed-ovoid, c 1 × 0.8 mm, included in the swollen, persistent perianth. Fig. 43/4–6.

Although very similar to *O. morifolia* Baker, which occurs on Madagascar, *O. ficifolia* is distinguished by the absence of stiff hairs on the petioles and inflorescence branches, which are a feature of *O. morifolia*.

Distr. ALDABRA: W; Rodrigues and Mauritius.

Notes. Only known on Aldabra from a few trees on the margins of the mixed scrub bordering the Casuarina Grove at Settlement; *Dupont* 127 (K); *Fryer* 17 (K); *Renvoize* 1357 (female) & 1358 (male) (both K). Leafless during the dry season; flowers during the wet season. Actively spreading; seeds locally dispersed by bulbul.

Vernac. 'figue marron'.

64. MORACEAE

Trees, shrubs or herbs; milky latex usually present. Leaves alternate, rarely opposite, simple, entire to variously toothed or lobed; stipules present. Flowers small or minute, unisexual, in spikes or heads with variously enlarged receptacles. Perianth said to be in 2 whorls of (2–)4(–6), reduced, scale-like segments, but these variously reduced and frequently seem to be only in one series of similar parts. Male flowers: stamens equal in number and opposite the perianth segments, erect. Female flowers: ovary 1, superior to inferior, 1-celled, ovules 1, pendulous; styles usually 2. Fruit drupaceous but variously united into multiples.

Large, mostly tropical family.

1. Leaves rolled in bud; stipules clasping; receptacle invaginated to form a sac, enclosing numerous achenes   **1. Ficus**
1. Leaves folded in bud; stipules not clasping; fruit globose, seeds 1
                                                   2. Maillardia

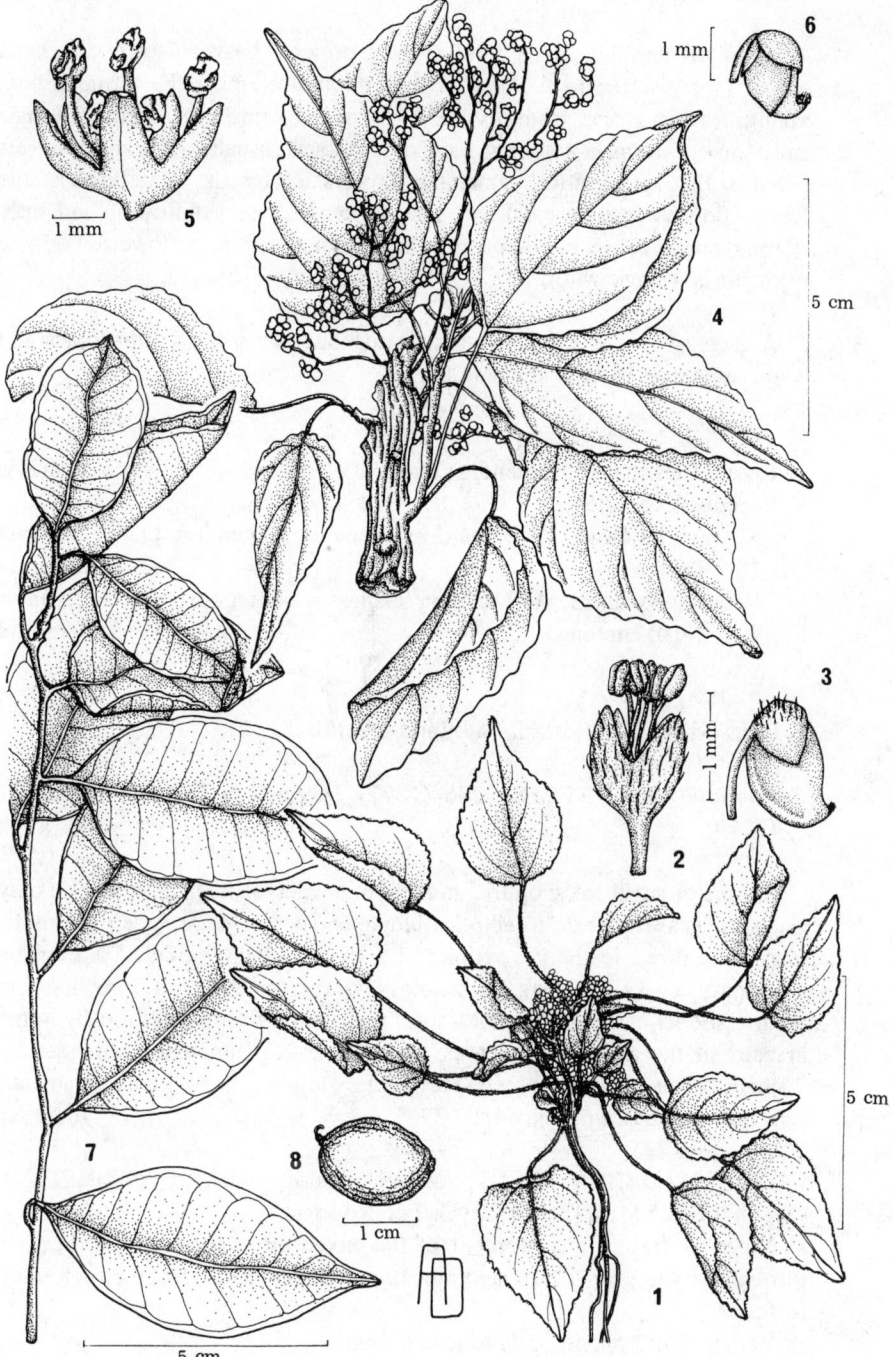

Fig. 43 1, *Laportea aestuans*, habit; 2, male flower; 3, female flower.
4, *Obetia ficifolia*, habit; 5, male flower; 6, female flower.
7, *Maillardia pendula*, habit; 8, fruit.

1. FICUS L.

Trees or climbing shrubs; milky latex present. Leaves usually alternate, entire, lobed or toothed; rolled in bud; stipules clasping the terminal bud, falling, leaving a scar around stem. Flowers tiny, unisexual, with both male and female flowers in a hollow, sac-like receptacle, usually a few male flowers towards the apical pore, remaining flowers all female, sometimes sterile female flowers (gall-flowers) may also be present, or gall-flowers and male flowers may occur on a separate plant from the female flowers. Perianth segments in a single whorl.

A large genus of c 800 species throughout the tropics, sub-tropics and warm temperate parts of the world.

1. Receptacles stalked, subtending bracts 2   **1. F. avi-avi**
1. Receptacles sessile
  2. Receptacles large, 10–18 mm in diameter, subtending bracts 3; leaves
    10–30 cm long   **2. F. nautarum**
  2. Receptacles small, 4–7 mm in diameter, subtending bracts 2; leaves
    4–6 (–10) cm long   **3. F. reflexa**

1. Ficus avi-avi Bl., Bijdr. Fl. Ned. Ind.: 446 (1825).

F. consimilis Baker, F.M.S.: 286 (1877); Hemsley in B.M.I.K. 1919: 131 (1919).

Shrubs or small trees, up to 5 m high; branchlets slender, pubescent, rarely glabrous. Leaves elliptic to elliptic-oblong, 6–14 × 3–6 cm, base subcordate, margins entire, leathery; petioles 1.5–5 cm long; stipules lanceolate-acuminate, 3–12 cm long on young vigorous growth, 1–1.5 cm long on mature shoots, glabrous or pubescent, falling early. Receptacles usually borne in pairs in the axils of the leaves, on stalks 5–13 mm long, subtended by 2 small, orbicular, lobed bracts; receptacles globose, up to 1 cm in diameter, glabrous or pubescent. Fig. 44/1.

Distr. ALDABRA: W, P, M, S, Esprit, Michel; ASSUMPTION; COSMOLEDO: M; ASTOVE; Seychelles, Rodrigues.
Notes. A frequent constituent of the mixed scrub communities. Flowers throughout the year if sufficient moisture available. Seed locally dispersed by tortoises.
Vernac. 'multipliant' or 'la fouche ti feuilles'.

2. Ficus nautarum Bak., F.M.S.: 285 (1877) & in B.M.I.K. 1899: 151 (1894); Schinz in A.S.N.G. 21: 83 (1897); Voeltzkow in A.S.N.G. 26: 549 (1902); Dupont, Report: 40 (1907); Hemsley in B.M.I.K. 1919: 131 (1919).

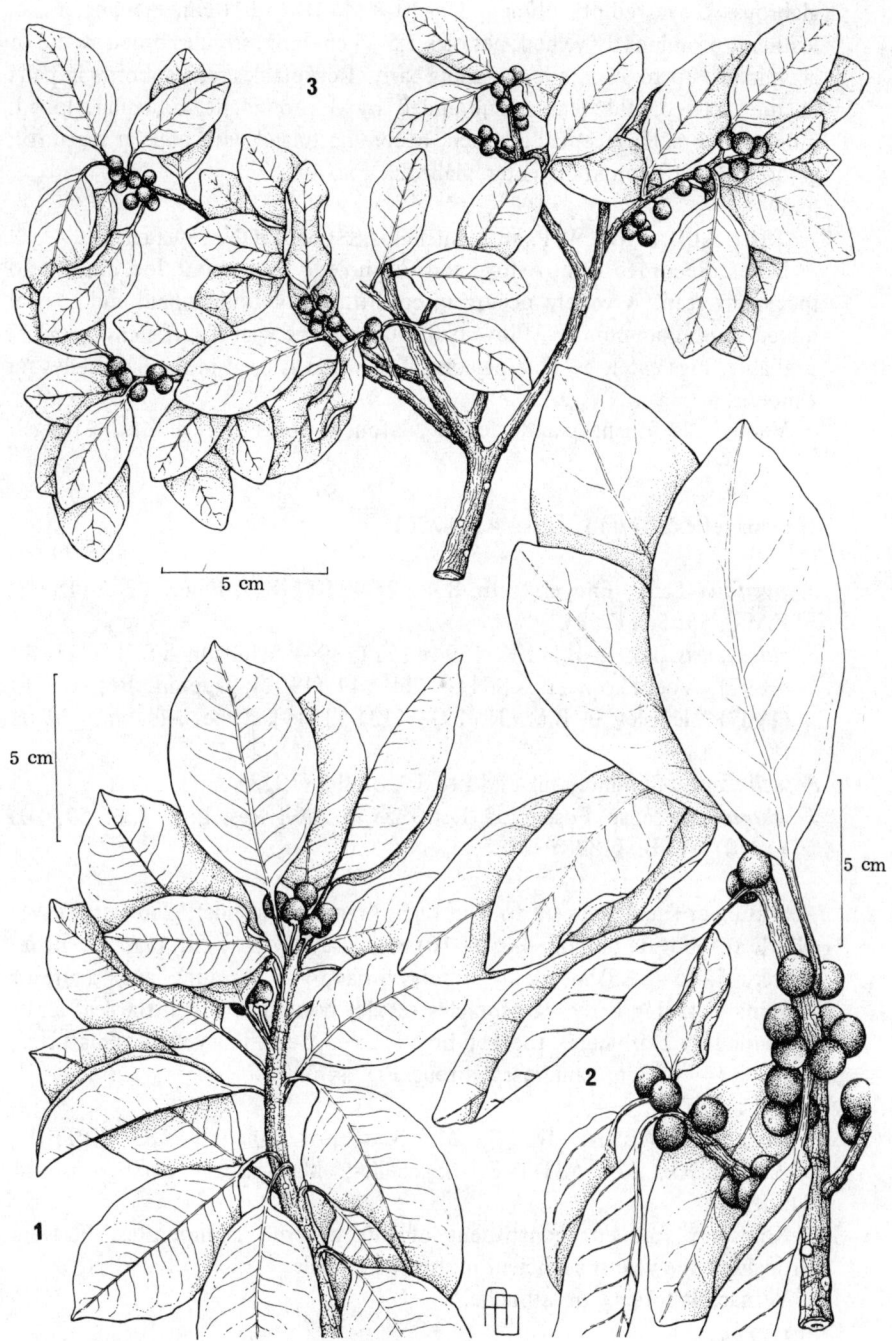

Fig. 44 1, *Ficus avi-avi*, habit. 2, *Ficus nautarum*, habit. 3, *Ficus reflexa*, habit.

Shrubs or spreading robust trees, up to 10 m high; branchlets stout, glabrous. Leaves elliptic-oblong, 10–30 × 4–10 (–14) cm, margins entire, leathery, prominently veined; petioles 1.5–5 cm long; stipules broad, acute or acuminate, up to 15 mm long, falling early. Receptacles sessile, borne in pairs in the axils of the leaves, subtended by 3 prominent, orbicular, lobed, pubescent bracts; receptacles green, becoming tinged with pink at maturity, globose, 10–18 mm in diameter, glabrous. Fig. 44/2.

Distr. ALDABRA: W, P, M, S, Michel; ASSUMPTION; Seychelles.

Notes. Recorded from Astove and Cosmoledo by Dupont, loc. cit. but no specimens seen. A widely occurring constituent of both inland and coastal mixed scrub communities. Flowers throughout the year if sufficient moisture available. Figs eaten by blue pigeons. The wood is used in the Seychelles for canoes.

Vernac. 'la fouche grande feuille', 'afouche rouge' or 'la fouche rouge'.

3. Ficus reflexa Thunb., Diss.: 41 (1786).

F. pyrifolia Lam., Encycl. Méth. Bot. 2: 497 (1788); F.M.S.: 285 (1877); F.M.C. 55: 56 (1952).

F. aldabrensis Bak. in B.M.I.K. 1894: 151 (1894); Schinz in A.S.N.G. 21: 83 (1897); Voeltzkow in A.S.N.G. 26: 549 (1902); Dupont, Report: 40 (1907); Hemsley in B.M.I.K. 1919: 131 (1919). Type: Aldabra, *Abbott* (K, holo.).

F. sechellarum Summerh. in B.M.I.K. 1928: 393 (1928).

F. thonningii sensu Fosberg & Renvoize in Atoll Res. Bull. 136: 59, 103 (1970), not Bl. (1838).

Shrubs or small trees up to 5 m high; branchlets slender, glabrous. Leaves elliptic to obovate-oblong, 3–5 (–10) × 1.5–2.5 (–5.5) cm, margins entire; petioles 5–15 (–25) mm long; stipules broad, 4–8 mm long, apex acute or acuminate, falling early. Receptacles sessile, borne in pairs in the leaf axils, subtended by 2 orbicular, toothed bracts; green, becoming yellow at maturity, globose, 4–7 mm in diameter, glabrous. Fig. 44/3.

Distr. ALDABRA: W, P, M, S, Esprit, Michel; ASSUMPTION; COSMOLEDO: M; ASTOVE; Seychelles, Madagascar, Rodrigues and Mauritius.

Notes. A frequent constituent of mixed scrub communites. Flowers throughout the year if sufficient moisture available.

Vernac. 'la fouche' or 'afouche'

2. MAILLARDIA Frap. & Duch.

Trees, milky latex present. Leaves alternate, entire, leathery, shiny; stipules linear-lanceolate or triangular-ovate, readily falling. Male and female flowers on separate trees. Male flowers: tiny, sessile, in dense, solitary spikes;

perianth of 4 segments; stamens 4. Female flowers: solitary, stalked, axillary; perianth closely applied to and enclosing the ovary, orifice 4-toothed. Fruit ovoid or globular, fleshy; seeds 1.

A genus of 6 species, mainly in Réunion and Madagascar.

Maillardia pendula Fosberg in K.B. 29: 266, fig. 2 (1974). Type: Aldabra, Takamaka Grove, *Whitton* 94, (K, holo., US, iso.,).*

Small, slender trees, up to 10 m tall; branches slender, pendulous. Leaves elliptic to elliptic-oblong, 6–9 × 2.5–3.5 cm, apex abruptly and narrowly elongated; petioles 8–10 mm long; stipules triangular-ovate, 1–1.5 mm long, readily falling. Male flowers: in dense, globose clusters up to 75 mm in diameter; perianth segments yellow-tinged, 1 mm long, membranous; anthers 1 mm long, filaments 1.5 mm long. Female flowers: not known. Fruit reddish-brown, ovoid, 12–15 × 7–8 mm, fruiting pedicel 3 mm long; seed 1, oblong-oval, 9 × 5 mm, strongly inrolled – grooved along 1 side. Fig. 43/7–8.

Distr. ALDABRA: S; endemic.
Notes. Only known from a few trees in Takamaka Grove, where it was locally common but due to recent vegetation changes may now be extinct. Few observations but apparently flowers during the wet season. Fruit eaten by blue pigeons.

65. CASUARINACEAE

Trees; branchlets cylindrical, jointed, striated, functioning as leaves; true leaves reduced to whorls to minute scales at the nodes. Flowers monoecious or dioecious, much reduced, arranged in catkins, wind pollinated. Perianth absent. Male catkins cylindrical, jointed. Female catkins developing into woody cone-like structures formed from thickened, hardened, floral bracts. Fruit a samara with a single wing.

A family of 1 genus (by some regarded as 2), principally Australasian.

CASUARINA L.**

Characters as for the family.

A small genus, principally Australian, with several species in New Caledonia, New Guinea and Malesia; 1 widespread in the Indo-Pacific region.

---

* Berg in Bull. Jard. Bot. Nat. Belg. 47: 376 (1977) regards *Maillardia pendula* as a synonym of *M. montana* Léandri from Madagascar, Comoros and Aldabra.
** Botanists differ as to whether *Casuarina* should be attributed to Linnaeus (1759) or Adanson (1763).

Casuarina equisetifolia L., Amoen. Acad. 4: 143 (1759); F.M.S.: 294 (1877); Schinz in A.S.N.G. 21: 81 (1897); Voeltzkow in A.S.N.G. 26: 549 (1902); Dupont, Report: 40 (1907); Hemsley in B.M.I.K. 1919: 131 (1919); Vesey-Fitzgerald in J. Ecol. 30: 11 (1942); Fosberg in P.T.R.S. B, 260: 221, 229 (1971); Renvoize in P.T.R.S. B, 260: 229, 230 (1971)*.

Tree, up to 15 m tall; heart-wood very hard and heavy; branchlets jointed, joints 1 cm long, 1 mm or less thick, with 6—8 striae. Leaves in whorls of 6 to 8. Flowers monoecious to dioecious. Male flowers in cylindrical elongated catkins. Female flowers in shorter, top-shaped catkins; styles maroon, slender. Fruiting catkins cylindrical to globose, c 2 × 1 cm; fruit oblong-oval, 7 × 2.5 mm, with a transparent wing. Fig. 39/3—5.

Distr. ALDABRA: W, P, M, S, Esprit, Michel, Moustique; ASSUMPTION; COSMOLEDO: W, M; ASTOVE; an Indo-Pacific strand plant.

Notes. Locally dominant along the coast. Flowers throughout the year if sufficient moisture available. Fruit eaten by turtle dove, white-eyes and fody, also water-dispersed (still viable after floating for more than 8 weeks).

A species well adapted to the rigorous and changing habitats on small islands, widespread in the western Indian Ocean Islands. It is difficult to know if it is native or introduced by man on a particular island. Because the oldest trees seem to be found in areas formerly occupied by man and to have given rise to spreading colonies of younger ones, with the smallest trees on the peripheries, we are inclined to regard this species as a human introduction in the Aldabra Group. Few of the native Aldabra plants seem able to thrive in its shade and its spread on the island is a matter of concern. A careful study of this question is needed.

Vernac. 'filao' or 'cêdre'.

---

* The correct name of this species, according to a strict interpretation of the International Code of Botanical Nomenclature, is *Casuarina litorea* L., as this name was effectively published in 1754, validated by reference to a previously effectively published description. However, this interpretation is not generally accepted, and its application would create havoc by replacing the names of a goodly number of well known plants. Hence, since there is the difference of opinion, it is my intention to propose the rejection of the Stickman Dissertation on Rumphius Herbarium Amboinense and Flora Ambinense for purposes of priority at the next International Botanical Congress. In anticipation of this we are using the later name, *Casuarina equisetifolia* L.   (F.R. Fosberg)

66. ORCHIDACEAE

Herbs, rarely scramblers or somewhat shrubby; roots often thickened, aerial roots, if present, with spongy, water-absorbent, outer layer; vegetative stems elongated, with several or many nodes, or strongly shortened and bulbous-thickened with several nodes, or reduced to 1 swollen internode (pseudobulb) bearing a leaf. Leaves usually alternate or basal, often 2-ranked, rarely a single, opposite pair or a whorl; petiole, if any, usually sheathing at base, sheath closed. Inflorescence various, flowers strongly irregular. Perianth in 2 series of 3 segments each; outer whorl usually of 3 parts, sepaloid or petaloid, equal or 1 larger; inner whorl of 2 petaloid parts, often similar to outer whorl, and 1 highly modified, often spurred, segment (lip or labellum). Stamens and ovary fused into a "column"; anthers 1, terminal on column or 2, subterminal, sessile, pollen usually coherent into masses (pollinia). Ovary inferior, 1-celled, placentation parietal, rarely 3-celled, placentation axile; style thickened, stigmas 3, or reduced to 2, with third modified into a projection (rostellum). Fruit a capsule, dehiscent, valves 3–6; seeds very numerous, very minute.

An enormous family, principally tropical but with some temperate and even arctic representatives, very largely epiphytic, but also many terrestrial genera.

Dupont, Report: 41 (1907) refers to 'orchids' on Aldabra. He collected 2 species on Aldabra but the reference is too vague to be included in the synonymy.

1. Spur present; leaves linear or oblanceolate.
  2. Flowers small, sessile; spur very short, conical   **1. Acampe**
  2. Flowers large, pedicelled, spur long, slender   **2. Angraecum**
1. Spur absent; leaves elliptic oblong.   **3. Hederorkis**

1. ACAMPE Lindl.

Epiphytes. Leaves thick, leathery. Flowers sessile, in clusters on a sparsely branched inflorescence. Sepals and petals free, entire; labellum 3-lobed, embracing the column, spur short, hairy inside, the opening lying at the base of the column. Column short, thick.

A genus of 15 species in tropical and southern Africa, Madagascar and tropical Asia.

Acampe rigida (Buch.-Ham. ex J.E.Sm.) P.F.Hunt in K.B. 24: 98 (1970); Seidenfaden in Bot. Mus. Leafl. Harv. Univ. 25: 60 (1977).

Aerides rigida Buch.-Ham. ex J.E.Sm. in Rees, Cyclop.: 39 (1819).

Acampe renschiana Reichenb.f. in Otia Bot. Hamb. 2: 77 (1878); Hemsley in
  B.M.I.K. 1919: 132 (1919).

A. pachyglossa Reichenb. subsp. *renschiana* (Reichenb.f.) Senghas in Die
  Orch. 15: 165 (1964).

Robust, erect or scrambling herb to 1.5 m; internodes often with long,
grey, flexuous or contorted, aerial roots up to 5 mm in diameter. Leaves
articulated at the base, leaving a persistent sheath, linear, 9–22 × 1–2
(–3) cm, apex obtuse and asymmetrical, leathery, shallowly folded. Flowers
fragrant, sessile, clustered in a short, stout, sparsely branched inflorescence
up to 7 cm long; bracts broadly ovate, 3 mm long, clasping the stem at the
base. Sepals and petals similar, yellow with transverse, dark red, blotchy
bands, obovate, 6–12 mm long, apex obtuse to acute, incurved, leathery;
labellum narrowly pouched, keeled, side lobes erect, terminal lobe lilac with
purple spots, deflexed downwards; spur very short, conical. Capsule brown,
spindle-shaped, 4–5 cm long, slightly 6-angled, remains of perianth long-
persistent. Fig. 45/1–2.

Distr. ALDABRA: W (south), W. Channel; Kenya to the Transvaal,
Madagascar, Comoros, Southeast Asia and China.
Notes. Found on West Island and islands in the West Channel, near
Polymnie and south of Esprit. Occurs in champignon fringing the mangroves,
also epiphytes on mangroves, *Pemphis* and *Sideroxylon*. Flowers mainly
during the mid dry season.
Vernac. 'fleur de putantil'.

2. ANGRAECUM Bory

Epiphytic herbs. Leaves fleshy, articulated at the base, leaving a persistent
sheath. Flowers fairly large, axillary, solitary or in clusters. Sepals and petals
free, labellum entire, free from the column, similar to the other petals or
larger, produced at the base into a long spur. Column very short and broad,
concave in front, without wings or a foot.

A large genus of over 200 species in tropical and southern Africa,
Mascarene Islands and Philippines.

Angraecum eburneum Bory, Voy. 1: 359, t. 19 (1804); F.M.S.: 356 (1877);
Hemsley in B.M.I.K. 1919: 132 (1919).

Robust sprawling herb up to 1.5 m high. Leaves oblanceolate, fleshy,
35–60 × 3–6 cm, apex rounded, fleshy, folded. Flowers fragrant, widely
spaced along stout, erect, stems; stems as long as the leaves, bracts dark
brown, broadly ovate, 1.5–2 cm long, clasping the stem and enclosing the
stout, twisted, 2–2.5 cm long pedicel. Sepals and petals similar, yellow-green,
ovate-lanceolate, 30 × 8 mm, apex acuminate, somewhat leathery, spreading

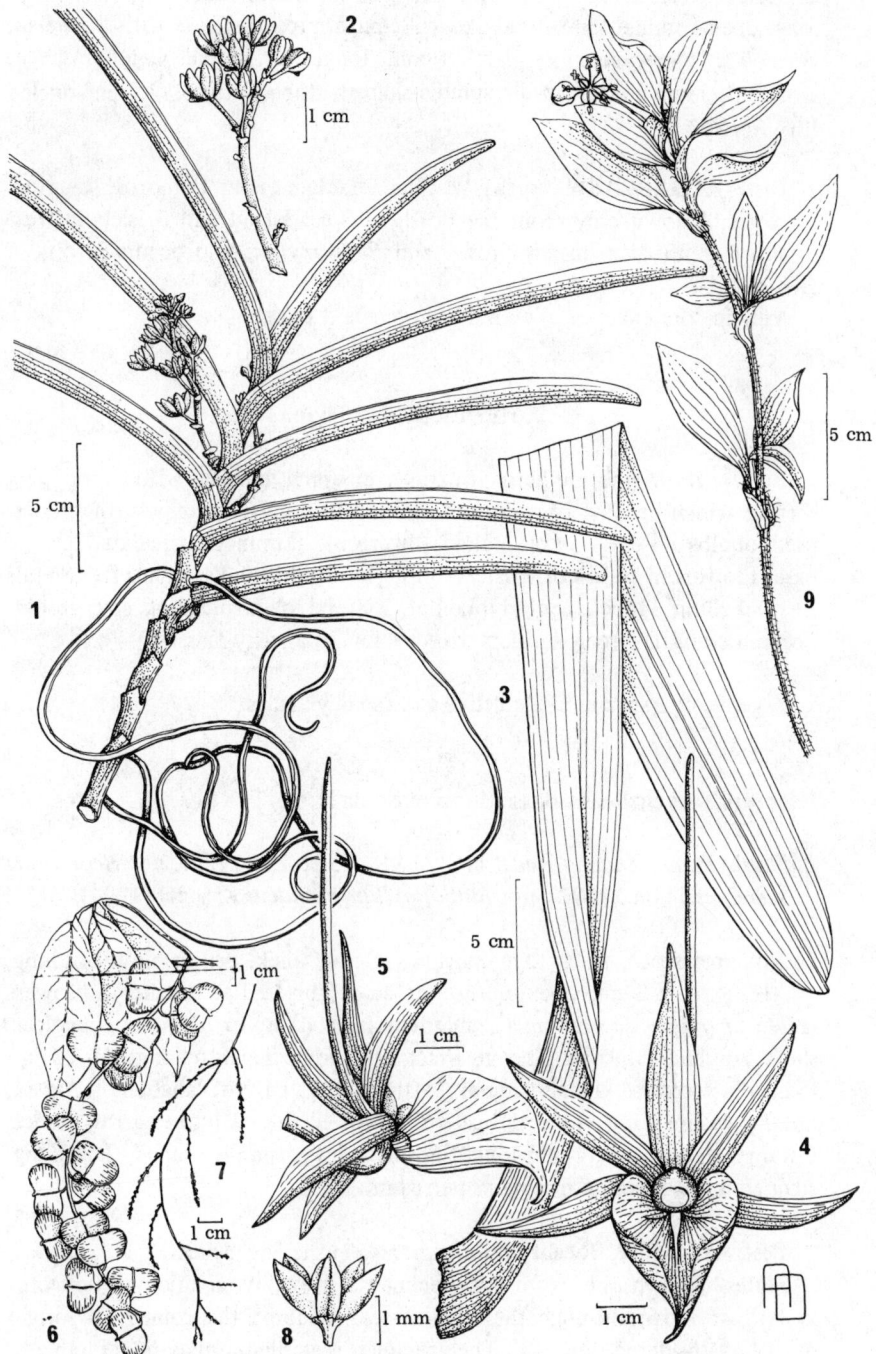

Fig. 45   1, *Acampe rigida*, habit; 2, flowers. 3, *Angraecum eburneum*, leaf; 4 & 5, flower. 6, *Dioscorea bemarivensis*, fruits; 7, flowering branch; 8, flower. 9, *Commelina benghalensis*, habit.

or reflexed; labellum white with pale yellow throat, heart-shaped, 30 mm long, apex rounded and hood-like with an abrupt tip, base with a median, lanceolate plate; spur slender, 70 mm long, horizontal. Column green, rounded, 5 mm long. Capsule spindle-shaped, ribbed 4.5–5 × 1.5 cm, angled. Fig. 45/3–5.

Distr. ALDABRA: W (north), W. Channel; Madagascar, Comoros, Réunion.
Notes. Known only from the north of West Island and 3 islets in West Channel. Epiphytes on mangroves and *Sideroxylon*; also occurs in pits in champignon.
Vernac. 'orchidée' or 'fleur paille-en-queue'.

3. HEDERORKIS Thou.

Epiphytic or epilithic herbs; rhizomes creeping, covered with overlapping sheaths which become fibrous with age and finally fall to expose the nodes; pseudobulbs absent. Leaves paired, divergent, terminal, articulated at the base. Flowers in spikes or clusters, borne on slender stalks. Sepals free. Petals free; labellum 3-lobed, central lobe flat, 2 lateral lobes spreading, spur absent. Column elongate, erect. Anthers with pollinia in 2 waxy masses.

A genus of 2 species in Mauritius and the Seychelles.

Hederorkis seychellensis Bosser in Adamsonia II, 16: 228 (1976).

Bulbophyllum scandens Rolfe in B.M.I.K. 1922:23 (1922), not *Hederorkis scandens* Thou. (1822), nor *Bulbophyllum scandens* Kraenzl (1904).

Rhizome stout, up to 2 m long, 8–10 mm thick. Leaves elliptic-oblong, 10–18 × 4–6.5 cm, apex obtuse or acute, blade flat, leathery. Flowers cream or purple, several, on a slender curving stalk up to 10 cm long; pedicels short, subtended by a small, ovate bract 2 mm long. Sepals oblong, 10–12 mm long, apex obtuse. Petals similar to the petals, curved; labellum recurved, oblong, 8 mm long, apex obtuse or indented, lateral lobes narrow, erect. Column club-shaped, 4 mm long. Capsule brown, spindle-shaped, 2 cm long, strongly ribbed, perianth remains persistent.

Dist. ALDABRA: locality unknown; also in the Seychelles.
Notes. Known only from the specimen at Kew, by an unknown collector in August 1916; although the label states 'Aldabra', the country of origin must be considered doubtful. The specimen is sterile but may be referable to the above species.

MUSACEAE

The banana (*Musa* sp.) has been recorded from Assumption by Dupont (1929), Report to the Seychelles Government, 'as growing in pits in the guano'. No specimens have been seen, neither has it been recorded in recent years.

67. DIOSCOREACEAE

Twining herbs, rarely shrubby; stems annual, arising from tubers or woody rootstocks. Leaves opposite or alternate, usually with strong veins, palmate from base and connected by a network of cross veins. Flowers in spikes, racemes or panicles; bisexual or unisexual and then usually dioecious. Perianth segments 6, in 2 similar series, usaully united basally. Stamens 6, inner 3 sometimes reduced to staminodes. Ovary inferior, 3-celled, ovules 2-numerous. Fruit usually a capsule, often 3-winged.

A small pantropical family with a few temperate members.

DIOSCOREA L.

Twining dioecious herbs; stems annual, arising from tubers. Leaves alternate or opposite, often ovate-cordate, entire, lobed or compound; stipules absent. Inflorescence spicate, male inflorescence axillary or terminal, female inflorescence axillary. Male flowers: perianth suberect or spreading; stamens 6, all fertile or 3 reduced to staminodes. Female flowers: perianth similar to male flowers; ovary oblong, 3-angled, 3-celled, ovules 2 per cell; styles 3. Capsule 3-lobed or triangular-ellipsoid, dehiscing into 3 valves; seeds 2 per cell, winged or wingless.

600 species distributed throughout the tropics and sub-tropics.

Dioscorea bemarivensis Jum. & Perr. in Ann. Col. Mus. Marseille, II, 8: 423 (1910); F.M.C. 44: 12, fig. 3 (1950).

D. nesiotis Hemsley in J. Bot. 55: 288 (1917) & in B.M.I.K. 1919: 132 (1919). Types: Aldabra, *Thomasset* 218 & 241 (both K, syn.).

Slender, scrambling, dioecious twiners; tubers small, globose, covered irregularly with small projections and occasionally having long, narrow protruberences; stems glabrescent, up to 4 m long. Leaves alternate, 3- or 5-foliolate; petiole 1.5–6 cm long; leaflets elliptic-ovate or lanceolate, 2–10 × 1–5 cm, margins entire, sparsely hairy or glabrous; petiolules

2–15 mm long. Flowers unisexual, very small, arranged alternately in simple, rarely branched, slender, solitary or clustered, axillary racemes up to 25 cm long; pedicels up to 3 mm long, slender in flower, stouter and reflexed in fruit; bracts 1–1.5 mm long, apex acuminate. Perianth segments cream, elliptic-oblong, 1–1.5 mm long, spreading. Male flowers: stamens 6, all fertile; filaments short, anthers small. Fruits 3-winged capsules in dense, pendulous clusters, each wing suborbicular, 1–1.5 cm broad, containing 1 or 2 seeds; seeds oblong, flattened 4 × 2 mm, with a brown, membranous, encircling wing, the whole disseminule being up to 1 cm in diameter. Fig. 45/6–8.

Fruiting specimens may be confused with *Gouania scandens* (Family 20/2), a tendril climber whose winged fruits have an ellipsoid central body.

Dist. ALDABRA: M (west), S; ASSUMPTION; ASTOVE. Madagascar.
Notes. An occasional constituent of mixed scrub and open champignon communities near the coast. The tuber is said to be eaten in times of famine. Vernac. 'cambare'.

68. LILIACEAE sens. lat*

Habit various, mainly herbaceous, frequently with bulbs or rhizomes; Inflorescences racemose, paniculate or umbelloid, or flowers solitary; bracts various. Perianth generally in 2 series of 3 parts each, series frequently similar, segments free or all 6 parts fused. Stamens usually 6, all fertile, rarely 3 sterile or absent. Ovary superior or inferior, 3-celled, with axile or basal placentation, rarely 1-celled with 3 parietal placentas; style usually 1, often with 3 branches or stigma-lobse. Fruit a capsule or berry; seeds few to usually many.

In this circumscription the family is practically world-wide, very large, but with few island genera, except for introductions. Most authors subdivide this family in various ways, almost everyone recognizing two large groups as families, the Liliaceae and Amaryllidaceae, variously distinguished and variously circumscribed. None of the characters, nor even the combinations of characters, used seem important enough to separate families. Under the present circumscription a great many tribes may be distinguished, but sub-families are difficult to separate convincingly. None of the schemes with which we are familiar seem very natural, so most of the liliacean families are here united.

1. Plants of tufted habit or with a short, stout stem; leaves thick, fleshy, arranged in a basal rosette

*Including the Amaryllidaceae and Agavaceae as well as various other small families recognised by Hutchinson, and in recent Floras of neighbouring areas.

2. Leaves very large, stout, straight, usually with a sharp, hardened tip; flowers white or cream, borne erect on horizontal branches of a pyramidal inflorescence at the apex of a long pole arising from the basal rosette
1. Agave
2. Leaves not very large, usually curling downwards with age; flowers reddish orange, borne horizontally on an erect spike-like inflorescence which may branch at the base to give 2 or 3 additional spikes and thereby having the appearance of candelabra
4. Lomatophyllum
1. Plants of bushy or arborescent habit; leaves not arranged in a basal rosette
3. Leaves reduced to scales, their photosynthetic function being taken over by small, linear, leaf-like shoots arranged in clusters and giving the plant a feathery appearance; flowers white, small, often in clusters
2. Asparagus
3. Leaves lanceolate, conspicuous, flaccid, usually at the apices of sparsely branching stems; flowers cream, in medium sized, branching inflorescences
3. Dracaena

1. AGAVE L.

Perennials; stems very short, rarely visible. Leaves numerous, arranged in a basal rosette, persistent, fleshy, stiff, armed with an apical spine. Inflorescence branched and pyramidal or elongated, borne at the end of a long, pole-like stalk. Perianth usually cream or yellow, 6-partite, sepals and petals similar, rather fleshy, fused at the base. Stamens 6, exerted beyond the perianth when mature. Ovary inferior, oblong, consisting of 3 carpels, each containing numerous ovules in 2 rows. Fruit an oblong to globose, dehiscent, capsule; seeds many.

These plants are remarkable for their mode of reproduction. Said to live up to 100 years, the plants are vegetative for most of their life and grow rather slowly. When they eventually flower the single inflorescence and its long stalk grow very rapidly. After the production of the inflorescence the plants die and perpetuation depends either on the seeds or on small bulbils which may develop on the inflorescence and replace the mature fruit.

A genus of c 300 species, native to tropical America, introduced elsewhere.

Agave sisalana (Perrine ex Engelm.) Drummond & Prain in Agric. Ledger 13, no. 7: 89 (1906).

A. rigida Northrop var. *sisalana* Perrine ex Engelm. in Trans. Acad. Sci. St. Louis 3:314, fig. 24 (1875).

Perennial plants with massive, fleshy leaves arranged in a dense rosette. Leaves linear-lanceolate, shallowly channelled above, slightly convex beneath, up to 150 × 15 cm, margins smooth or armed with prickles, apex usually with hard, brown spine up to 2.5 cm long. Inflorescence a dense cluster of erect flowers borne on horizontal branches near the apex of a stout pole 7 m high. Flowers 4.5–5.5 cm long from base of the ovary. Perianth segments cream, 3.5–4 cm long, lower third fused. Anthers linear, 2.5 cm long, exserted at maturity, filaments 3.5–6 cm long. Style 3–6 cm long at maturity. Capsules rarely produced on Aldabra, reproduction mainly by bulbils developing after flowering in the axils of the bracts. Fig. 46/1

Distr. ALDABRA: W, P, M, S (west) Michel; COSMOLEDO: W; ASTOVE; native of Mexico now widely cultivated throughout the tropics and naturalized.

Notes: Introduced around settlement sites for cultivation for sisal, which is manufactured from the leaf fibres. Now more or less naturalized, but being eradicated wherever possible.

Vernac. 'agave' or 'sisal'.

2. ASPARAGUS L

Herbs or shrubs; rhizomes perennial; stems annual or perennial, erect or decumbent, often green and much branched. Leaves reduced to membranous scales or spines, with solitary or clusters of small, linear, green, "leaf-like" shoots in their axils giving the plants a feathery appearance. Flowers small, solitary or clustered, terminal, in the axils of the green shoots or on simple branches. Perianth in 2 series of 3, white or cream, similar segments, free at the base. Stamens 6, fused to the perianth segments near the base. Ovary globose, 3 celled, ovules several per cell; style divided into 3 parts for much of its length or at the apex only. Fruit an indehiscent, globose berry; seeds 1-several.

A genus of c 300 species restricted to the Old World.

Asparagus umbellulatus Bresler, Diss. Gen. Aspar. Hist. Nat. Med. n. 24 (1826); F.M.S.: 377 (1877); Baker in B.M.I.K. 1894:151 (1894); Schinz in A.S.N.G. 21:83 (1897); Voeltzkow in A.S.N.G. 26:549 (1902); Dupont, Report: 41 (1907). [as *umbellatus*]; Hemsley in B.M.I.K. 1919: 132 (1919) – not to be confused with *A. umbellatus* Link from the Canary Is.

Asparagopsis umbellulata (Bresler) Kunth, Enum. 5:81 (1850).

Erect, decumbent or scrambling, glabrous, feathery shrub; stems slender, perennial, up to 2 m long, becoming branched upwards, branches often deflexed. Leaves reduced to small papery scales; leaf-like axillary shoots readily falling, 1–2.5 cm long, in clusters of 1–3. Flowers on pedicels 1–2 mm long, jointed in the middle, usually in dense clusters borne in the axils of the small shoots which are often lost by the time of flowering. Perianth segments cream, ovate-lanceolate, 3–4 mm long, glabrous rather papery, spreading at maturity. Stamens as long as the perianth segments; anthers oblong, 0.5–0.75 mm long. Berry pale orange-brown, globose, 5–7 mm in diameter; seeds (2–) 3(–4), black, 3–angled, 2.5 mm in diameter. Fig 46/2.

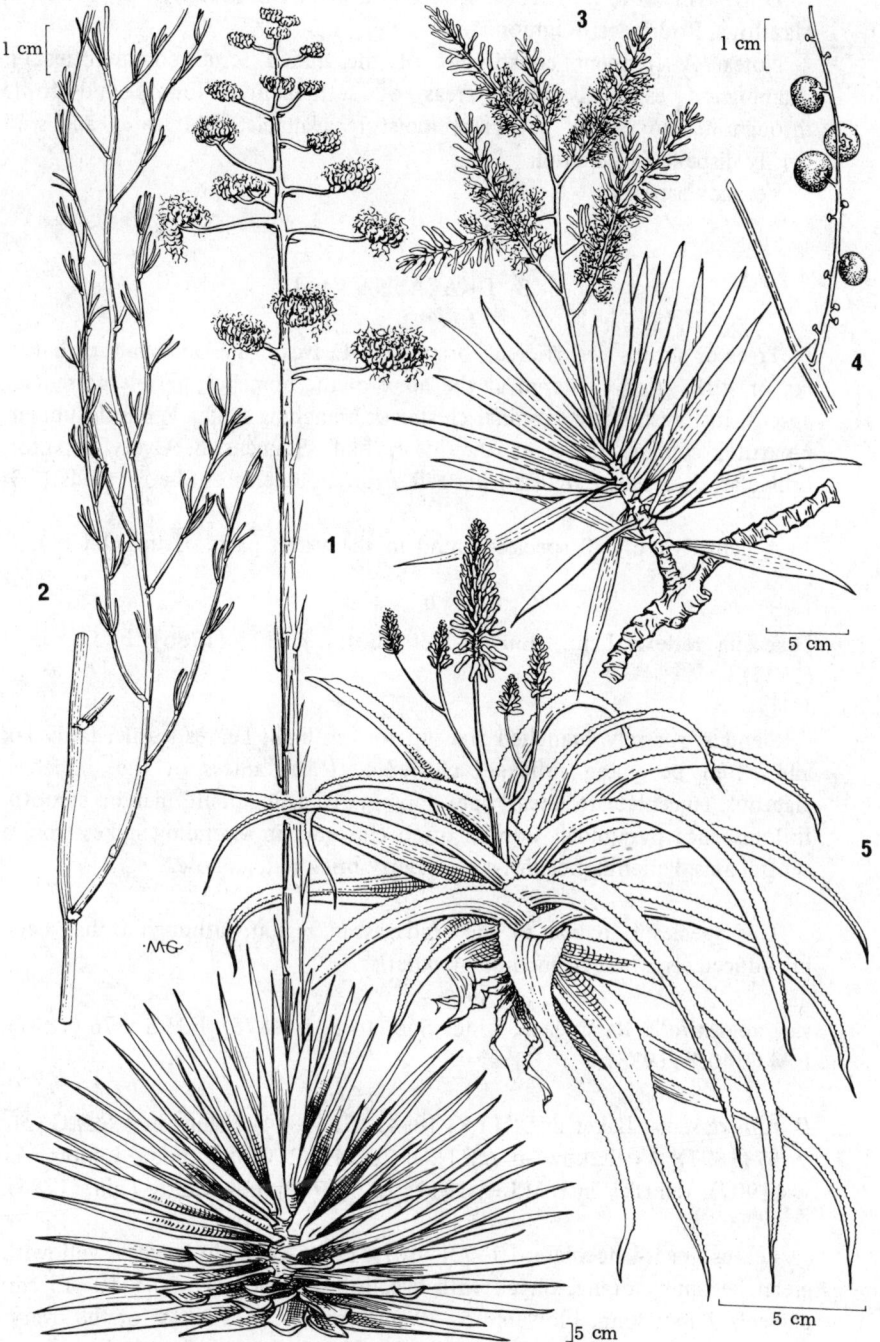

Fig. 46 1, *Agave sisalana*, habit. 2, *Asparagus umbellulatus*, habit.
3, *Dracaena reflexa* var. *angustifolia*, habit; 4, fruits.
5, *Lomatophyllum aldabrense*, habit.

Distr. ALDABRA: W, P, M, S, lagoon islets; COSMOLEDO:M; ASTOVE; Mauritius, Rodrigues, Réunion.

Notes. A frequent constituent of the mixed scrub communities on champignon especially near areas of settlement. Flowers and fruits throughout the year if sufficient moisture available. Fruit eaten and seed locally dispersed by bulbul.

Vernac: 'asperge'.

3. DRACAENA Vand.

Trees or shrubs; stems erect, branching. Leaves ovate-lanceolate or linear, lax or stiff, often clustered at the apex of the branches; petiole present or absent. Inflorescence a terminal cluster or branching spike.. Perianth tubular, 6-partite, similar, fused in the lower half. Stamens 6. Ovary superior, orbicular, 2–3 celled, ovules 1 per cell. Fruit an indehiscent berry; seeds 1–3.

A genus of c 150 species, found in the warm parts of the Old World.

Dracaena reflexa Lam., Encycl. Méth. Bot.   2: 324 (1786); F.M.S.: 375 (1877).

Slender, sparsely branched tree up to 5 m high. Leaves sessile, fairly lax and often becoming reflexed, clustered at the apices of the branches, glabrous, lanceolate to linear, apex tapering to a fine point, margins smooth. Inflorescence pyramidal, flowers often clustered in ascending spikes from a central axis. Perianth greenish-white. Berry brick red, globose.

This species is native to the Madagascan region, although it has been introduced elsewhere; it includes 4 varieties.

var. **angustifolia** Baker in J. Linn. Soc. 14:531 (1875); F.M.S.:376 (1877); F.M.C. 40:8 (1938).

D. reflexa sensu Baker in B.M.I.K. 1894: 151 (1894); Schinz in A.S.N.G. 21: 87 (1897); Voeltzkow in A.S.N.G. 26:549 (1902); Dupont, Report: 41 (1907); Hemsley in B.M.I.K. 1919: 132 (1919), not strictly Lam. (1786).

Leaves linear-lanceolate, 10–23(–40) x 1–1.5 cm, light green or yellowish green becoming orange-tinged with age. Inflorescence 20–30 x 10–15 cm; pedicels 2 mm long. Flowers 15–20 mm long from the base of the ovary. Perianth segments greenish-white, linear, 15 mm long, reflexed. Stamens as long as the perianth; anthers oblong, 15–20 mm long. Stigma as long as the perianth. Berry brick red, globose, 10 mm in diameter; seeds 2, orange-brown, subglobose, 4 mm in diameter. Fig. 46/3–4.

Distr. ALDABRA: W, P, M, S; ASSUMPTION; Madagascar, Réunion, Rodrigues and Mauritius.

Notes: Collections by Abbott and Dupont from Aldabra cited by Hemsley have not been traced. Dupont, loc-cit. records *Dracaena* from Cosmoledo and Astove but no specimens seen. Occurs occasionally throughout the mixed scrub communities. Flowers mainly during the early to mid dry season. New leaves produced in mid to late wet season. Ripe fruit may remain on the tree for many months.

Vernac: 'bois de chandelle' or 'bois de chandelle rouge'.

4. LOMATOPHYLLUM Willd.

Robust, perennial, 'Aloe' –like plants; stems very short and inconspicuous or fairly long and usually erect. Leaves thick, fleshy, clustered in a terminal rosette. Inflorescence of solitary or branching, erect spikes. Flowers tubular, perianth 6-partite, segments similar, fused at the base. Stamens 6, not greatly exserted when mature. Ovary superior, orbicular, 3-celled, ovules several per cell. Fruit an indehiscent, globose berry.

A genus of c 10 species, restricted to the Madagascan and Mascarene Regions.

Lomatophyllum aldabrense Marais in K.B. 29: 722 (1974). Type: Aldabra, West Is., *Stoddart* 920 (K, holo., US, iso.).

L. borbonicum sensu Baker in B.M.I.K. 1894: 151 (1894); Schinz in A.S.N.G.
   21: 82 (1897); Voeltzkow in A.S.N.G. 26: 549 (1902); Dupont, Report:
   41 (1907); Hemsley in B.M.I.K. 1919: 132 (1919), not Willd. (1811).

Perennial plants with fleshy leaves clustered at the apex of a stout, erect or decumbent stem up to 2 m long, stemless when young. Leaves linear-lanceolate, up to 100 × 10 cm, broad at the base, tapering to a pointed apex, margins usually horny and weakly spiny, upper surface concave, the whole leaf becoming reflexed with age, dark green often tinged with orange or red, glabrous. Peduncle up to 4–5 cm long, flattened, with 3–5 branches, pedicels 15–25 mm long; bracts ovate to triangular, 2.5–4.5 mm long. Flowers 20–25 mm long from base of ovary. Perianth bright reddish-orange outside, yellowish inside with a prominent dark vein down each segment. Anthers oblong, 2 mm long. Style 15–20 mm long. Berry purplish-red, oblong-globose, 12 mm in diameter; seeds few, black, 3-angled, 3 × 3 mm. Fig. 46/5.

Distr. ALDABRA: W, P, M, S; ASSUMPTION; ASTOVE; endemic.

Notes. Occurs frequently in the inland mixed scrub, on coastal champignon, and under *Casuarina*, its fiery red spikes making it one of the most conspicuous of all the indigenous plants. Flowers mainly during the mid wet to early dry season but unseasonal rain can extend the flowering period. Leaves turn from green to copper with increasing drought. Green fruit available throughout most of the year. Seeds locally dispersed by bulbul and blue pigeon.

Vernac. 'zanana mowo'

69 COMMELINACEAE

Herbs, rarely shrubs. Leaves, alternate or densely spirally arranged, simple, entire. Flowers in scorpioid cymes, often with conspicuous bracts. Calyx and corolla strongly differentiated, each of 3 segments, petals all equal or 1 smaller, all ephemeral and deliquescent. Stamens 6 or 3 (rarely 1), filaments usually with copious, long, bead-like hairs. Ovary superior, usually 3-celled, ovules 1—few, placentation axile; style 1. Fruit usually a loculicidal capsule, rarely indehiscent; seeds large, usually with a sculptured testa.

Pantropical and subtropical, rarely temperate.

COMMELINA L.

Small to medium sized, succulent and delicate herbs. Leaves membranous, with a distinctive sheathing base. Flowers irregular, bisexual, clustered in a branched inflorescence which is enclosed, or almost so, by a membranous, heart-shaped, folded bract. Sepals 3, free, oblong to ovate, smaller than the petals. Petals 3, free, clawed, 2 equal, 1 smaller. Stamens 6, 3 fertile, 3 sterile, filaments without long hairs. Ovary 3-celled, 1 cell containing 1 ovule, remaining 2 cells containing 1—2 ovules. Fruit a small, partly dehiscent capsule, the chamber containing 1 ovule being indehiscent; seeds with a sculptured testa.

A genus of c 230 species distributed throughout the tropics and sub-tropics.

Commelina benghalensis L., Sp. Pl.: 41 (1753); F.M.S.: 324 (1877); Hemsley in B.M.I.K. 1919: 132 (1919); F.M.C. 37: 15 (1938).

Decumbent or semi-erect, annual herbs; stems jointed, succulent, trailing, pubescent, up to 1 m long. Leaf ovate to ovate-lanceolate, 3—5(—10) × 1.5—3(—5.5) cm, apex acute, base tapering abruptly to a petiole, shortly pubescent; petiole expands to form a pale, papery sheath around the stem, margins with few, long, stiff, reddish hairs. Inflorescence a small cluster of flowers exserted from or enclosed by a small, leaf-like, pubescent bract, its basal margins united. Sepals pale blue, broadly ovate or elliptic, 2—4 mm long. Petals bright blue, roughly orbicular, 2 dorsal petals 5—7 mm long, reflexed, ventral petal smaller, not reflexed. Stamens 3 fertile, 3 sterile and shorter; stamens projecting forward above the ventral petal. Ovary superior; style projecting forward above the ventral petal. Capsule oblong, 4—5 × 3 mm, valves 2, dehiscing; seeds 2 per valve, oblong, 1.5—2.5 × 1.5—2 mm, wrinkled. Fig. 45/9.

Distr. ASSUMPTION; pantropical weed.

Notes. Only known from 1 collection, *Dupont* 253 (K). Hemsley cites a :ollection by *Fryer*, possibly in error since no specimen has been found.

Vernac. 'herbe aux cochons' or 'herbe cochon'.

70. PALMAE

Trees, shrubs or lianas, often very large; stems with scattered vascular)undles, very densely distributed in the outer portion where they may form ιn extremely hard wood. Leaves folded fan-like, often gigantic, spirally ιrranged in a dense rosette, or, more rarely, scattered, or still more rarely :rowded in 2 or 3 ranks, pinnately compound or palmately veined and cut;)etiole present, the base in some genera forming a sheath around the terminal)ud. Flowers in a panicle (spadix), rarely spicate, spadices usually enclosed in or more bracts (spathes); flowers bisexual or more usually unisexual and hen usually monoecious, rarely dioecious, 3-merous but parts often eduplicated. Stamens (3–)6 or more. Ovary 1–3-celled, ovules 1 per cell, ιsually only 1 developing. Fruit a drupe or berry, often fibrous, sometimes overed by overlapping scales, endosperm oily, fleshy or bony.

A large, mostly tropical family, many species being of economic mportance.

. Pinnae folded so that in cross-section the midrib is at the apex of an
   inverted 'V'; fruit globose-ovoid, 20–30 cm long   **1. Cocos**
. Pinnae folded so that in cross-section the midrib is at the base of an erect
   'V'; fruit oblong-ovoid, 4–7 cm long   **2. Phoenix**

The Seychelles endemic 'coco de mer' or 'double coconut', *Lodoicea ιaldivica* (Gmel.) Pers., is reported in the Seychelles Archives in a despatch y the Governor as having been planted on Aldabra in 1900. No specimens ollected nor any further records of the plant.

I. COCOS L.

Monoecious trees; trunks stout, unbranched, bearing a terminal crown of :aves. Leaves 25–35, large, pinnate. Flowers unisexual, in dense panicles of ιany male flowers and few female flowers, each panicle subtended by one nall and one large, striate, woody bract. Perianth of 6 segments in 2 'horls of 3, female flowers subtended by 2 bracteoles. Stamens 6. Ovary ιrge, 3-celled; stigmas 3. Fruit a large, 3-sided drupe with a tough smooth, uter skin, thick, fibrous mesocarp and a hard, dark-brown, inner shell; seeds 1.

The endosperm of the seed forms a white, oily, solid layer inside the shell, leaving a cavity filled with liquid.

A pantropical, cultivated, monotypic genus.

Cocos nucifera L., Sp. Pl.: 118 (1753).

Palms, up to 30 m tall; trunk columnar, up to 40 cm in diameter, thicker near base, light greyish brown in colour. Leaves pinnate, 4.5–6 m long, pinnae 200–250 per leaf, linear, 50–120 × 1.5–5 cm; petiole stout, with a net-like, sheathing base. Panicle large, 1–2 m long, flowers fragrant, densely clustered, male flowers numerous, mostly towards the apex, female flowers solitary or few, towards the base. Male flowers: small, pale yellow, outer 3 perianth segments cuff-like, 3 mm long, inner 3 segments ovate, 13 mm long, tough and shell-like, enclosing the stamens and a much reduced ovary. Female flowers: large, globose, 2–3 cm in diameter, bracteoles and perianth segments brownish-green, of equal size, sub-orbicular, tough, papery. Fruit large, elliptic or ovoid, 20–30 cm long, 3-sided. Fig. 47/1.

In our area the plants are all of the characteristic small-seeded Seychelles type, the seed being ½ the total length of the fruit. In Pacific plants the seed occupies a much greater proportion of the fruit.

Distr.   ALDABRA: W, P, M, S,   Esprit,   Michel;   ASSUMPTION; COSMOLEDO: W, M; ASTOVE; pantropical strand plant, said by some to be native in the Seychelles but this is controversial.

Notes. Found on coastal sands, often as plantations, sometimes (?) locally dispersed and naturalized. New leaves produced during the wet season. Flowers throughout the year but chiefly during the wet season; fruits least abundant during the late dry season. Fruits sea-dispersed – found germinating on strand line in Aldabra

Vernac. 'cocotier' or 'coconut'.

2. PHOENIX L.

Dioecious tree; trunk stout, covered with old leaf bases, usually unbranched, bearing a terminal crown of leaves. Leaves pinnate; leaflets folded upwards from the midrib. Inflorescences unisexual, flowers borne singly in a spiral on the inflorescence branches, each flower subtended by 1 bracteole. Male flowers: calyx cup-shaped, lobes 3, short, triangular; corolla shortly tubular, lobes 3, exceeding the calyx; stamens (3–)6(–9), ovary reduced or absent. Female flowers: similar to the male: staminodes 6; carpels 3, free, elongated, stigmas short. Fruit with a smooth or waxy epicarp, fleshy mesocarp and thin membranous endocarp; seeds 1, longitudinally grooved.

17 species in the Old World tropics and subtropics.

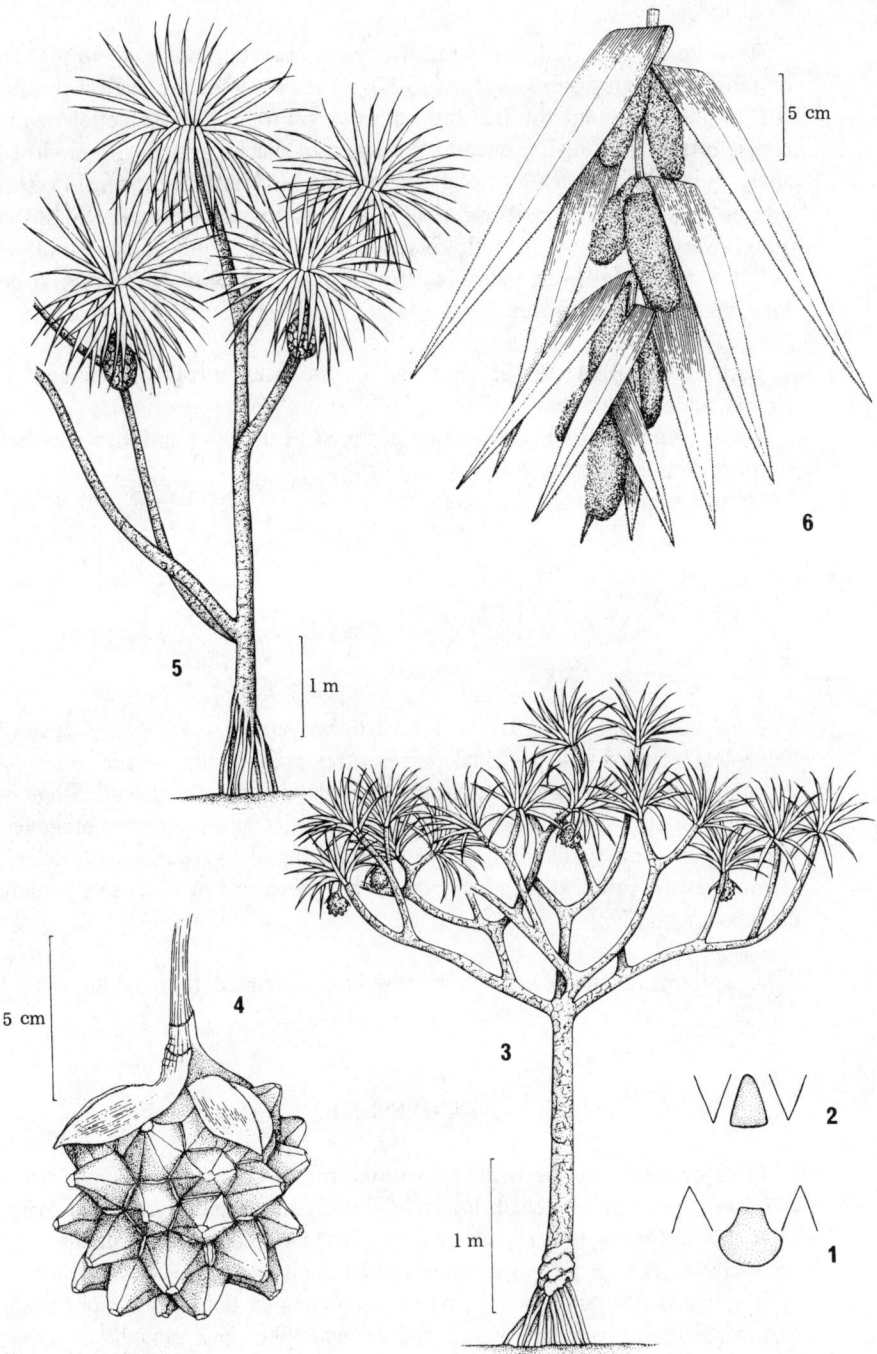

Fig. 47   1, *Cocos nucifera*, diagram of leaflet arrangement. 2, *Phoenix dactylifera*, diagram of leaflet arrangement. 3, *Pandanus aldabraensis*, habit; 4, fruit. 5, *Pandanus tectorius*, habit; 6, male inflorescence.

Phoenix dactylifera L., Sp. Pl.: 118 (1753).

Tree up to 20 m tall; trunk massive, erect. Leaves pinnate up to 300 cm or more long, grey-green, leaflets c. 80 on each side of the central rhachis, stiff, tending to point towards the leaf apex, up to 30×2 cm; petiole with a coarse brown, sacking-like sheath; basal part of rhachis armed with modified, sharp pointed, lower leaflets. Male flowers: in large inflorescence, flowers in groups or solitary; calyx 2 mm long, 3-toothed; petals cream, fleshy, 8 mm long; ovary minute. Female flowers globose; calyx 2 mm long, 3-toothed; petals 4 mm long. Fruit yellow to brown or almost black, oblong, 4–7 cm long, mesocarp varying from thick and juicy to thin and dry. Fig. 47/2.

Distr. ALDABRA: W, S; unknown in the wild, possibly originated in Arabia, widely cultivated.

Notes. Known from 1 immature plant at Settlement and also reported south of Bras Takamaka.

Vernac. 'dattier', 'datt' or 'date'.

71. PANDANACEAE

Tree, shrubs and vines. Leaves crowded, in 3 spiral ranks, strap-shaped or lanceolate margins and principal nerves often spiny, veins parallel. Flowers dioecious, borne in bracteate heads or spikes. Perianth absent. Stamens crowded on club-shaped spikes or in naked clusters on inforescence branches. Carpels 1 or usually coherent in bundles (phalanges), ovaries 1–many-celled; stigmas sessile. Fruit either a berry or a drupe consisting of 1 or usually many carpels.

A widespread Old-World family not closely related to anything else; 3 genera.

PANDANUS L.

Trees or rarely shrubs with prop roots; trunks and branches ringed with leaf scars, sparingly branched, branches thick, compressed at base and some-what clasping trunk or branches with bundles scattered in pith, abundant and coalescing to form a porous woody cylinder in cortical area. Leaves crowded, in 3 spiral ranks, lanceolate or narrowly oblong to linear or strap-shaped, narrowing to an angular, awl-shaped or whip-like apex, angular in cross section, base clasping, blades stiff, leathery, 1 to several times folded fan-wise, veins parallel, close-set, usually with prickles on margins and midribs. Flowers dioecious, reduced to naked stamens or carpels, these variously solitary or usually fused into fascicles (stamens) or phalanges (ovaries). Male flowers: on branchlets of an alternately branching, 2-ranked or 1-sided, bracteate inflorescence; stamen-filaments awl-shaped, connective often prolonged into an apical appendage. Female flowers: collected into heads (often incorrectly called syncarps), heads solitary or spicately arranged on an axis, subtended by leaf-like bracts; phalanges of 1–many carpels, fused for most of their length, divided transversally into 3 sections, a basal, fleshy, fibrous mesocarp, a central, bony endocarp containing 1 seed per cell, and an apical mesocarp, the upper part of which is conic or pyramidal, or more usually divided into nipple-shaped, conical, or pyramidal lobes (enormously thickened styles?), bearing at their apices the stigmas, each representing a carpel; stigmas kidney-shaped or orbicular. Seeds oily, ellipsoid to cylindical, testa very thin.

An enormous Old World tropical genus with possibly many hundreds of species, the number recognized depending on different species concepts.

1. Well-developed leaves mostly 30–60 cm long, weakly prickly; phalanges with 1–5, closely crowded stigmas on apex of a single pyramidal process
              1. P. aldabraensis
1. Well-developed leaves over 1 m long, notably prickly; phalanges with 8–17 stigmas, each on a separate lobe or process.   **2. P. tectorius**

1. **Pandanus aldabraensis** St. John in Pac. Sci. 28: 83, fig. 328 (1974); Fosberg in K.B. 31: 838 (1977). Type: Aldabra, Takamaka area, *Fosberg 49058* (BISH, holo., K, US, iso.).

P. vandermeeschii sensu Schinz in A.S.N.G. 21: 81 (1897); Voeltzkow in
   A.S.N.G. 26: 549 (1902), not Balf. f. in F.M.S.: 398 (1877) nor probably
   Dupont, Report: 41 (1907) & Hemsley in B.M.I.K. 1919: 133 (1919).
Pandanus sp., St. John in Proc. Sci. 28: 99, fig. 336 (1974).

Small, little-branched tree, 2–2.5 m tall, trunk and branches almost smooth, prickles very few, branches relatively soft, 2.5–3 cm thick, leaf-scars and leaves crowded. Leaves linear, up to 63 × 4 cm, tapering to a whip-like, prickly tip, margins and midrib with small prickles, directed upwards, closely appressed, those on midrib only on distal half, blade minutely yellow-dotted beneath, c 30 veins per side. Male panicle with at least 8 major bracts and several reduced bracts, slender ramifications covered distally with stamens and with pimple-like protuberances 2–3 mm long, covered with stamens; anthers linear, c. 2 mm long, twisted, filaments about as long or shorter. Peduncle of female inflorescence 3-angled, with several pairs of bracts up to 18 cm long, keeled, leaf-like, pouched, subtending a single head or bracts separated and each subtending a small head, heads (when mature or almost so) globose, c 8 × 8 cm, each consisting of 30–50 phalanges, the free parts of these c 10 mm high, pyramidal, each with 1–5 stigmas crowded together at the summit, loop-like, open side almost closed, convex sides mostly radiating outward and toward the distal end of the head, endocarp very irregular c 13–17 mm long, to 15 mm wide, with 1 or more cells, basal mesocarp 14–16 mm long; seed broadly ellipsoid, 6 mm long. Fig. 47/3–4.

A very fragmentary and immature specimen of what is probably this species, *Voeltzkow* 1a, in the Zurich herbarium, seen by us and also studied and illustrated by St. John, was with some hesitation called by Schinz *P. vandermeeschii* Balf. f. The latter is a rather similar species endemic to Mauritius. Schinz's record gave rise to the idea that the common Aldabra *Pandanus* belonged to *P. vandermeeschii*.

Distr. ALDABRA: M(west), S(east); endemic.

Notes. Only known from the Gionnet Channel and Takamaka areas, where it occurs in mixed and tall scrub communities. Flowering not observed but many fruits seen during the later dry season. Fruits eaten and seeds locally dispersed by tortoises.

Vernac. 'vaqua' or 'vaquois'.

2. Pandanus tectorius Park., Journ. Voy. S. Seas: 46 (1773); Fosberg in K.B. 31: 840 (1977).

P. alloios St. John in Pac. Sci. 28: 86, fig. 329 (1974). Type: Aldabra, South Is., Cinq Cases, *Fosberg* 49233 (BISH, holo., K, US, iso.).

P. chelyon St. John in Pac. Sci. 28: 88, fig. 330 (1974). Type: Aldabra, South Is., Cinq Cases, *Fosberg* 49196 (BISH, holo., K, US, iso.).

P. impar St. John in Pac. Sci. 28: 90, figs, 331—333 (1974). Type: Aldabra, South Is., Cinq Cases, *Fosberg* 48997 (BISH, holo., K, US, iso.).

P. intraconicus St. John in Pac. Sci. 28: 96, fig. 334 (1974). Type: Aldabra, South Is., Cinq Cases, *Fosberg* 49236 (BISH, holo., K, US, iso.).

P. subcubicus St. John in Pac. Sci. 28: 97, fig. 335 (1974). Type: Aldabra, South Is., Takamaka, *Fosberg* 49347 (BISH, holo., K, US, iso.).

P. vandermeeschii sensu Dupont, Report: 41 (1907). Hemsley in B.M.I.K. 1919: 133 (1919) probably, not Balf. f. (1877).

Tree up to 8 m tall, sparingly branched, cone of stilt-roots tending to be dense, roots forked 1 to 3 or 4 times, well-developed. Leaves 75—135 cm long, usually less, tips long, whip-like, prickly, 2-folded, spines closely forward appressed on margins, close together, few and more or less backward pointing on midrib, c 45—50 parallel veins per side. Male inflorescence pendent, rachis strongly undulating, branches subtended by large, white bracts, bearing numerous erect structures, each bearing a clump of 45—90 stamens. Female heads solitary, pedunculate, subtended by white bracts exceeding head, phalanges 26—38, each with 8—17 cells, lobes and stigmas; stigmas with convex side out. Fruiting head vertically pendent, cylindrical, c 25 × 12 cm, with 26—38, broadly top-shaped, angular phalanges, 5—7 × 4—6 cm, spirally arranged in 5—6 rows on a club-shaped receptacle, apical surface varying from almost flat to deeply lobed. Fig. 47/5—6.

This species, here as in other parts of its vast range, is notoriously polymorphic; 5 segregate species having been described from the Aldabra population alone, differing in details of shape of the mature phalanges and in the sculpturing of their distal ends. They are here treated as individuals of a single variable population.

Distr. ALDABRA: W, P, M, S; and Indio-Pacific strand species.

Notes. Fairly generally distributed in mixed scrub and very common, sometimes dominant in *Guettarda* scrub on coastal champignon ridges, locally especially luxuriant around water holes, but by no means correlated in distribution with water holes, as has been previously suggested. Flowers mainly during the mid wet season. Fruit may remain on the plant until late in the dry season. Fruits remain viable after floating in the sea for more than 8 weeks. Fruits eaten and seeds locally dispersed by tortoises.

Vernac. 'vaqua' or 'vaquois'.

72. NAJADACEAE

Elongated, branched, slender, weak, submerged aquatics. Leaves opposite or whorled, linear, sheathing at base, margins entire or toothed. Flowers minute, sessile in branch of leaf axils, unisexual, dioecious or monoecious. Male flowers: spathe tubular to inflated along 1 side; perianth very thin, sac-like; anther 1, sessile. Female flowers: perianth absent or apparently so; ovary 1-celled, ovule 1, basal; style 1, stigmas 2–4, linear. Fruit an achene, pericarp thin; seeds 1, testa hard.

A family of 1 genus, widespread in temperate and tropical waters, both fresh and saline.

NAJAS [NAIAS] L.

Generic description as for family.

Najas graminea Delile, Fl. Egypt.: 282, fig. 50 (1813–1814).

Stems slender, much-branched, delicate. Leaves opposite, narrowly linear, 15–25 × 0.4–0.6 mm, margins and apex with short spine-like teeth, base expanded into a sharply lobed sheath, lobes acute or acuminate. Flowers monoecious. Female flowers: sessile, style, including branches, at least twice as long as ovary, branches somewhat exceeding united portion. Fruit cylindrical-ellipsoid, 1.75–2 mm long, beak long, persisting almost to maturity, areolae more or less isodiametric. Fig. 48/1–3.

Distr. ALDABRA: S (east); widespread through the tropics of the Old World.

Notes. Only 2 collections from somewhat brackish pools south west of Cinq Cases, *Fosberg* 49030 & 49035 (both K, US).

73. POTAMOGETONACEAE

Plants adapted for a submerged aquatic existence; stems elongated, diffuse or rhizome-like with erect, elongate or condensed branches. Leaves alternate, in 2 ranks, sheathing at base, usually thin. Flowers variously reduced, perianth (or pseudoperianth) in 1 whorl or absent, 3–6-lobed, lobes touching, not overlapping. Stamens 1–4. Ovary 1-celled, ovules 1, pendulous, apical or parietal placentation. Fruit an indehiscent nut or drupe.

A practically cosmopolitan family found in fresh and salt water, if accepted in the broad sense adopted here.

1. Stems delicate, thread-like, branched; rhizomes absent　　　　**3. Ruppia**
1. Rhizomes and erect stems cord-like, coarse
  2. Roots and leafy branches arising from every 4th node
　　　　　　　　　　　　　　　　　　　　　　5. Thalassodendron
  2. Roots and leafy branches arising from nodes, either every node or irregularly distributed
    3. Leaves round in cross-section　　　　　　**4. Syringodium**
    3. Leaves flat, grass-like
      4. Leaves with 3 nerves, the lateral nerves faint　　**2. Halodule**
      4. Leaves with more than 3 nerves　　　　**1. Cymodocea**

1. CYMODOCEA Koen.

Rhizome prostrate, herbaceous, monopodial, many vascular bundles in cortex, roots 1–5, and a short, erect, leafy branch per node. Leaves sheathing at the base, blades linear, soon falling, sheaths more persistent, compressed, leaving open or encircling leaf scar, leaf tips obtuse to indented, toothed or subentire, nerves numerous, 7–17. Flowers dioecious, solitary, terminal, enclosed in a leaf sheath. Perianth absent. Male flowers: peduncle present; anthers 2, joined dorsally at same height on peduncle, with awl-shaped appendage. Female flowers: sessile or almost so; carpels or ovaries 2, naked, each with short style divided into 2 awl-shaped stigmas. Fruit a nut, some-what compressed, dorsally ridged, beaked.

4 species in the Old World tropics and Mediterranean.

1. Leaf scars completely encircling short leafy branches; leaf tips subentire or roughly and minutely serrate at high magnifications　　**1. C. rotundata**
1. Leaf scars with a small gap; leaf tips noticeably toothed, ciliate, even at low magnifications　　　　　　　　　　**2. C. serrulata**

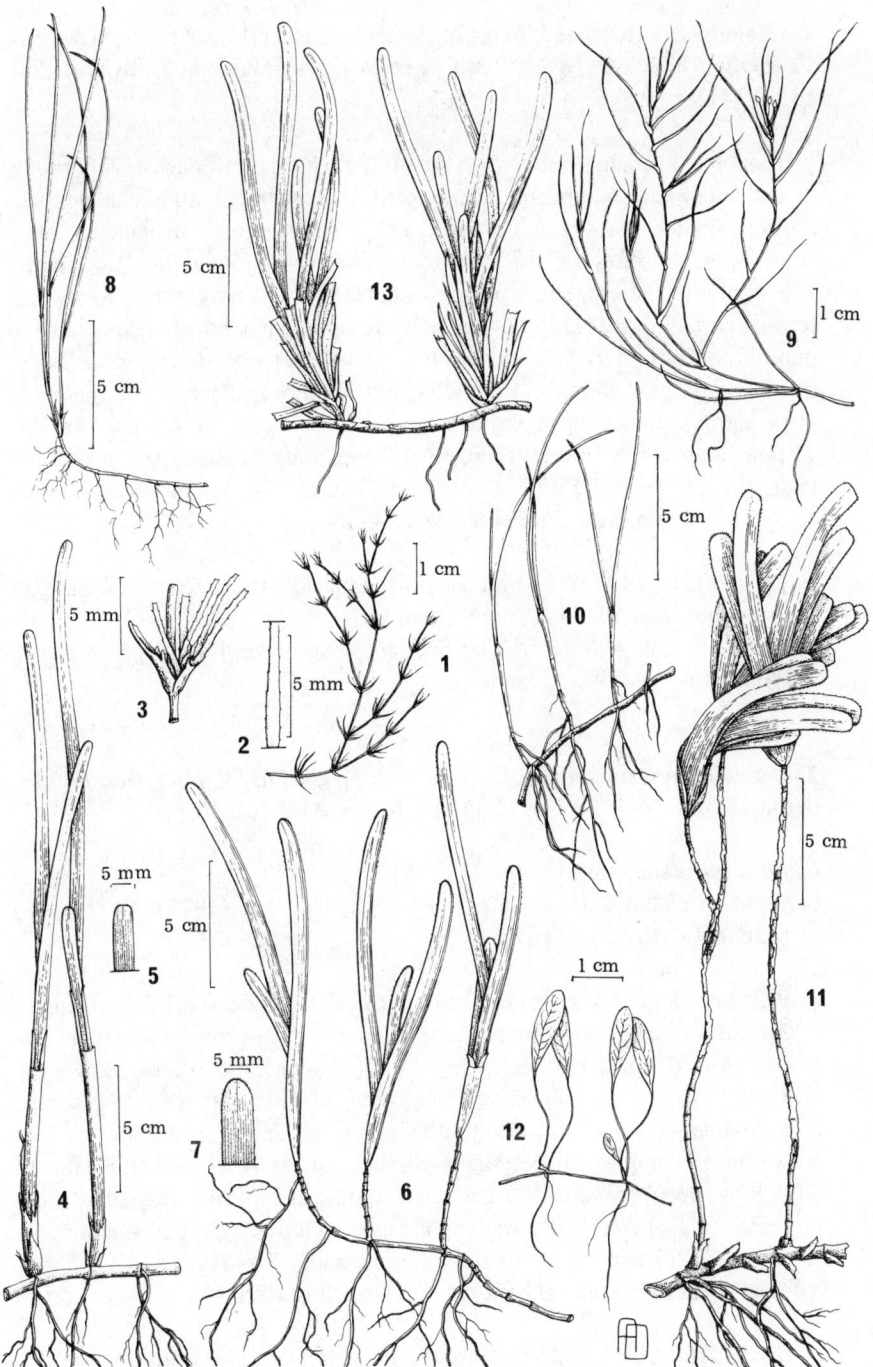

Fig. 48   1, *Najas graminea*, habit; 2, part of leaf; 3, nodal leaf cluster.
4, *Cymodocea rotundata*, habit; 5, leaf apex. 6, *Cymodocea serrulata*, habit; 7, leaf apex. 8, *Halodule uninervis*, habit. 9, *Ruppia maritima*, habit. 10, *Syringodium isoetifolium*, habit.
11, *Thalassodendron ciliatum*, habit. 12, *Halophila minor*, habit.
13, *Thalassia hemprichii*, habit.

1. **Cymodocea rotundata** Ehrenb. & Hempr. ex Aschers. in Sitz.-Ber. Ges. Nat. Fr. Berlin 1870: 84 (1870); F.M.C. 21: 14 (1950); Price in P.T.R.S. B, 260: 135 (1971).

Rhizome 2–3 mm thick when dried, branched, internodes 1–4 cm long, nodes somewhat prominent, mostly with 1–3, strong, cord-like, sometimes sparsely branched roots and a short, erect, leafy branch, arising alternately from opposite sides of the rhizome; branches to 5 cm long, internodes 1–5 mm long, alternately congested and lax, completely encircled by leaf scars. Leaves linear, blade, 3–4 mm wide, apex rounded, margins entire to minutely serrate, nerves 9–15, usually closely beset with short, oblong cysto-liths, sheath pale purplish, thin, dry, membranous, margins contiguous at base, flaring toward apex, lobes rounded to subacute or ? acute. Anthers 11 mm long. Ovary and style together 5 mm long, stigmas 30 mm or more long, spirally coiled. Fruits 1–2, sessile, subglobose, laterally compressed, 10 × 6 × 1.5 mm, beak 2 mm long. Fig. 48/4–5.

Distr. ALDABRA: W; an Indo-Pacific marine aquatic, to Carolines, Bismark Archipelago, Queensland and New Caledonia.

Notes. Forming closed mat on fine sands exposed at lowest spring tides at the southern end of West Island.

2. **Cymodocea serrulata** (R.Br.) Aschers & Magnus in Sitz.-Ber. Ges. Nat. Fr. Berlin 1870: 84 (1870); F.M.C. 21: 14, fig. 1/10 (1950).

Caulinia serrulata R.Br., Prodr.: 339 (1810).
Cymodocea ciliata sensu Price in P.T.R.S. B, 260: 135 (1971) not (Forssk.) Ehrenb. ex Aschers. (1867).

Rhizome slender, c 2 mm thick when dried, internodes variable in length, 2–3 roots and erect leafy branches sometimes on successive nodes, branches to at least 10 cm long; leaf scars 1–5 mm apart, not quite completely encircling stem. Leaves linear, c 5 mm wide, blades to at least 15 cm long, apex rounded, strongly ciliolate-toothed, margins for a short distance back of apex remotely appressed spiny-hairy-ciliolate, nerves 12–15(–17); sheaths to 3 cm long, open, margins 2–3 mm apart, flaring at top, with acuminate lobes on each side. Female flowers sessile, ovary elliptic, style 2–4 mm long, stigmas 23–27 mm long. Fruit compressed, 7–9 × 3.75–4.5 × 2 mm, 3 dorsal blunt ridges, apical beak c 1 mm long. Fig. 48/6–7.

Distr. ALDABRA: M; an Indo-Pacific marine aquatic to Ryuku Is., Carolines, Queensland and New Caledonia.

Notes. Only known from 1 collection, *Rhyne* 1188 (US) from shallow water channels in the mangrove area but possibly more common than appears from the single collection available.

2. HALODULE Endl.

Rhizome slender, herbaceous, 2 vascular bundles in cortical layer, each node bearing roots and a short, erect, leafy branch. Leaves with sheath narrow, linear, blade narrowly linear, falling before the sheath, apex variously toothed; nerves 3, midnerve stronger, lateral nerves close to the margin. Flowers dioecious, reduced to naked stamens or carpels, terminal, enclosed in or exserted from a leaf-sheath. Male flowers: pedunculate; anthers 2, joined dorsally in part but not at exactly the same level on the peduncle. Female flowers: subsessile, carpels or ovaries 2, free, with long undivided styles. Fruit a globose to ovoid, indehiscent nut, stylar beak short, pericarp hard.

Probably what is here referred to as a "flower" is a much-reduced inflorescence, therefore the term peduncle is used, rather than pedicel.

The taxonomy of this genus is poorly understood, largely due to the difficulty in obtaining fertile material.

Pantropical, with 2 or 3, or at most a very few, widespread species found in shallow salt water.

1. Leaf tip with 3 teeth; blade 0.25–3.5 mm wide   **1. H. uninervis**
1. Leaf tip with 2 lateral teeth; blade 0.3–1 mm wide   **2. H. wrightii**

1. Halodule uninervis (Forssk.) Aschers. in Boiss., Fl. Orient. 5: 24 (1882); Price in P.T.R.S. B, 260: 137 (1971).

Zostera uninervis Forssk., Fl. Aegypt.-Arab.: cxx 157, (1775).
Diplanthera uninervis (Forssk.) Aschers. in Engler & Prantl, Nat. Pflanzenf. Nacht. 1: 37 (1897); F.M.C. 21: 16 (1950).

Rhizome less than 2 mm thick, often much thinner, nodes prominent; roots 1–6(–10) per node. Leaves thread-like to linear, 0.2–3.5 mm wide, tips with 2 lateral projections and a middle one that may be shorter than or equal to (rarely longer than) the lateral ones. Male flowers: peduncle 6–20 mm long; anthers 2–3 mm long, red. Female flowers: ovary 1 mm long, style 28–42 mm long. Fruit broadly ovoid to globose, 2–2.5 mm long, apical beak up to 1 mm long. Fig. 48/8.

Distr. ALDABRA: W, M; COSMOLEDO: M; an Indo-Pacific marine aquatic, eastwards to Japan, Tonga and Queensland.
Notes. Occurs in shallow salt-water on reef flats and just below beaches, growing in sand or silt.

2. Halodule wrightii Aschers. in Sitz.-Ber. Ges. Nat. Fr. Berlin 1868: 19, 24 (1868) & in Bot. Zeit. 26: 511 (1868); Price in P.T.R.S. B, 260: 137 (1971).

Rhizome less than 2 mm thick, nodes prominent; roots 2–5 per node. Leaves thread-like to linear, 0.3–1 mm wide, tips with 2 lateral projections, rarely with a central projection of the midrib that easily breaks off. Male flowers: peduncle 12.5–23 mm long; anthers 3.5–5 mm long, red. Female flowers: ovary 1.5–2 mm long, style 10–28 mm long. Fruit ovoid or globose, 1.5–2 mm long, beak apical or lateral short.

Distr. ALDABRA: W; COSMOLEDO: M; Caribbean, east Atlantic and Indo-Pacific marine aquatic.

Notes. Occurs in shallow salt-water on reef flats and just below beaches, growing in sand or silt, apparently more frequent than *H. uninervis*.

3. RUPPIA L.

Rhizome very slender, branched, tending to zig-zag, bearing a leaf and 1 or more roots at most nodes. Leaves alternate, sheath open, blade thread-like, apparently with 1 nerve (consisting of 3 vascular strands when examined anatomically). Flowers bisexual, in 2-flowered spikes, much reduced. Perianth absent. Stamens 2, anthers sessile. Carpels 4 or more, sessile, ovules 1, pendulous; style short or absent, stigmas attached within the margin. Ovaries in fruit becoming drupelets on long, straight stalks.

A genus of very few, ill-defined species of saline habitats.

Ruppia maritima L., Sp. Pl.: 127 (1753); F.M.C. 21: 10 (1950).

R. rostellata Koch ex Reichenb., Icon. Pl. Crit. 2: 66, fig. 174/306 (1823); F.M.S.: 392 (1877).
R. maritima L. subsp. *rostellata* (Koch ex Reichenb.) Aschers. & Graeb. in Engler, Planzenr. IV, 11: 144 (1907); F.M.C. 21: 11, fig. 1/6–7 (1950).

Stems long, weak, slender, forking. Leaves linear-thread-like, to 10 cm long. Peduncles short, elongating after flowering, curved. Fruits ovoid, 2–3 mm long, more or less asymmetric, beaked, on stalks to 3 cm long. Fig. 48/9.

Distr. ALDABRA: S (east); cosmopolitan aquatic of slightly brackish to extremely saline waters.

Notes. Only known from 1 fragmentary collection from a pool in the Takamaka area, *Fosberg & McKenzie* 49293 (US).

4. SYRINGODIUM Kütz.

Rhizome prostrate, herbaceous, monopodial; many vascular bundles in cortex; roots 1 or more and a short erect leafy branch per node. Leaves circular in cross-section, with a central vascular strand, air channels and smaller, outer vascular bundles, blade soon falling, sheath lobed, persistent.

Flowers dioecious, cymose, each flower enclosed by a reduced leaf or bract with inflated sheath. Perianth absent. Male flowers: pedicel present, anthers 2, dorsally joined in lower part, attached at same level. Female flowers: sessile; carpels or ovaries 2, free, each with short style divided into 2 short, stigmatic branches. Fruit a 4-sided nut, with a dorsal ridge and short, 2-fid beak.

Pantropical, 2 species only.

Syringodium isoetifolium (Aschers.) Dandy in J. Bot. 77: 116 (1939); Price in P.T.R.S. B, 260: 135 (1971).

Cymodocea isoetifolia Aschers. in Sitz.-Ber. Ges. Nat. Fr. Berlin 1867: 3 (1867); F.M.C. 21: 15 (1950).

Rhizome slender; roots 1–3, unbranched or scarcely branched; stems erect, with 2–3 leaves. Leaves round in cross-section; sheaths to 4 cm long, blades up to 30 × 0.2 cm. Cymes up to 12 cm or more long, branched. Male flowers: pedicel to 7 mm long; anthers 4 mm long. Female flowers: sessile; ovary ellipsoid, 3–4 mm long, style branches 4–8 mm long. Fruit obliquely ellipsoid, to 4 × 2 × 1.5 mm, beak 2 mm long. Fig. 48/10.

Distr. ALDABRA: W (south); an Indo-Pacific marine aquatic, eastwards to the Ryukyu Is., New Caledonia, Tonga and Fiji.
Notes. Locally common on sandy bottoms in shallow water at or some-what below low-tide level at Passe de Femme, *Fosberg* 49442 (K, US) & *Rhyne* 1194 (US).

5. THALASSODENDRON den Hartog

Rhizome prostrate, to 5 mm thick, hard, sympodial, bearing roots on every 4th internode and an erect leafy branch and a bud on the internode following that with roots, cortex sclerenchymatic, each node bearing a sheathing scale, falling early. Leaves in 2-ranked clusters at ends of erect branches, linear, apex toothed, nerves many, with oblique cross veins, sheath compressed, open, ligulate, lobed. Flowers dioecious, solitary on dwarf branches, lateral on leafy shoots, enclosed by 4 bracts, outer bract usually lacking blade, inner 3 bracts resembling strongly reduced leaves, innermost lacking a ligule. Male flowers: anthers 2, dorsally united, almost sessile, with short, terminal appendages. Female flowers: carpels 2, free; ovary short, style short, divided above into 2, slender, awl-shaped stigmas. Fruit of 2 fertilised ovaries, 1 usually abortive, with inner bract persistent; germination frequently while still attached to mother plant; seedling free floating.

An Indo-Pacific genus of 2 species.

Thalassodendron ciliatum (Forssk.) den Hartog, Sea-grasses: 188, fig. 52 (1970).

Zostera ciliata Forssk., Fl. Aegypt.-Arab.: 157 (1775).

Cymodocea ciliata (Forssk.) Ehrenb. ex Aschers. in Sitz. Ber. Ges. Nat. Fr.
    Berlin 1867: 3 (1867); Hemsley in B.M.I.K. 1919: 133 (1919); F.M.C. 21:
    15 (1950).

Rhizome extensively branched, forming large patches, scales ovate, up to
15 mm long, tending to become fibrous before being shed, apex drying somewhat twisted; internodes 1–3 cm long; roots 1–5, branching, the smaller
branches, twisted; leafy branches 1(–2), near the distal end of every 4th
internode, usually 10–30(–40) cm or more long, simple or rarely branched,
internodes up to 1 cm long, nodes mostly crowded, leaf-scars encircling nodes,
prominent. Leaves dark purplish-green, broadly linear, 10–15 × 0.6–1.3 cm,
saddle-shaped, apex rounded to almost blunt or slightly indented, margins
sparsely spiny-toothed, especially near apex, nerves many, connected by
mostly oblique cross-veins especially near apex, teeth often forked, those near
the middle much more widely separated, sheath pinkish cream to light purple,
1.5–3 cm long, obovate, shed with blade attached, margins widely separated
above, closed at base, lobes not prominent, ligule prominent, entire, to
2.5 mm long. Anthers 6–7 mm long. Ovary and style c 6 mm long, stigmatic
branches up to 20 mm long. Detached fruiting branchlets up to 5 cm long.
Fig. 48/11.

Distr. ALDABRA: W, M (east), S; COSMOLEDO: M; an Indo-Pacific
marine aquatic extending from the Western Indian Ocean to the Solomon Is.
and Queensland.

Notes. Abundant, forming a dense mat on reef-flats, purplish-green when
viewed from above.

74. HYDROCHARITACEAE

Herbs adapted to aquatic conditions. Leaves all basal or scattered along an
elongate stem. Flowers regular, bisexual or unisexual, solitary and then
monoecious or dioecious, or in umbels subtended by 1 or 2 bracts, sessile or
peduncles long. Perianth of 1 or 2 whorls of free parts, inner often petaloid.
Stamens 3–many. Ovary inferior, 1-celled, ovules numerous, placentation
parietal. Fruit berry-like, indehiscent or somewhat irregularly splitting into
valves.

Found throughout tropical and temperate regions in both fresh and salt
waters; genera several.

1. Leaves opposite, elliptic or oblong, relatively broad   **1. Halophila**
1. Leaves alternate, linear, without ligules; cross nerves abundant **2. Thalassia**

1. HALOPHILA Thouars

Rhizome slender, creeping, buried or rarely exposed; each node producing 2 scales, a bud or usually a dwarf, leafy shoot and 1 or more, unbranched, sometimes sparsely hairy roots. Leaves in pairs, pseudo-whorls, or 2-ranked, and opposite, linear to elliptic or oblong, venation pinnate with a marginal or submarginal vein; petiole present or absent. Inflorescence terminal on lateral branch or branchlet, with 1 or 2 flowers enclosed in a spathe of 2 overlapping bracts; flowers unisexual, monoecious or dioecious. Male flowers: pedicellate; stamens 3, alternating with 3 perianth segments, anthers 2- or 4-celled. Female flowers: sessile; perianth tubular, lobes 3, reduced; ovary inferior, 1-celled, styles 3—5, slender. Fruit beaked, pericarp membraneous; seeds few to many.

Pantropical marine genus of 8 species.

Halophila minor (Zoll. ex Miq.) den Hartog in Fl. Males. I, 5: 410, fig. 17b (1957); Price in P.T.R.S. B, 260: 136 (1971); Sachet & Fosberg in Taxon 22: 441 (1973).

Lemnopsis minor Zoll. ex Miq. in Zoll., Syst. Verz. 1:75 (1854).
Halophila ovata sensu F.M.S.: 393 (1877); F.M.C. 26: 4, fig. 1/3—5 (1946), not Gaud. (1826).

Rhizome thin, fragile, 1 root per node; scales transparent. Leaves opposite, on dwarf branches, these scarcely evident, blades oblong-elliptic or narrowly obovate, up to 1.5 × 0.6 cm, apex usually obtuse, base obtuse to wedge-shaped, thin, veins usually 5—8, widely spreading; spathe obovate to sub-orbicular, keeled, 2—3 mm long, apex acute or acuminate; petiole 0.5—2 cm long. Flowers dioecious. Male flowers: pedicel 2 cm long; perianth segments erect, obtuse, 3 mm long; anthers 2 mm long. Female flowers: hypanthium 2—5 mm long, ovary to 3 mm long, styles 3, to 15 mm long. Fruit ellipsoid, ovoid or globose, 2—4 mm long, beak 2—6 mm long; seeds yellow-brown, subglobose, c 20. Fig. 48/12.

Taylor's collections have oblong-elliptic leaves with (6—)7—9(—10) pairs of veins at an angle of about 60—45° from midrib, submarginal vein rather close to margin. They have been seen by den Hartog who did not object to their determination.

Distr. ALDABRA; M (Pass Gionnet), S; an Indo-Pacific marine aquatic, from East Africa to the Marianas and New Caledonia.
Notes. Found in shallow water on sandy or muddy bottoms in Passe Gionnet, Gros Ilot, *Taylor* (BM) and Bras Takamaka, *Taylor* (BM).

2. THALASSIA Banks ex König

Rhizome prostrate, branching, monopodial, cortex with no vascular strands, internodes short, nodes not prominent, scales at each node membranous, thin, dry, falling, scale scars encircling nodes, very narrow, roots potentially 1 per node, often irregularly absent, unbranched, the proximal part, at least, densely white woolly-hairy, short, erect, leafy branchlets appearing irregularly, lateral, opposite the scales at nodes, internodes of branchlets covered by a dense mass of thin, dry, overlapping, persistent leaf-sheaths. Leaves 2-ranked, green, broadly linear, apex rounded, nerves 9−17, parallel, connected by occasional transverse cross-nerves, blade changing abruptly into a thin, dry, membranous sheath, flat above, compressed and sheathing below, closed at base, ligule absent, blade falling, sheath long-persistent. Flowers dioecious; peduncle present. Perianth segments 3. Male flowers: pedicel present, flowers 2; spathe of 2 bracts joined along 1 side; stamens 3−12, anthers oblong, erect, 4-celled, pollen in a gelatinous mass. Female flowers: subsessile, flowers 1; spathe of 2 bracts joined along both sides; ovary covered with minute, spiky projections, styles 6−8, each branched into thread-like stigmas. Fruit globose, covered with stout, blunt prickles, splitting into several spreading valves; seeds few.

A tropical genus of 2 species, 1 Caribbean, 1 Indo-West Pacific. Growing in shallow marine habitats.

Thalassia hemprichii (Ehrenb.) Aschers. in Peterm., Geogr. Mitt. 17: 242 (1871); Price in P.T.R.S. B, 260: 138 (1971).

Schizotheca hemprichii Ehrenb. in Abh. Berl. Akad. Wiss. 1852: 429 (1834). *Cynodocea rotundata* sensu Hemsley in B.M.I.K. 1919: 133 (1919), not Ehrenb. & Hempr. ex Aschers. (1870).

Rhizome 2−5 mm thick, internodes short, 4−7 mm long, with irregularly diagonal, very narrow, scale scars; leafy branchlets occasional on rhizome, at irregular intervals. Leaves up to 40 × 0.4−1.1 cm, apex jagged to subentire, margins with sparsely appressed, minute, spiny teeth near apex; sheath to 7 cm long, persistent, fraying into a papery-fibrous mass with age. Spathe leaves oblong to lanceolate, up to 25 mm long. Perianth segments elliptic, 7−8 × 3 mm. Male flowers: stamens mostly 6−9, anthers 7−11 × 1−1.5 mm. Female flowers: hypanthium 2−3 cm long; ovary 1 cm long, styles 6, 5−7 mm long, stigmatic branches 10−15 mm long. Fruit on pedicel up to 10 mm long, ovoid-globose, 2−2.5 × 1.3−3.3 cm, shortly beaked, densely covered with minute, spinose pimples, splitting into 8−20 valves; seeds 3−9, greenish-brown, conical, 8 × 8 mm. Fig. 48/13.

Distr. ALDABRA: W (south), M (east); COSMOLEDO: M; an Indo-Pacific marine aquatic, from the African coast to New Caledonia and the Marshall Is. and Samoa.

Notes. Occurs in extensive meadows on sandy and muddy bottoms.

75. CYPERACEAE

Mostly herbs, sometimes vine-like, rarely shrubs; stems (culms) solid, triangular to cylindrical. Leaves basal or in 3-ranks on stems, differentiated into a sheath and blade, or sheaths bladeless, sheath, at least to begin with, closed around stem. Flowers (florets) very much reduced, in 2-ranks or spirally arranged, in spikelets, each floret subtended by a scale-like bract. Floret composed of 1–6 stamens and/or an ovary with a 2- or 3- (or more-) branched style. Ovules 1, erect; perianth represented by bristles or scales or these absent. In some genera a bracteole (prophyll) subtends or surrounds the ovary and develops into a sac (perigynium) surrounding the fruit. Fruit a 3-angled or lens-shaped nut (achene); seed 1, free, erect.

A large world-wide family.

1. Scales of spikelets clearly in 2 ranks   **2. Cyperus**
1. Scales spirally arranged
  2. Summit of leaf sheath with ear-like appendage and with a tuft of long hairs; blade noticeably hairy   **1. Bulbostylis**
  2. Summit of leaf sheath not appendaged, without a tuft of hairs; blade not conspicuously hairy   **3. Fimbristylis**

1. BULBOSTYLIS Kunth

Perennial or annual; stemless to unbranched stems, rarely branched. Leaves usually in a dense rosette, often hairy; sheath with ear-shaped appendages at summit and a tuft of long hairs; blade slender. Spikelets umbellate, capitate or solitary, pedunculate, on slender culms or basal with the inner foliage leaves of rosette also serving as floral bracts, scales overlapping, usually spirally arranged, rarely 2-ranked. Stamens 1 to 3. Style branches 3, base of style usually bulbous, usually persistent on fruit. Achenes usually 3-angled.

Widespread tropical and subtropical genus, (by some united with *Fimbristylis*, to which it is very close); often found in savannas, sandy, or rocky open places.

1. Florets and achenes basal, in axils of reduced foliage leaves; plant a tiny tuft of leaves with no culm evident   **1. B. basalis**
1. Florets borne in spikelets solitary or usually paniculate, at summits of obvious culms; plant tufted with culms 10–30 cm tall   **2. B. hirta**

Both the Aldabra species of *Bulbostylis* lack the persistent bulbous style-base on the achene, having enlarged but scarcely bulbous bases which fall with the styles. This makes them intermediate in the technical distinction between *Bulbostylis* and *Fimbristylis*. Because their habits are that of *Bulbostylis* they are placed in this genus, although their existence weakens the distinction between the 2 genera.

1. Bulbostylis basalis Fosberg in K.B. 31: 829 (1977). Type: Aldabra, West Is., Bassin Cabri, *Merton* 7068 (US, holo., K, iso.).

Diminutive, densely tufted plant, consisting of crowded rosettes, culms so reduced that plants are essentially stemless, axis apparently not more than 5 mm long, covered by crowded leaf sheaths. Leaves needle-like, up to 7.5 cm long, usually much shorter, c 0.3 mm thick above sheath, strongly channelled above, crescent shaped in transverse section; minute, bristly hairs sparse but noticeable; sheaths brown, broadly lanceolate-ovate, c 5 mm long, expanded, open, margins very thin, translucent, ciliate on outer leaves, glabrous except at summit on inner leaf sheaths which have ear-shaped appendages and very long, tangled hairs. Floral bracts (scales) not sharply distinct from the leaves; outer scales like reduced leaves, blades shorter; blades of inner bracts progressively shortened to prominent spinose tips, sheaths less or not at all appendaged, less to not at all ciliate at summit; fertile scales lanceolate-ovate, 3–5 mm long, apex acuminate to a green or brown, blunt, spine-like tip, 3 glossy brown, strong nerves close together dorsally, expanded part reddish tan. Stamens 1 per floret, filaments elongate, flattened, very thin, translucent; anthers c 1.7–1.8 mm long, broadly linear, apex with spinose tip. Style not at all fringed, somewhat enlarged at base but not bulbous, entirely falling from achene, entire part c 2 mm long, branches 3, very elongated, very slender. Achenes pale brown to dull medium brown, obovate in outline, 1.0–1.2 mm long, apex rounded, narrowed to an almost shortly stalked base, extremely variable in this respect, broadly 3-angled to almost plano-convex with faint keel, strongly and sharply transversely wrinkled. Fig. 49/1.

Distr. ALDABRA: W.P.S; endemic.
Notes. A tiny tufted annual, behaves as an ephemeral, green at first, turning orange with drought. Locally common on very thin fine silty soil, in places flooded during rainy season, also in crevices in limestone; heavily grazed and seed locally dispersed by tortoises.

2. Bulbostylis hirta (Thunb.) Svenson in Contrib. Ocas. Mus. Hist. Nat. Coll. La Salle, Havana 4: 11 (1946).

Cyperus hirtus Thunb. in Hoffm., Phytogr. Blaetter 1: 6 (1803).
Fimbristylis exilis (H.B.K.) Roem. & Schult., Syst. 2: 98 (1817); Turrill in B.M.I.K. 1919: 140 (1919).

Densely tufted sedge; stems to 30 cm or more tall, striate, finely shaggy-haired. Leaves coarsely needle-like, shorter than stems, striate, softly hairy; sheaths with broad, brown, thin, dry margins, sparsely to copiously long-hairy at summits. Inflorescence of 1 sessile spikelet subtended by a bract, usually with 1–3 ray peduncles 2–3 mm long, each subtended by a bract equalling or exceeding the spikelet, each ray bearing a spikelet subtended by a short, awl-shaped bract; base of whorl with long, white hairs. Spikelets narrowly elliptic to narrowly oval, 5–10 × 1.5–2 mm, scales ovate, chestnut brown, rarely very dark, paler towards the margins, minutely rough to the

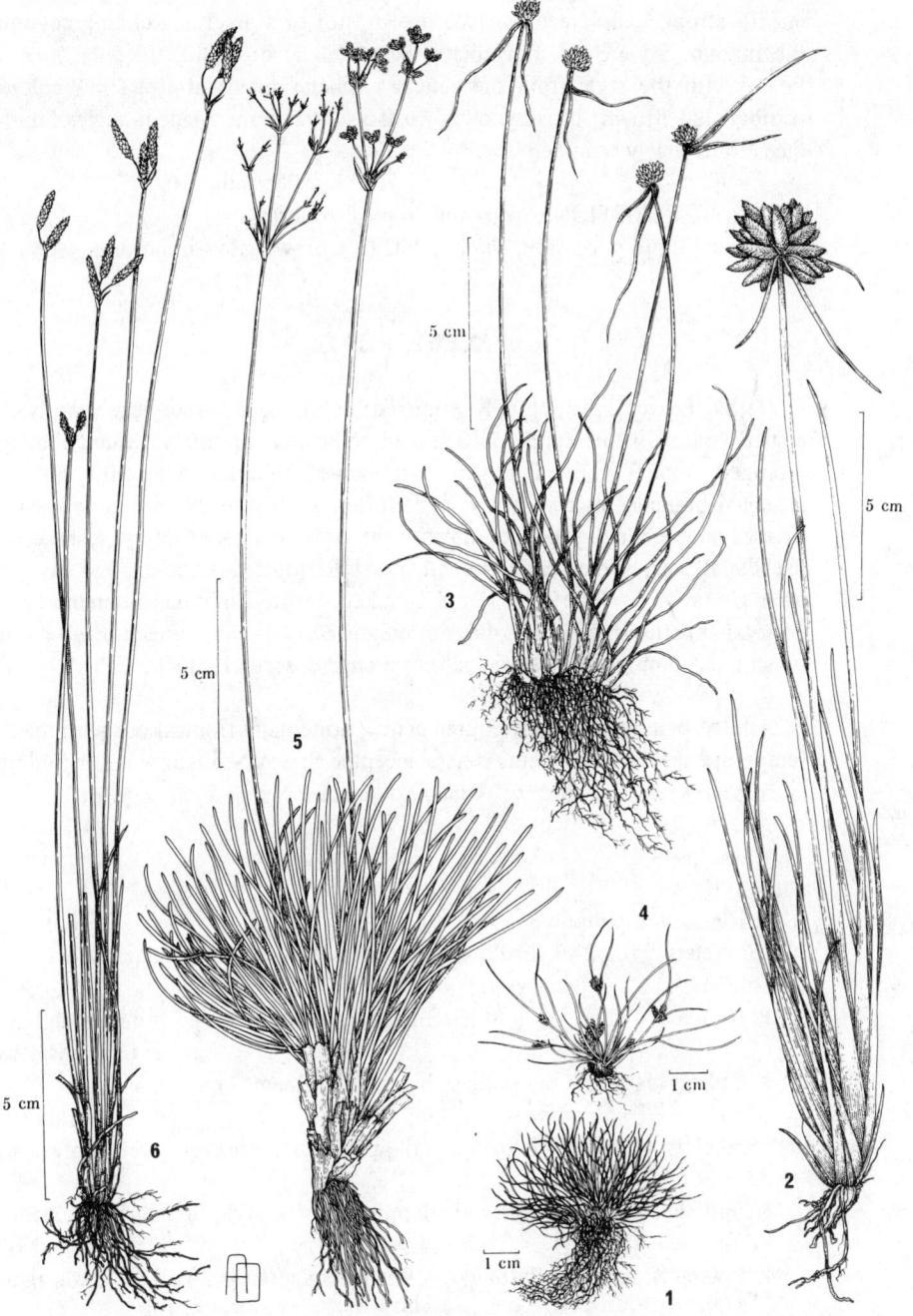

Fig. 49   1, *Bulbostylis basalis*, habit. 2, *Cyperus conglomeratus*, habit.
3, *Cyperus dubius*, habit. 4, *Cyperus pumilus*, habit.
5, *Fimbristylis cymosa*, habit. 6, *Fimbristylis ferrugineus*, habit.

touch, not or scarcely pouched (in Assumption and most African specimens), midrib strong, pale green, or pale brown, not or scarcely extending beyond the margin. Style 3-fid, not ciliate, base conical or ovoid, the enlargement falling with the style from the achene. Achene dirty white or straw colour turning dull brown, broadly ovoid, c 0.5 × 0.5 mm, 3-sided, angles blunt, faces transversely wrinkled.

Distr. ASSUMPTION; Africa and tropical America.
Notes. Found once only, *Dupont* 102 (K), growing in wet ground.

2. CYPERUS L.

Herbs. Leaves basal, spirally arranged; culms rounded to very sharply 3-angled, solid. Inflorescence of 2-ranked spikelets, capitately to umbellately arranged, umbel formed of an extremely condensed axis with spirally arranged branches or rays, leafy bracts often 2–3 or more, subtending heads or rays and becoming reduced upward on axis; rays with prophylls; spikelets usually at least somewhat flattened, rhachilla jointed at base or at base of each glume or not jointed. Florets 2-ranked, mostly bisexual. Stamens 1–3. Ovary 1, bristles 0; style 2–3-fid, not enlarged at base nor fringed. Fruit a nut or achene, compressed (lens-shaped) or 3-angled; seed 1, erect.

A large practically cosmopolitan genus, principally tropical but with many temperate species. The genus is here accepted in a very broad sense, including such segregates as, *Pycreus* Beauv., *Mariscus* Vahl, *Kyllinga* Rottb., and others.

1. Inflorescence capitate
  2. Spikelets with 2–3 fertile scales; florets 1–2; style branches 2; achenes biconvex
    3. Culms crowded on a thick rhizome, tending to be bulbous at base
                                                        1. C. aromaticus
    3. Culms scattered on slender branched rhizome, not at all bulbous at base   **2. C. bigibbosus**
  2. Spikelets with more than 3, overlapping, fertile scales; florets more than 2
    4. Spikelets rounded; scales overlapping, 6–10; style branches 3; achenes 3-angled   **5. C. dubius**
    4. Spikelets strongly flattened; scales conspicuously 2-ranked, more than 10
      5. Diminutive plants; spikelets in loose heads, linear, dark brown; style branches 2; achene oblong to almost orbicular   **8. C. pumilus**
      5. Plants 15 cm or more tall; spikelets in tight heads, ovate to elliptic, whitish; style branches 3; achenes 3-angled
        6. Culms round in cross-section, stiff, glaucous, 2–4 mm thick; leaves stiff, appearing awl-shaped   **4. C. conglomeratus**

6. Culms weak, not conspicuously rounded or glaucous; leaves flat, slender, not at all stiff **7. C. niveus**
1. Inflorescence umbelloid, spicate, or paniculate
7. Culm bases conspicuously bulbous; spikelets in simple spikes or spikes somewhat branched at base; leaves weak, very slender **3. C. bulbosus**
7. Culms bases not at all bulbous; spikelets in umbels or panicles
8. Diminutive plants; spikelets compressed, linear, scales with apices somewhat curved outward; style branches 2 **8. C. pumilus**
8. Robust plants; spikelets scarcely compressed, lanceolate to narrowly oblong; scales appressed, even at tips; style branches 3 **6. C. ligularis**

Cyperus articulatus L. cited by Turrill in B.M.I.K. 1919: 133 (1919) is not known from the flora area. The specimen cited, *Dupont* 295 (K), is from Mahé.

1. Cyperus aromaticus (Ridley) Mattf. & Kük. in Engler, Pflanzenr. 101, IV, 20: 581 (1936).

Kyllinga aromatica Ridley in Trans. Linn. Soc. II, Bot. 2: 145 (1895).
K. polyphylla Willd. ex Kunth, Enum. 2: 134 (1837): F.M.S.: 415 (1877), not *Cyperus polyphylla* Vahl (1806).

Culms several to many, erect, strong, striate, to 50 (−80) cm tall, congested on a thick, hard, horizontal rhizome, bases thickened and covered by strong, stiff, purple, scale-like, overlapping, reduced leaves, outer scales c 1 cm long, inner c 10 cm long, bladeless, then well developed. Leaves with blades up to ½ or more as long as culms, blades flat, 2.5−5 mm wide, apex somewhat acuminate, sheath truncate on side opposite blade. Spikelets in dense, broadly ovoid to globose heads, these tending to be compound, e.g. with smaller ones crowded at base, to 1 cm long; involucral bracts 7, unequal, longest up to 20 cm or more, leaf-like. Spikelets straw-coloured, oblique, lanceolate, to 4 mm; scales 3, 2 sterile, 1 fertile: outermost scale smooth, about one third as long as spikelet, subacute; second almost as long as spikelet, with 5−7 widely spaced nerves, keel folded and somewhat pouched, with sparse and minute spines, ending in a stout spinose tip originating below the 2-fid apex; fertile scale slightly longer and straighter but otherwise similar. Stamens 3−6, anthers linear, 2 mm long, apex with an abrupt spinose tip, base slightly arrow-shaped. Style 2-fid. Achene glossy, dark brown, somewhat obliquely ovoid or oblong, c 1.2 mm long, somewhat compressed, with white, isodiametric cellular reticulation.

A widespread African species. The variety *aromaticus* is not known from Aldabra.

var. **elatus** (Steud.) Kük. in Engler, Pflanzenr. 101, IV, 20: 582 (1936).

Kyllinga elata Steud., Syn. Pl. Glum. 2: 70 (1855).

As described above, characterized by its thick congested rhizomes, more numerous subsidiary heads and spikelets broadly lanceolate rather than ovate, up to 4 mm long. Fig. 50/1.

Distr. ALDABRA: W; East Africa and Madagascan regions, Comoros and Seychelles.

Notes. Only known from Aldabra by the collections behind Settlement, *Fosberg* 48811 and *Renvoize* 1404 (both K, US).

2. Cyperus bigibbosus Fosberg in K.B. 31: 832 (1977). Type: Aldabra, South Is., Takamaka (Wilson's) Well, *Fosberg* 49333 (US, holo., K, iso.).

Loosely tufted; rhizomes ascending, branched, rather slender, with fibrous roots; culms slender, erect or ascending, to c 40 cm tall, usually shorter, not noticeably bulbous at base, bases closely sheathed by almost bladeless, purplish sheaths, 2–3 cm long, very thin, truncate at apex, tipped with a rudimentary blade, inner sheaths with progressively developed blades, innermost with blades almost or quite equalling culms, 1.5–2 mm wide, apex long-acuminate, margins minutely rough to the touch. Spikelets white, sometimes, when old, turning brown on drying, in globose heads 5–10 mm in diameter; whorl of bracts 3–5, unequal, longest to 25 cm. Spiklets strongly compressed, strongly oblique, 2–3.5 × 1.2 mm; scales 5, 2 sterile, 3 fertile; outer sterile scale boat-shaped, strongly acuminate, one third to half the length of spikelet; inner sterile scale broader, thinner, about a quarter to one third the length of spikelet, acute to sub-obtuse, very thin, scarcely keeled; lowest fertile scale subtending a bisexual floret, very broadly ovate or elliptic, 3/4 or more as long as spikelet, apex acute, alightly hooded, with or without several, short, pointed bristles closely investing spikelet, scarcely keeled, somewhat pouched just above middle; second and third fertile scales very thin, very broadly ovate enclosing a male floret, apical portions inrolled forming a tubular tip through which stamens and style are exserted. Stamens 3 per floret, anthers linear (1.0–) 2.0 mm, blunt at both ends. Styles 2, branched, branches white and slender. Achene dark brown, strongly compressed, biconvex, oblong-elliptic, 1.4–1.8 mm long, ends rounded, minutely, longitudinally, cellular pitted, showing irregularly distributed, oblong, darker drown markings when moistened. Fig. 50/2.

Distr. ALDABRA: S (east); endemic.

Notes. So far as known this species is endemic to Aldabra where it is found on 'platin' or flat-lying, slabby limestone, or on rough limestone, in open to closed mixed scrub vegetation on the eastern part of South Island.

3. Cyperus bulbosus Vahl, Enum. 2: 342 (1806).

Culms slender, 5–20 cm tall; roots fibrous, very fragrant, from base of hard, ovoid, blackish bulbs, 6–9 mm long; base of culms invested by fibrous remains of leaf-sheaths. Leaves few to numerous, equalling or shorter than

Fig. 50   1, *Cyperus aromaticus* var. *elatus*, habit. 2, *Cyperus bigibbosus*,
habit. 3, *Cyperus bulbosus*, habit. 4, *Cyperus ligularis*, habit.
5, *Cyperus niveus* var. *leucocephalus*, habit.

culms, very slender, lower parts to 2 mm wide, inrolled, especially distally, appearing almost needle-like, tapering to very slender tips. Spikelets loosely arranged in simple or, near base, somewhat branched spikes, lower spikelets in fascicles of 2–3 (–4), with 2–4, unequal, leaf-like 'involucral' bracts, these scattered on the rachis, 1 for each branch or fascicle of spikelets. Spikelets maroon or deep chestnut to greenish-brown, linear or lanceolate-linear, 5–15 × 1–2 mm, not strongly compressed, scales 6–20, ovate, c 2.5–3 mm long, apex sub acute to rounded, sometimes almost spine-tipped, completely covering rhachilla, noticeably striated, midrib greenish, margin thin, very narrow, not well-marked. Anthers linear-oblong. Style branches 3, elongated, tending to coil. Achene grey, c 1–1.2 mm long, 3-angled, angles blunt, cellular reticulate, cells weakly hexagonal. (Description mostly from extra-Aldabra material). Fig. 50/3.

Distr. ASSUMPTION; Widespread in the Old World tropics from West Africa to Malesia and Australia.

Notes. Two sterile collections from Assumption, probably adventive, as it occurs between the rail tracks in the guano works, *Frazier* 763 & 784 (both US).

4. Cyperus conglomeratus Rottb., Descr. et Icon. Nov. Pl. 21, fig. 15/7 (1773).

C. maritimus Poir. in Lam., Encycl. Méth. Bot.   7: 240 (1806), not *C. maritimus* Bojer (1837).

Stiff, densely tufted plant, 20 cm to almost 1 m tall, glaucous, glabrous; culms 2–4 mm thick, weakly striate. Leaves usually shorter than culms, stiff, flat but strongly inrolled and appearing awl-shaped, up to 75 cm long, tapering gradually to a rather blunt point, base strongly expanded, margins scarcely rough; sheath open, brown, strongly striate without. Spikelets many, in very compact, globose heads 1.5 to 3.5 cm in diameter, subtended by 2–4, strongly unequal, leaf-like bracts up to 2–3 cm or more long. Spikelets dull white or pale brown, compressed, ovate, up to 5 × 15 mm, acute; scales up to 12 per side, closely overlapping, elliptic, c 4 mm long, apex subobtuse and keeled, concave. Stamens 3, filaments flat, anthers oblong. Style 3-fid. Achene subglossy to glossy, dark brown, slightly obovate to oval, 1 × 1.5 mm, apex at base rounded, plump, 1 side obscurely angled.

This is a very complex species, here accepted in a very broad sense, compared with the treatment by Kükenthal in Engler, Pflanzenr. 101, IV, 20: 267–276 (1936). Detailed study over its entire African, Asiatic, and Indian Ocean range may possibly support the recognition of more entities, but so far the available treatments do not seem to fit the array of specimens examined. Fig. 49/2.

Distr. ALDABRA: W, S (west), Esprit; widespread through the tropics of the Old World.

Notes. Found along the strand line of sandy beaches. Occurs near the Research Station on West Island, the extreme west of South Island and on Ile Esprit. Flowers mainly during the mid wet season but occasional rain can stimulate flowering.

Vernac. 'laiche'.

5. Cyperus dubius Rottb., Descr. et Icon. Nov. Pl.: 20 fig. 4/5 (1773); F.M.S.: 409 (1877).

Mariscus dregeanus sensu Hemsley in B.M.I.K. 1919: 133 (1919), not Kunth (1837).

Tufted, clumps with few to many, weak culms, variable in length, up to 30 or more cm, striate when dry; base somewhat bulbous, basal 3—5 cm invested by a persistent, brown, striated, leaf sheath. Leaf blades thin, equalling or exceeding culms, 1—2 (—3) mm wide, margins inrolled. Spikelets in white, ovoid heads 0.5—1 cm or more long and wide, subtended by 2—3 (—4) leaf-like, unequal, usually reflexed bracts. Spikelets scarcely compressed, narrowly ovoid, 2.5—5 mm long, apex somewhat 2-lobed, base oblique; scales 6—10, whitish to slightly yellowish, ovate, 2 mm long, apex rounded but with a thick, glandular tip, prominently striate. Anthers oblong, 0.7—0.8 mm long, connective reddish brown. Style 3-branched. Achene chestnut brown, obovoid or ovoid, 1 × 0.5 mm, apex at base subtruncate strongly 3-sided, angles not sharp, smooth. Fig. 49/3.

Distr. ALDABRA: W, M, S, Esprit; ASSUMPTION; widespread in tropical Africa and Asia.

Notes. Fairly common on rough limestone, sand, and platin. Flowers from mid to late wet season.

Vernac. 'laiche' or 'oignon'.

6. Cyperus ligularis L., Pl. Jam., Pugil. 3: (1759).

Mariscus rufus H.B.K., Nov. Gen. Sp. Pl. 1: 126, fig. 67 (1815); Turrill in B.M.I.K. 1919: 133 (1919).

Tufts of several to many triangular culms up, to 1.3 m tall, spreading, prostrate to erect. Leaves grey-green, basal or nearly so, rather numerous, shorter than to as long as, or even slightly longer than culms, margins minutely and remotely rough, gradually dilated below into sheaths which are ruptured and open when mature, outer leaves reduced to bladeless sheaths a few cm long. Inflorescence a compound umbel with as many as 10—12, strongly unequal rays up to 10—15 cm long, strongly divergent, the distal portion branched, spikelets crowded in broadly cylindrical to ovate or globose heads. Spikelets dark red-brown, or bright rusty brown, fading to dull brown, lanceolate to narrowly oblong, 3.5—7 m long, apex acute, base often slightly curved, disarticulating at base from persistent rachis, scarcely compressed; scales 4–6, strongly alternating, closely overlapping, broadly ovate, apex obtuse, margins translucent, concave, 11–13-nerved, weakly keeled. Stamens 3, anthers broadly linear. Style slender, 3-fid. Achene dark blood-red to brownish, somewhat obovate 1.5 × 0.8 mm, strongly 3-angled, smooth. Fig. 50/4.

Distr. ALDABRA: W, M, S, Esprit; ASSUMPTION; COSMOLEDO: M; ASTOVE; pantropical.

Notes. Flowers mainly during the wet season but unseasonal rains can stimulate flowering. Plant eaten and seeds locally dispersed by tortoises.

Vernac. 'laiche'.

7. Cyperus niveus Retz., Obs. Bot. 5: 12 (1789).

var. **leucocephalus** (Kunth) Fosberg in K.B. 31: 835 (1977).

C. obtusiflorus Vahl, Enum. 2: 307 (1806).

C. sphaerocephalus Vahl var. *leucocephalus* Kunth, Enum. 2: 45 (1837).

C. compactus Lam., Tabl. Encycl. Méth. Bot.1: 44 (1791); Baker in B.M.I.K. 1894: 151 (1894); Schinz in A.S.N.G. 21: 82 (1897); Voeltzkow in A.S.N.G. 26: 549 (1902); Dupont, Report: 41 (1907); Turrill in B.M.I.K. 1919: 133 (1919), not *C. compactus* Retz. (1788).

C. sp. near *C. debilissimus* Baker, Turrill in B.M.I.K. 1919: 133 (1919).

C. compressus sensu Turrill in B.M.I.K. 1919: 133 (1919), not L. (1753).

Tufted herb with few to many, weak, rather slender culms, 10–50 cm tall; quite leafy at base. Leaves shorter than to exceeding culms, flat, 3 mm wide near base, tapering gradually to a slender apex, glabrous except for very minutely roughened margins, leaf bases very dark brown or purplish. Spikelets white to dull pale brown, flat, ovate, to 15 (–20) × 8 mm, apex acute to obtuse. Inflorescence a head of 5–10 (–12) spikelets subtended by 4, unequal, leaf-like bracts, 3–12 cm or even 6–30 cm long in some inflorescence, scales white, strongly overlapping, ovate, 4–4.5 × 3 mm, apex subobtuse, flattened, keeled and distally somewhat crested. Stamens 3. Style branches 3, drying reddish or brownish. Achenes grey, oblong-ovoid, c 1.2 × 1 mm, strongly 3-angled, angles black. Fig. 50/5.

This plant has generally been known as *C. compactus* Lam. or *C. obtusiflorus* Vahl, but scarcely differs from the Asiatic *C. niveus* Retz.; var. *niveus* from south Asia with pale, brownish white, ovate-lanceolate spikelets and 2 involucral bracts has not been recorded from Aldabra.

Distr. ALDABRA: W, P, M, S; tropical Africa, Madagascar.

Notes. Fairly common on limestone in both open and closed scrub communities. Flowers mainly during the wet season. Achenes eaten and locally dispersed by fody.

8. Cyperus pumilus L., Amoen. Acad. 4: 302 (1759) & Sp. Pl. ed. 2: 69 (1762).

Tufted glabrous annual; culms slender, to 25 cm tall (Aldabra material much smaller). Leaves usually shorter than culms, or exceeding them in dwarfed specimens, very narrow, at most 1.5 mm wide, usually much narrower. Bracts 1–3, greatly exceeding inflorescence, spikelets in loosely globose heads, compound in large specimens. Spikelets very flat, linear-oblong, to 15 × 1 mm: scales many, sharply keeled, median area green to brown, extending into a strong, blunt, slightly out-curved spiny tip, sides with 2 or 3 nerves, margins translucent, abruptly expanded, overlapping in 2 ranks, divergent when mature, into broad wing. Stamens 2, anthers broadly elliptic. Style branches 2. Achene dull brown to greyish black, or white (possibly fully formed but immature), rather plump, broadly oblong to almost as broad as long, c 0.5 × 0.3–0.4 mm, apex subtruncate to slightly notched, style base persisting. Fig. 49/4.

Distr. ALDABRA: S (east); pantropical.

Notes. Found in areas heavily grazed by tortoises near Cinq Cases; *Hnatiuk* 731514 (US) & 731548 (K). The Aldabra plants are of an extremely dwarfed, stemless form with the inflorescence sessile among the basal leaves, both leaves and bracts thread-like. Flowers during the mid wet season. Achenes locally dispersed by tortoises.

3. FIMBRISTYLIS Vahl

Leaves basal, scapes naked, except for a bract subtending the inflorescence. Spikelets solitary, capitate or cymose; scales and subtended florets arranged spirally, or rarely in 2, irregular ranks; bristles absent. Stamens 3, hypogynous filaments flattened, anthers or anthers with filaments shed. Style flattened, fringed on 2 edges and basal part of 2–3 (rarely more) branches, or not, style falling. Achene plano-convex or 3-angled, without persistent style base.

Pantropical and extending to some extent into warm temperate regions.

The species often referred to as *Fimbristylis hispidula* Kunth or *F. exilis* (H.B.K.) Roem. & Schultes is here regarded as a *Bulbostylis* and is treated in that genus as *B. hirta* (Thunb.) Svenson.

1. Culms hard, arising from usually clumped rosettes of linear, well-developed, usually stiff leaves; style glabrous; achenes black or brown   **1. F. cymosa**
1. Culms hollow, leaves 2–4 at base of each, in tufted clumps; styles fringed; achenes white, yellow, or reddish tan   **2. F. ferruginea**

1. Fimbristylis cymosa R.Br., Prodr.: 22 (1810).

F. obtusifolia sensu Baker in B.M.I.K. 1894: 151 (1894); Dupont, Report, 41
(1907); Turrill in B.M.I.K. 1919: 134 (1919), not (Lam.) Kunth (1837).
F. spathacea Roth, Nov. Pl. Sp.: 24 (1821); Turrill in B.M.I.K. 1919: 134
(1919).
F. glomerata (Retz.) Nees in Linnaea 9: 290 (1834); Turrill in B.M.I.K. 1919:
134 (1919).

Usually tufted plant, variable in habit, responding vegetatively to moisture
differences by changes in stiffness, size, and luxuriance; roots with a pleasant
spicy odour. Leaves in dense rosettes, narrowly linear, scarcely or not
tapering, apex blunt, flattened, tending to be stiffish. Culms tending to be
stiff, erect to spreading; cymes subcapitate to open, branched or umbelloid,
inflorescence bracts reduced. Spikelets few to many, ovoid to broadly
lanceolate-ovoid, apex rather obtuse; scales closely overlapping, rusty brown
with translucent margins, broadly ovate-obtuse. Stamens 2–3, filaments
flattened. Style flattish, margins not fringed, glabrous, 2- or 3-(rarely more-)
branched, branches recurved. Achenes black or dark brown, obovate,
flattened or somewhat 3-angled, surface covered with wart-like projections
to smooth. Fig. 49/5.

This is a pantropical species, usually but by no means always coastal,
extremely variable and with a number of synonyms, mostly representing
variations in habit, inflorescence branching, number of style branches, shape
and surface of achenes. Those Aldabra plants with 2-fid styles and plano-
convex, minutely pimply achenes would fall in what is often called
F. spathacea Roth or *F. cymosa* var. *spathacea* (Roth) Koyama.
The Aldabra populations of this species do not seem to fit well into the
varieties recognized elsewhere, as the style branches may be either 2 or 3,
often on the same plant, and the achenes likewise lens-shaped or 3-sided and
much less papillate than usual for the extra-Pacific forms.
A diminutive form of this species (*Hnatiuk* 731550) occurs on Aldabra,
identical with the larger ones in every respect except size. It is very slender
and only a few cm tall, with almost thread-like leaves, spikelets 1–4,
clustered, clusters with 1–2 rays bearing smaller clusters, style branches 2–3,
achenes medium brown to dark brown. It seems to be a mere habitat-induced
fluctuation, but retained its small size when a turf obtained from Cinq Cases
was raised in a garden at the Station and protected for a year from grazing by
tortoises; unfortunately there is no information as to how much it was
watered. We are at present of the opinion that it has no genetic basis and we
are therefore not naming it as a distinct taxon.

Distr. ALDABRA: W, M, S, Esprit; ASSUMPTION; COSMOLEDO: W, M;
ASTOVE; pantorpical.
Notes. Extremely common, generally in open places, especially where the
soil is thin or almost absent. Flowers throughout the wet season.
Vernac. 'herbe oignon' or 'laiche'.

2. Fimbristylis ferruginea (L.) Vahl, Enum. Pl. 2: 291 (1806); F.M.S.: 419 (1877); Dupont, Report: 41 (1907); Turrill in B.M.I.K. 1919: 134 (1919).

Scirpus ferruginea L., Sp. Pl.: 50 (1853).

Densely tufted plant; rhizomes congested; culms erect or ascending, hollow to 1.5 mm thick, to 70 cm tall or more, with 2-4 brown leaf sheaths at base, lower sheath spathe-like, upper tubular, with reduced, linear-awl-shaped blades, 1–3 (–4) cm long. Inflorescence cymose-umbelloid, with up to 7 spikelets on pedicels of varied length, subtended by awl-shaped bracts up to 15 mm long, central spikelet usually sessile or subsessile. Spikelets narrowly cylindrical-ovoid, lanceolate in outline, varying greatly in size, 4–27 mm long; scales closely overlapping, broadly ovate, apex subobtuse, strongly spine-tipped, midrib strong, marked with chestnut or rusty-brown, parallel lines, central portion slightly grey, margins tending to be very thin, translucent. Stamens 2, filaments flat, very short to exceeding length of scale, anthers linear, included and sterile to well-exserted and fertile. Style flat, strongly fringed-ciliate distally, branches 2, basally ciliate, exserted or variously reduced, stigmatic portions erect on reduced styles, strongly divergent, recurved or even coiled when exserted. Achenes white to pale reddish tan or dull yellow, broadly obovoid to suborbicular, 1 × 0.8 mm, smooth. Fig. 49/6.

In the same spikelet, stamens, styles and achenes can be found in various stages of development.

Distr. ALDABRA: S (east); tropical and warm temperate countries nearly throughout the world.

Notes. So far only collected from the eastern end of South Island, where it is locally abundant on plantin, especially near freshwater pools. Flowers throughout the wet season. Achenes locally dispersed by tortoises.

76. GRAMINEAE

Annual or perennial herbs; stems jointed, usually hollow at the internodes and solid at the nodes. Leaves solitary, arranged alternately at the nodes, consisting of sheath, ligule and blade; sheath surrounding the stem with margins free and overlapping or joined, sometimes with small protuberances (auricles) at the mouth; ligule a membranous flap or fringe of hairs at the junction of the sheath and blade; blade continuous with the sheath (rarely with a small, false petiole) usually linear to lanceolate but frequently ovate, wiry or needle-shaped. Inflorescence a collection of spikelets in a panicle or in spikes or racemes, usually solitary, digitate or arranged along a central axis. Spikelets consisting of bracts arranged in 2 ranks on a slender axis (rhachilla); lower 2 bracts (glumes) empty, the succeeding 1–many bracts (lemmas) each enclosing a flower and opposed by a translucent scale (palea); lemma, palea and flower together termed a floret; glumes or lemmas often bearing 1 or more stiff bristles (awns). Flowers usually bisexual, sometimes unisexual, perianth represented by 2, rarely 3, minute, translucent or fleshy scales (lodicules). Stamens usually 3, rarely 6. Ovary 1-celled; styles $(1-)2(-3)$, stigmas plumose. Fruit a caryopsis, pericarp thin, united to the seed and often combined with various parts of the spikelet to form a false fruit.

An enormous family of c 9,000 species, distributed throughout the world.

1. Plants with tall, woody, persistent stems   **1. Bambusa**
1. Plants with herbaceous stems
  2. Plants distinctly aromatic when crushed   **3. Cymbopogon**
  2. Plants not disntinctly aromatic when crushed
    3. Some or all spikelets with a conspicuous awn 4–20 mm long
      4. Lower glume wrinkled (in our area), sessile spikelets only with a bent
        / awn   **11. Ischaemum**
      4. Lower glume smooth, all spikelets with a straight awn
        5. Racemes digitate; decumbent annual   **5. Daknopholis**
        5. Racemes solitary; erect perennial   **8. Enteropogon**
    3. Spikelets awnless or with a short awn point, at most 2 mm long
      6. Spikelets sunk in the main axis
        7. Main axis fragile, readily breaking up at maturity   **12. Lepturus**
        7. Main axis tough, not breaking up at maturity   **18. Stenotaphrum**
      6. Spikelets not sunk in the main axis
        8. Spikelets all sessile on the rhachis or main axis; inflorescence of 1 or
          more racemes terminating the stem
          9. Spikelets surrounded by a whorl of bristles
            10. Bristles spinose, flattened, fused at the base   **2. Cenchrus**
            10. Bristles thread-like, free at the base   **15. Pennisetum**
          9. Spikelets without a whorl of bristles, arranged along 1 side of the
            rhachis only
            11. Robust, densley tufted plants; leaf-blades erect, rigid, circular in
              cross-section   **16. Sclerodactylon**
            11. Loosely tufted, decumbent or spreading plants
              12. Rhachis of raceme projecting as a small point beyond the
                uppermost spikelets   **4. Dactyloctenium**
              12. Rhachis of spike or raceme terminating in a spikelet
                13. Inflorescence of 2 spikes   **14. Paspalum**
                13. Inflorescence of more than 2 spikes   **7. Eleusine**
        8. Spikelets, or some of them, distinctly pedicelled; inflorescence a
          panicle or of racemes scattered along a central axis
          14. Inflorescence of 2 different kinds; terminal (male), consisting of
            several, spreading racemes; axillary (female), a stout spike
            (cultivated)   **19. Zea**
          14. Inflorescences all alike
            15. Inflorescence an effuse, branching panicle or spike-like, with the
              branches contracted around the central axis

16. Spikelets many-flowered   **9. Eragrostis**
16. Spikelets 1- or 2-flowered
   17. Lemma soft, spikelets 1-flowered   **17. Sporobolus**
   17. Upper lemma hard, shining or wrinkled, shell-like, spikelets
      2-flowered   **13. Panicum**
15. Inflorescence of several, narrow, digitate or subdigitate racemes
   18. Spikelets with a small bead-like swelling at the base
                                   10. Eriochloa
   18. Spikelets without a bead-like swelling at the base   **6. Digitaria**

1. BAMBUSA Schreb.

Perennials; stems stout, woody, persistent. Leaf-blades flat, with transverse and logitudinal venation. Inflorescence paniculate or racemose. Spikelets 1-many-flowered, all florets usually bisexual. Glumes 2, membranous and unequal, lemmas longer than the glumes. Lodicules 3, ovate or lanceolate, ciliate. Stamens 3, rarely 6.

About 50 most Asiatic species but a few also in North and South America and South Africa.

Bambusa vulgaris Schrad. ex Wendl., Coll. Pl. 2: 26, fig. 47 (1810).

Erect bamboo, 3–5 m high; stems and young branches often black. Culm sheaths auriculate, leaving a distinct ridge below the nodes. Leaf-blades linear-lanceolate to lanceolate, 8–20 × 1.5–3.5 cm, apex finely acuminate.

Distr. ALDABRA: W; native of tropical Asia, introduced into most tropical countries and sometimes naturalized.
Notes. Occurs on waste ground behind Settlement. No flowers seen on Aldabra plants.
Vernac. 'bambou'.

2. CENCHRUS L.

Annuals or perennials. Leaf-blades flat, linear-lanceolate. Inflorescence a solitary, false spike. Spikelets 2-flowered, sub-sessile, awnless, in clusters of 2–7 surrounded by a burr-like whorl of flattened, spiny bristles fused at the base into a cup, bristles and spikelets falling together at maturity. Glumes translucent or membranous, upper glume longer than the lower. Lower floret male or sterile, lemma membranous; upper floret bisexual, lemma leathery.

25 species throughout the tropical and warm temperate parts of the World.

Cenchrus echinatus L., Sp. Pl.: 1050 (1753); F.M.S.: 440 (1877).

Erect tufted or stoloniferous annual, 25—50 cm high. Leaf-blades flat, linear-lanceolate, 6—8 × 0.4—0.8 cm, apex narrowly acute, upper surface hairy. Inflorescence 4—9 cm long, burrs of 2—6 spikelets loosely or densely arranged on the main axis. Whorl of bristles barbed backwards, pilose towards the base, enclosing spikelets 4.5—6 mm long. Fig. 51/1—2.

Distr. ASTOVE; a native of tropical America, now widespread throughout the tropics and subtropics.
Notes. Only found growing in the low mixed scrub zone and on the dunes between the coconut grove and lagoon on Astove.
Vernac. 'herbe à cateaux'.

3. CYMBOPOGON Spreng.

Robust perennials, rarely annuals. Leaf-blades broadly-linear to wiry, aromatic. Panicle often large, complexly branched, composed of paired racemes enclosed by boat-shaped spatheoles. Spikelets paired, lower spikelet 2-flowered, hermaphrodite, awned or awnless, upper spikelet 2-flowered, male or sterile, awnless.

40 species in the Old World tropics and subtropics, introduced elsewhere. Several species are cultivated for their essential oils.

Cymbopogon citratus (DC.) Stapf in B.M.I.K. 1906: 357 (1906).

Andropogon citratus DC., Cat. Hort. Monsp.: 78 (1813).

Tufted perennial to c 1 m high. Spikelets awnless (flowering specimens not seen).

Distr. ALDABRA: W; probably native in tropical Asia, now cultivated throughout the tropics.
Notes. Grown near the Manager's House at Settlement. Collections required to confirm identity.
Vernac. 'citronelle' or 'lemon grass'.

4. DACTYLOCTENIUM Willd.

Annuals or perennials. Leaf-blades flat, linear to lanceolate. Inflorescence of 1-several, terminal, digitate racemes. Spikelets laterally compressed, 3-5-flowered, arranged in 2 rows along 1 side of the rhachis which projects as a small point beyond the uppermost spikelets; florets all bisexual. Glumes unequal, keeled, 1-nerved; lower glume apex acute, upper with the nerve produced into a short spiny point. Lemmas ovate, apex acuminate to shortly awned, keeled, 3-nerved.

10 species in the tropics, sub-tropics and warm temperate regions of the world.

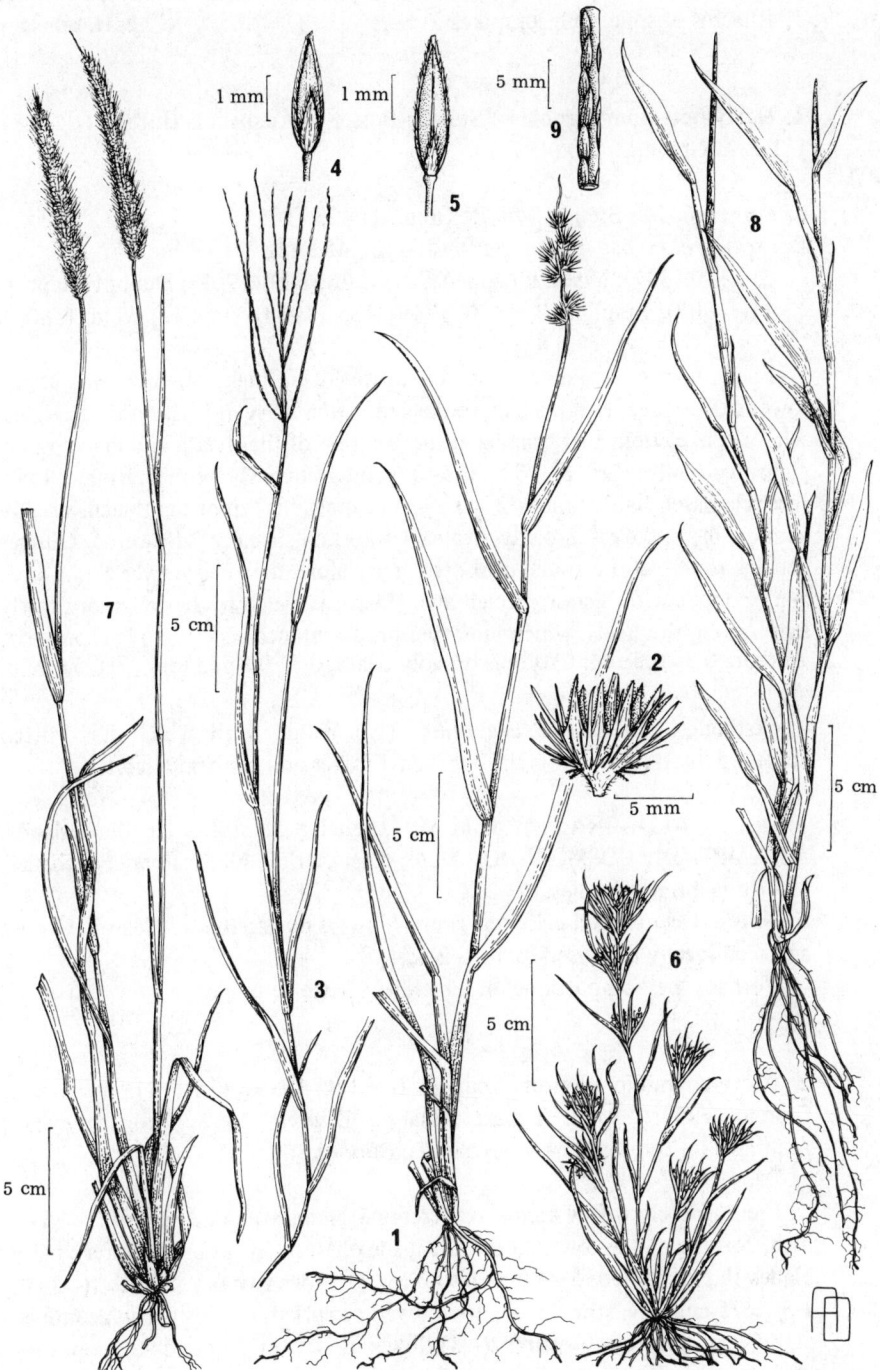

Fig. 51 1, *Cenchrus echinatus*, habit; 2, spikelet. 3, *Digitaria horizontalis*,
  habit; 4, spikelet. 5, *Digitaria setigera*, spikelet. 6, *Panicum
  assumptionis*, habit. 7, *Pennisetum polystachion*, habit.
  8, *Stenotaphrum micranthum*, habit; 9, detail of spike.

1. Rhachis of spikes glabrous, spikes usually more than 2   **1. D. ctenoides**
1. Rhachis of spikes pilose, spikes 2   **2. D. pilosum**

1. **Dactyloctenium ctenoides** (Steud.) Bosser in Adansonia II, 8: 516 (1968); F.T.E.A. Gramin.: 253 (1974).

Chloris ctenoides Steud., Syn. Pl. Glum. 1: 423 (1855).
Dactyloctenium aegyptium sensu F.M.S.: 452 (1877); Schinz in A.S.N.G. 21: 81 (1897); Voeltzkow in A.S.N.G. 26: 549 (1902); Dupont, Report: 41 (1907); Stapf in B.M.I.K. 1919: 135 (1919), not (L.) Willd. (1809).

Erect or semidecumbent annuals, up to 80 cm high, often rooting at the lower nodes. Leaf-sheaths slightly keeled, often hairy at least on the margins, with a conspicuous tuft of hairs at the junction of the sheath and blade, rarely glabrous; blades flat, 3–16 × 0.3–0.6 cm, frequently with scattered, long, tubercle-based hairs. Spikes 2–8, 2–3 cm long, the tip of the rhachis usually acuminate. Spikelets broadly ovate, 4 × 3 mm, usually 3-flowered. Glumes usually rough to the touch or shortly hairy along the keels; lower glume apex acute, upper with a short, rough awn. Lemmas rough to the touch or shortly hairy along the keels, which are often produced into a short, rough spiny tip. Anthers 0.3–0.5 mm. Caryopis broadly ovate, 0.7–0.8 mm long. Fig. 52/1–3.

Distinguished from *D. aegyptium* (L.) Willd., with which it is often confused, by its granular grain, short stiff spikes and prostrate habit.

Distr. ALDABRA: W, P, M, S, Michel, Mentor, lagoon islands; ASSUMPTION; COSMOLEDO: W, M; East Africa, Madagascar, Seychelles, Mauritius. Strand species.
Notes. Occurs on the coastal sands. Flowers during the wet season. Grazed and seed locally dispersed by tortoises.
Vernac. "herbe bourrique' or 'chiendent patte de poule'.

2. **Dactyloctenium pilosum** Stapf in B.M.I.K. 1919: 135 (1919); F.T.E.A. Gramin.: 254 (1974). Types: Aldabra, *Dupont* 70 & *Thomasset* 228; Assumption, *Dupont* 254 & Seychelles, *Dupont* 30 (all K, syn.).

Erect, loosely tufted annual or perennial; stems semi-decumbent, 5–25 cm high, rooting at the lower nodes. Leaf-sheaths usually glabrous, rarely hairy; blades flat or folded, 3–9 × 0.1–0.2 cm, glabrous to hairy. Spikes 2(–3–4), 1.5–3.75 cm long, straight, slightly curved or curled, rhachis hairy, acuminate at the tip. Spikelets usually 2(–3)-flowered. Lower glume 2–2.5 mm long, keel shortly ciliate; upper glume 2 mm long, keel smooth or only slightly rough. Anthers 0.75 mm long. Caryopsis elliptic, 0.8 × 0.6 mm, laterally compressed. Fig. 52/4–5.

Distr. ALDABRA: W, P, M, S; ASSUMPTION; Kenya Coast, Coetivy, Agalega and Seychelles.

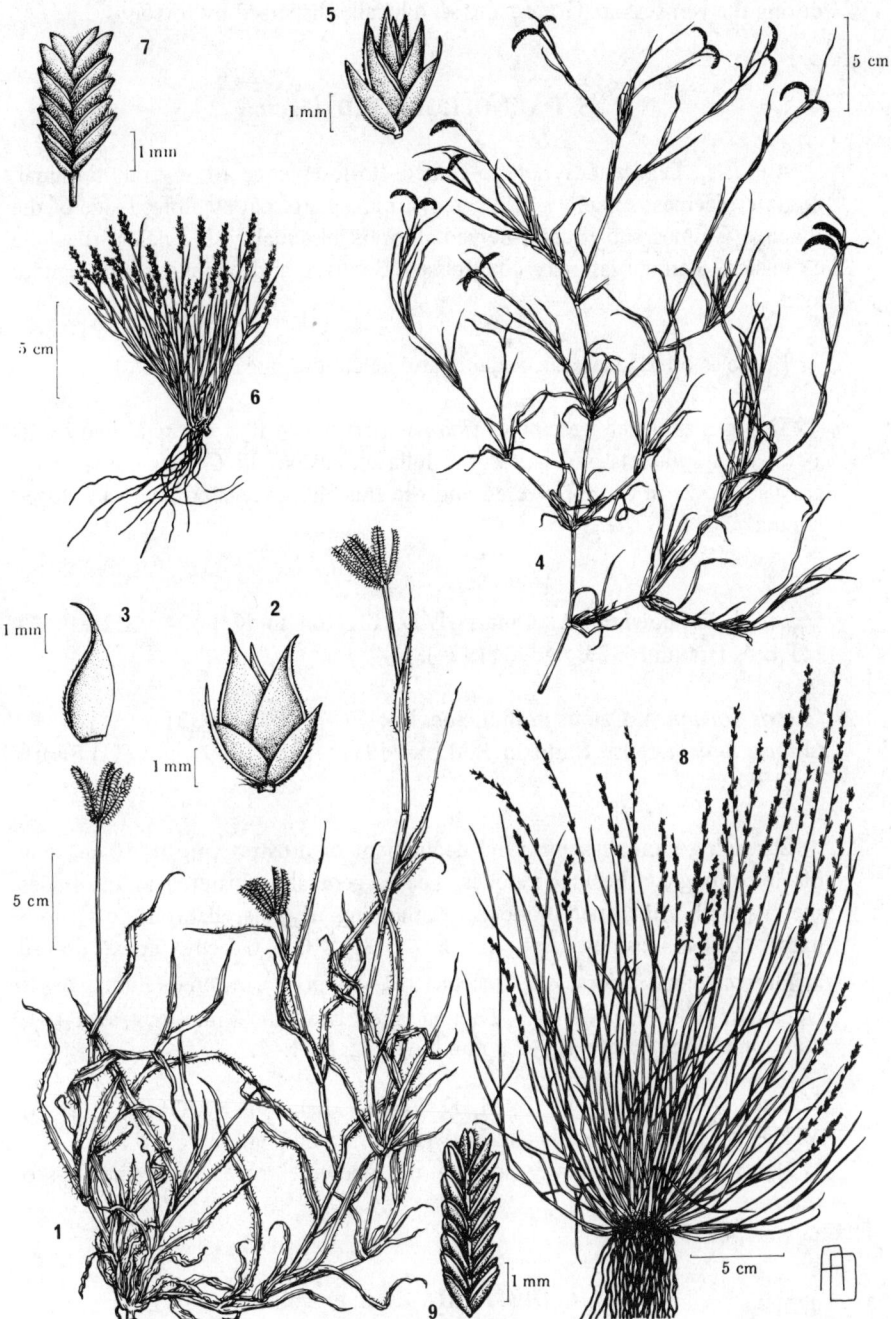

Fig. 52 1, *Dactyloctenium ctenoides*, habit; 2, spikelet; 3, lemma, side view.
4, *Dactyloctenium pilosum*, habit; 5, spikelet; 6, *Eragrostis decumbens*, habit; 7, spikelet. 8, *Eragrostis subaequiglumis*, habit;
9, spikelet.

Notes. Grows in open situations in the mixed scrub. Flowers mainly during the wet season. Grazed and seed locally dispersed by tortoises.

5. DAKNOPHOLIS W.D. Clayton

Annuals. Leaf-blades flat or folded. Inflorescence of several, terminal, digitate racemes; spikelets 1-flowered, arranged alternatley along 1 side of the rhachis. Glumes sub-equal 1-nerved. Florets bisexual with a naked rhachilla extension. Lemma laterally compressed, 3-nerved and bearing a long slender awn.

1 species on Madagascar, Aldabra and neighbouring islands.

This genus closely resembles *Chloris* from which it is distinguished by its 1-flowered spikelets and naked rhachilla extension. In *Chloris* the spikelets are usually 2- or more-flowered and the rhachilla extension bears a reduced lemma.

Daknopholis boivinii (A. Camus) W.D. Clayton in K.B. 21: 102 (1967); F.T.E.A. Gramin.: 323, fig. 90 (1974).

Chloris boivinii A. Camus in Bull. Soc. Bot. Fr. 79: 845 (1933).
Chloris radiata sensu Stapf in B.M.I.K. 1919: 135 (1919); not (L.) Swartz (1788).

Spreading annual; stems semi decumbent or prostrate, up to 50 cm long, often rooting at the lower nodes. Leaves generally clustered at the nodes; sheaths with thin, membranous, overlapping margins, glabrous to sparsely hairy; blades usually flat, linear, 1–3.5 × 0.1–0.4 cm, apex obtuse. Inflorescence of 2–4 slender racemes, (3–)4–6(–7)cm long. Glumes ovate-lanceolate, 0.8–1.2 mm long. Lemma lanceolate, 1.5–2 mm long, awn 4–10 mm long. Floret ovate 1.5–2.5 mm long. Fig. 53/1–3.

Distr. ALDABRA: W, P, M, S, Esprit, Michel; COSMOLEDO: W, M; ASTOVE; Kenya, Tanzania and Madagascar, strand plant.
Notes. Occurs on sandy soils in the coconut groves or at the tops of beaches.

6. DIGITARIA Heist. ex Fabr.

Annuals or perennials. Leaf-blades flat, linear-lanceolate. Inflorescence of narrow, digitate or distant racemes on a common axis. Spikelets 2-flowered, lanceolate to elliptic-oblong, abaxial (lower glume away from the rhachis), awnless, arranged in 2–4 rows, closely appressed, on 1-side of the rhachis. Glumes unequal, lower minute or absent; upper one eighth to almost as long as the spikelet. Lower floret male or sterile, lemma as long as the spikelet, palea absent; upper floret bisexual, lemma and palea both thinly hardened.

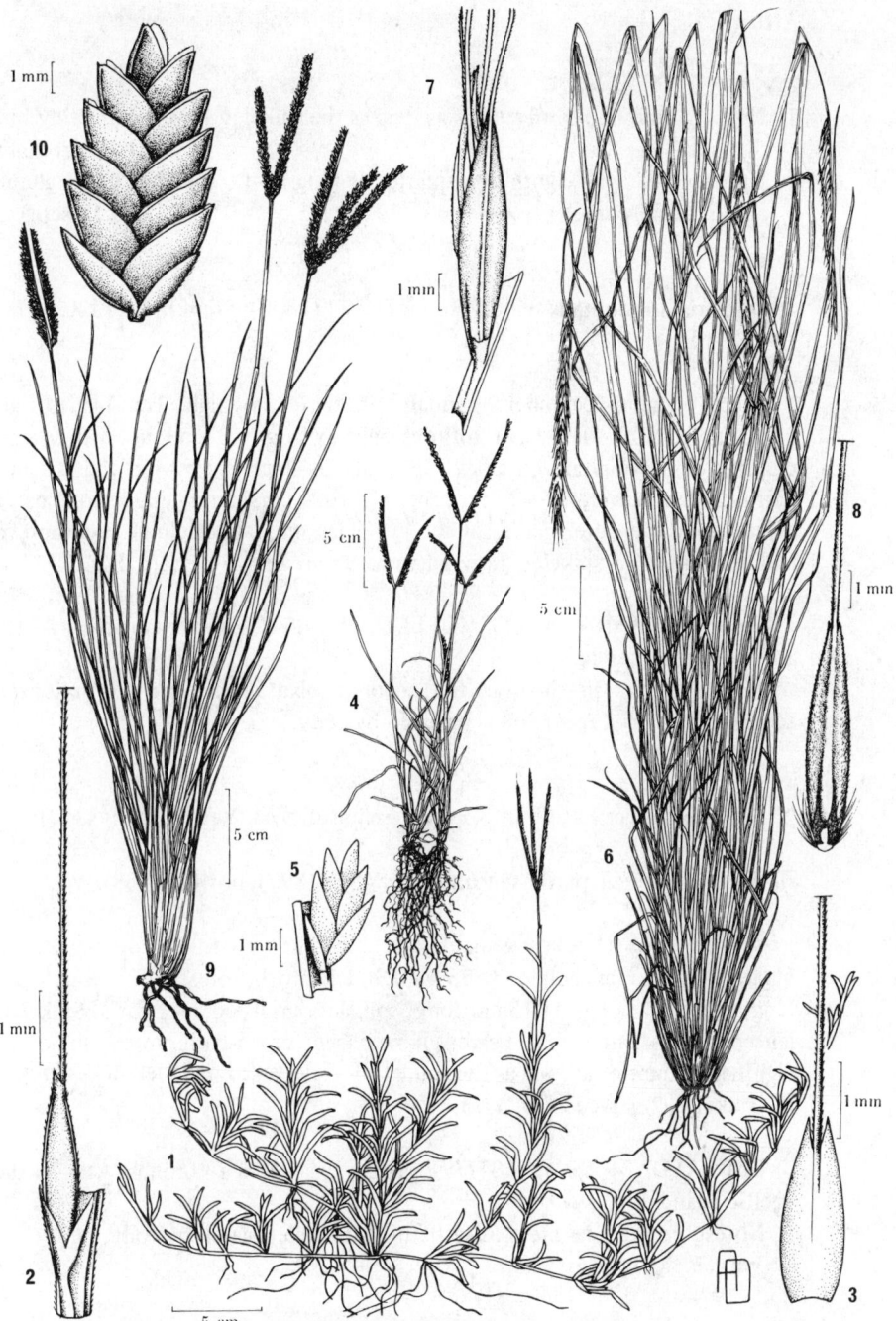

Fig. 53   1, *Daknopholis boivinii*, habit; 2, spikelet; 3, lemma. 4, *Eleusine indica* subsp. *indica*, habit; 5, spikelet. 6, *Enteropogon sechellensis*, habit; 7, spikelet; 8, lemma; 9, *Sclerodactylon macrostachyum*, habit; 10, spikelet.

Over 300 species in the tropics and sub-tropics of the World, mainly in Africa.

1. Upper glume one third to half as long as the spikelet, lower glume tiny but distinct   **1. D. horizontalis**
1. Upper glume one eighth to a quarter as long as the spikelet, lower glume absent or obscure   **2. D. setigera**

1. Digitaria horizontalis Willd., Enum. Pl.: 92 (1809); Stapf in B.M.I.K. 1919: 134 (1919).

Erect or semi-decumbent annuals, up to 60 cm high. Leaf-blades flat, linear, up to 20 × 0.3–1 cm. Inflorescence axis up to 3 cm long, with several digitate or sub-digitate racemes. Spikelets in 2 or 3 rows on the rhachis, narrowly lanceolate, 2.5–3 mm long, glabrous or sparsely hairy to densely ciliate along the outer nerves of the lower lemma. Upper glume one third to half the length of spikelet. Lower lemma 3 mm long, 5-nerved. Fig. 51/3–4.

Distr. ALDABRA: W; COSMOLEDO: M; ASTOVE; west tropical Africa and tropical America.
Notes. Occurs in or near the coconut plantations where it has been accidentally introduced. Spikelets eaten by fody.

2. Digitaria setigera Roth in Roem. & Schultes, Syst. Veg. 2: 474 (1817).

Panicum sanguinale partly sensu F.M.S.: 435 (1877), not L. (1753).

Annual grasses; culms geniculately ascending, up to 80 cm high. Leaf-blades linear to lanceolate, 3–25 × 0.3–1.2 cm. Inflorescence axis 1–6 cm long; racemes, 3–15, 4–15 cm long, spikelets arranged in 2 rows. Spikelets lanceolate, 2–3 mm long. Lower glume absent or obscure, upper glume one eighth to a quarter as long as the spiklet, 0–3-nerved. Lower lemma as long as the spikelet, 7-nerved. Fig. 51/5.

Distr. ALDABRA: W; ASTOVE; from East Africa through Asia to the Pacific Islands.
Notes. Found near areas of settlement. Spikelets eaten by fody.
Vernac. 'herbe gros gazon'.

7. ELEUSINE Gaertn.

Tufted annual or perennial grasses. Leaf-blades linear, often folded. Inflorescence of several terminal, digitate spikes. Spikelets many-flowered, ovate or oblong, bilaterally compressed, arranged in 2 ranks on 1-side of the spikes. Glumes 1—several-nerved, keeled, persistent. Lemmas 3-nerved, apex obtuse or acute, awnless, keeled, disarticulating at maturity.

A genus of 9 species, mainly tropical African.

Eleusine indica (L.) Gaertn., Fruct. 1: 8 (1788);F.M.S.: 451 (1877);F.T.E.A. Gramin.: 262 (1974).

Cynosurus indicus L., Sp. Pl.: 72 (1753).

Slender or robust, erect or ascending annuals, up to 85 cm high. Leaf-blades often folded, 5—35 × 0.25—0.6 cm. Inflorescence up to 10, slender, digitate or sub-digitate spikes, 3—15 cm long. Spikelets 3—9-flowered, 4.5—7.5 mm long. Glumes lanceolate 1—3 mm long, apex acute. Lemmas lanceolate, 2.5—5 mm long, apex acute. Grain reddish, elliptic or oblong, 1—1.5 mm long, prominently sculptured.

subsp. **indica**; F.T.E.A. Gramin.: 263 (1974).

Spikes 3—5.5 mm wide. Lower glume 1—2.5 mm long, 1-nerved. Lemmas 2.5—3.5 mm long. Grain elliptic, 1—1.3 mm long. Fig. 53/4—5.

Distr. ALDABRA: W; COSMOLEDO: M; pantropical weed.
Notes. Found near areas of settlement.
Vernac. 'chiendent patte de poule' or 'herbe patte de poule'.

8. ENTEROPOGON Nees

Tufted perennials with wiry culms. Leaf-blades flat, linear or inrolled and wiry. Inflorescence a solitary, terminal, 1-sided raceme. Spiklets 2—3-flowered, lanceolate-elliptic, arranged on the rhachis in 2 rows. Glumes lanceolate, membranous, terminating in a short awn point. Lower floret bisexual, lemma long-awned from the sinus of a bidentate apex; middle floret, when present, male or sometimes bisexual, slightly smaller than the lower floret; upper floret male.

6 species in the Old World tropics.

Enteropogon sechellensis (Baker) Dur. & Schinz, Consp. Fl. Afr. 5: 859 (1884); F.T.E.A. Gramin.: 333, fig. 94 (1974).

Ctenium sechellense Baker, F.M.S.: 452 (1877).
Enteropogon melicoides sensu Stapf in B.M.I.K. 1919: 135 (1919), not (Rottler) Nees (1836).

Erect tufted perennial with wiry culms up to 65 cm high. Leaf-sheaths strongly keeled, margins sparsely to densely hairy; blades up to

30 × 0.1–0.2 cm, flat and folded or inrolled. Inflorescence 7–16 cm long. Spikelets 3-flowered, 5 mm long excluding the awn, which is 10–20 mm long in the lowermost floret. Fig. 53/6–8.

Closely related to *E. monostachyos* (Vahl) K. Schum. from which it differs in having a longer awn to the fertile floret.

Distr. ALDABRA: W, P, M, S (south), lagoon islets; ASSUMPTION; COSMOLEDO: W; ASTOVE; coastal regions of East Africa, Madagascar and the Seychelles.

Notes. Occurs in shade of *Casuarina* and coconuts. Flowers during the wet season; seed may not be shed until late in the dry season.

Vernac. 'gazon'.

9. ERAGROSTIS Wolf

Annuals or perennials. Leaf-blades linear. Inflorescence an effuse or contracted panicle, rarely spicate. Spikelets 2–many-flowered, laterally compressed, rhachis tough or disarticulating, florets generally breaking up at maturity. Glumes unequal or sub-equal. Florets all bisexual. Lemmas membranous or leathery, keeled, 3-nerved. Paleas translucent, 2-keeled. Stamens 3 or 2.

300 species throughout the tropics and sub-tropics of the World.

1. Spikelets broadly oblong, 2.5–8 × 1.5–3 mm   **1. E. decumbens**
1. Spikelets linear oblong, 3–6 × 0.5–1.25 mm   **E. subaequiglumis**

1. Eragrostis decumbens Renvoize in K.B. 25: 418 (1971). Type: Aldabra, South Is., Cinq Cases, *Renvoize* 806 (K, holo., US, iso.).

Perennial, culms wiry, prostrate or erect, up to 10(–30) cm high, often forming a rosette. Leaf-blades very narrow, inrolled, 1–4(–6) cm long. Inflorescence a contracted panicle, almost spicate, with very short branches closely appressed to the main axis, never spreading. Spikelets broadly oblong, 2.5–8 × 1.5–3 mm, on short pedicels, 1–4 mm long. Glumes sub-equal. Paleas conspicuously ciliate on the keels. Anthers 2, 0.2 mm long. Caryopsis ovate or elliptic, 0.5 mm long. Fig. 52/6–7.

Distr. ALDABRA; W, P, M, S, Esprit, Michel; ASSUMPTION; COSMOLEDO: W, M; ASTOVE; endemic.

Notes. Occurs on exposed platin or in small holes and crevices in champignon. Flowers during the mid wet season. Grazed and seeds locally dispersed by tortoises.

2. Eragrostis subaequiglumis Renvoize in K.B. 25: 419 (1971). Type: Aldabra, Ile Michel, *Renvoize* 1039 (K, holo., US, iso.).

Tufted annual or perennial; culms wiry, erect, 10–45 cm high. Leaf-blades very narrow, inrolled, stiff or flexuous, 4–14 cm long, rarely flat, then 0.5–2 mm wide. Inflorescence a contracted panicle, almost spicate, with branches being closely appressed to the main axis, never spreading. Spikelets linear-oblong, 3–6 × 0.5–1.25 mm. Glumes sub-equal. Paleas ciliate along the keels. Anthers 2, 0.3 mm long. Caryopsis elliptic, 0.5 mm long. Fig. 52/8–9.

This species is closely related to *E. tenella* (L.) P.Beauv. ex Roem. & Schult. which has an effuse panicle and 3 anthers.

Dist. ALDABRA: W, P, M, S, Esprit, Michel; ASSUMPTION; COSMOLEDO: W, M; ASTOVE; Seychelles.

Notes. Occurs in a variety of habitats, from sandy soils in coconut plantations to rocky crevices in the mixed scrub communities. Flowers mainly during the late wet season but unseasonal rains can stimulate plants that have not succumbed to the dry season. Grazed and seeds locally dispersed by tortoises.

10. ERIOCHLOA Kunth

Annuals or perennials. Leaf-blades linear to lanceolate. Inflorescence of several, non-branching racemes along a central axis, occasionally branching in *E. meyerana*. Spikelets 2-flowered, pedicelled, ovate-lanceolate, apex acute to acuminate or with a stiff bristle-like awn with a small bead-like swelling at the base formed from the lowest rhachilla internode and adherent lower glume. Glumes very unequal, lower glume reduced to a small translucent scale which in *E. meyerana* extends as a small flap above the bead; upper as long as the spikelet, ovate-lanceolate, membranous. Lower floret male or sterile; lemma membranous, similar in size and shape to the upper glume; palea membranous or translucent or absent. Upper floret bisexual; lemma and palea hard, thin and brittle.

About 30 species throughout the tropics and sub-tropics.

1. Perennial up to 1 m high; leaf-blade linear or linear-lanceolate; spikelets sparsely hairy with an acute apex, not with a distinct awn point
   1. E. meyerana
1. Delicate annual up to 70 cm high; leaf-blade narrowly lanceolate; spikelets hairy with a distinct awn point   **2. E. subulifera**

1. Eriochloa meyerana (Nees) Pilger in Engler & Prantl, Pflanzenfam. ed. 2, 14e: 56 (1940).

Panicum meyeranum Nees, Fl. Afr. Austr.: 32 (1841); Stapf in B.M.I.K. 1919: 134 (1919).

Loosely tufted, rhizomatous perennial, up to 1 m high. Leaf-blades linear-lanceolate to linear, flat or folded, 6–16 × 0.3–0.6 cm. Inflorescence resembling a racemose panicle, 8–16 cm long, but spikelets frequently on short secondary branches. Spikelets ovate-acute, 3–3.5 mm long. Lower glume a small, translucent, persistent scale, 0.2–0.3 mm long; upper glume and lower lemma sparsely hairy or glabrous, 3.5 mm long. Fig. 54/1.

Because of the presence of a lower glume this species was formerly placed in the genus *Panicum*,but the small bead at the base of the spikelet indicates that it is more closely related to species of *Eriochloa*.

Distr. ALDABRA; ASSUMPTION; tropical and subtropical Africa.

Notes. Only known on Aldabra from 3 early collections, *Dupont* 99 & 293 and *Fryer* 23 (all K), all without location, suggesting that it may no longer be extant. The 2 collections from Assumption, *Dupont* 75 (K) and *Stoddart* 1063 (K, US), indicate its continued presence there.

2. **Eriochloa subulifera** Stapf in B.M.I.K. 1919: 141 (1919). Types: Assumption, *Dupont* 258 & 261 (both K, syn.).

Delicate annuals, 30–70 cm high. Leaf-blades flat or folded, narrowly lanceolate to linear-lanceolate, 2–15 × 0.3–0.5 cm. Inflorescence a racemose panicle, 4–12 cm long. Spikelets ovate-lanceolate, 1.5–2 mm long, (excluding awn point), pedicels 2–4 mm long. Lower glume completely appressed to the bead-like swelling, upper glume and lower lemma densely hairy, up to 2 mm long. Fig. 54/2–3.

This species is very similar to the widespread African species *E. fatmensis* (Hochst. & Steud.) Clayton, from which it differs in having a more delicate panicle and smaller spikelets.

Distr. ASSUMPTION; ASTOVE; northwest Madagascar.

11. ISCHAEMUM L.

Annuals or perennials. Leaf-blades flat, linear to linear-lanceolate. Inflorescence of paired, digitate or clustered racemes. Spiklets paired, 1 sessile, the other pedicelled, arranged more or less alternately on a fragile, jointed rhachis; spikelets 2-flowered. Sessile spikelet: glumes equal or subequal, enclosing the lemmas and paleas; lower glume rounded or flat on the back, occasionally wrinkled, leathery: lower floret male, lemma membranous or translucent, awnless; upper floret bisexual, lemma membranous or translucent, usually 2-lobed, awned. Pedicelled spikelet: awnless, otherwise similar to sessile spikelet or male or sterile.

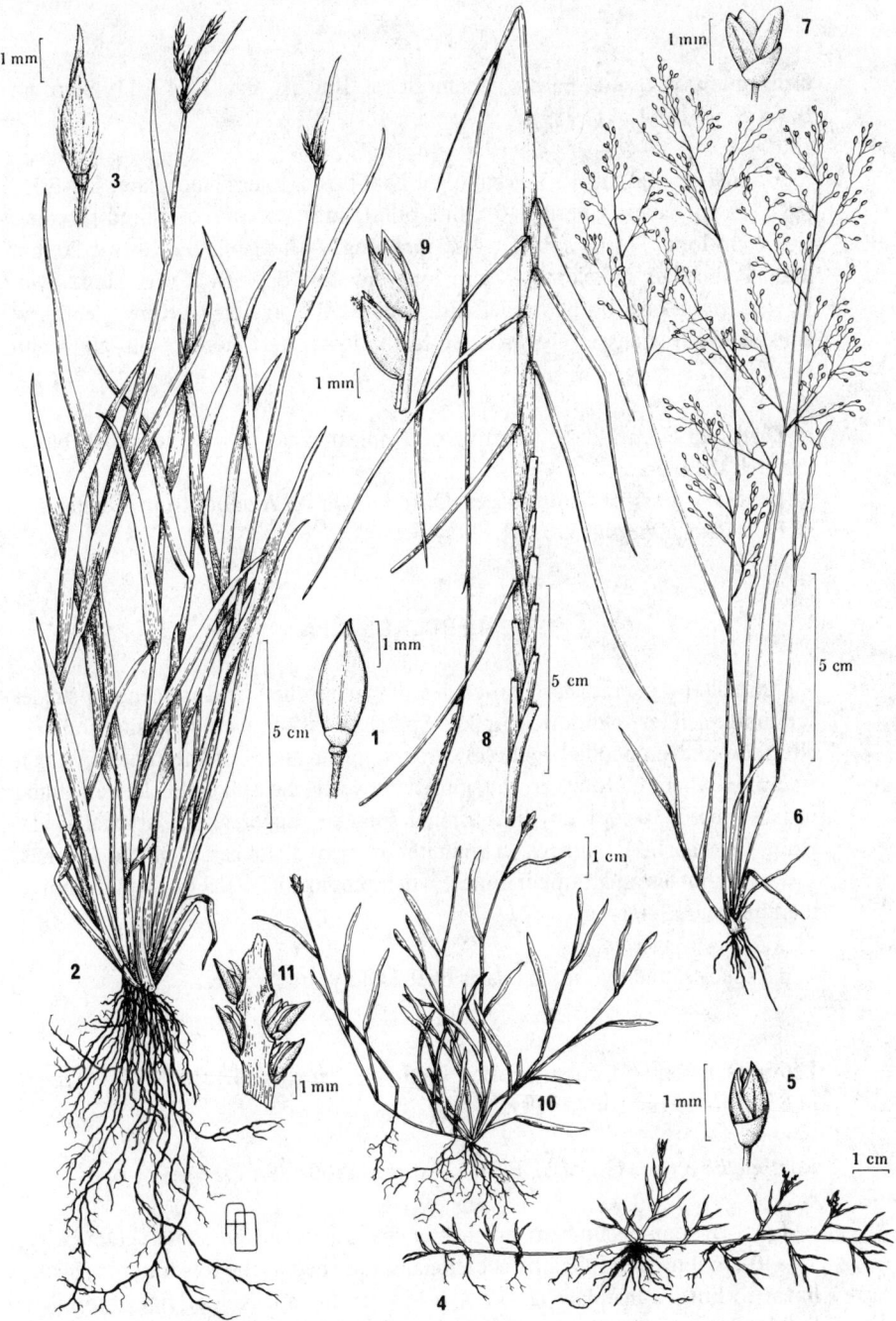

Fig. 54   1, *Eriochloa meyerana*, spikelet. 2, *Eriochloa subulifera*, habit;
3, spikelet. 4, *Panicum aldabrensis*, habit; 5, spikelet. 6, *Panicum*
voeltzkowii, habit; 7, spikelet. 8, *Paspalum distichum*, habit;
9, detail of spikelets. 10, *Stenotaphrum clavigerum*, habit;
11, detail of spike.

About 50 tropical species mainly in the Old World.

Ischaemum rugosum Salisb., Icon. Stirp. Rar. 1, fig. 1 (1791); Stapf in B.M.I.K. 1919: 134 (1919).

Robust annual, up to 120 cm high. Leaf-blades linear-lanceolate, 15–30 × 0.8–1.5 cm, apex tapering to a fine point. Inflorescence of paired racemes, 6–12 cm long. Sessile spikelet 4–5 mm long, with a geniculate awn c 20 mm long. Pedicelled spikelet 2–3 mm long, awnless or with a very short awn. Lower glume of both sessile and pedicelled spikelets, yellow, leathery, conspicuously transversely wrinkled in the lower part, herbaceous, green and veined at the apex.

Distr. ALDABRA: S; a native of tropical Asia, now widely distributed through the tropics.

Notes. A grass of damp places. Only known on Aldabra from 1 collection at Takamaka, *Dupont* 230 (K).

12. LEPTURUS R.Br.

Annuals or perennials. Leaf-blades flat or inrolled. Inflorescence a single, terminal, slender, cylindrical spike. Spikelets 1–2-flowered; sessile, solitary, alternate and embedded in hollows on opposite sides of the rhachis, which fractures when mature at the joints between the spikelets. Lower glume usually absent except on the terminal spikelet; upper glume abaxial (away from the rhachis), narrow, acuminate or awned, longer than the lemmas. Lower floret bisexual; upper floret, when present, bisexual or sterile, lemmas membranous, 3-nerved.

12 species, mainly on sea coasts of the Old World tropics.

Lepturus repens (G.Forst.) R.Br., Prodr. Fl. Nov. Holl.: 207 (1810); Stapf in B.M.I.K. 1919: 136 (1919).

Rottboellia repens G.Forst., Fl. Ins. Austral. Prodr.: 9 (1786).

Erect or semi-decumbent much branched, glaucous perennial; culms wiry, 15–40 cm long. Leaf-sheaths occasionally bunched at the lower nodes; blades linear to linear-lanceolate, 2–12 × 0.15–0.5 cm, apex acute, flat or inrolled. Spikes 5–10 × 0.1–0.15 cm. Spikelets 1(–2)-flowered. Upper glume ovate-lanceolate, 5–10 mm long or longer in the terminal spikelet, apex acute to finely acuminate, herbaceous. Lemma and palea ovate to ovate-oblong, 4–5 mm long, translucent, completely covered by the upper glume. Anthers 1–2 mm long. Fig. 55/1–2.

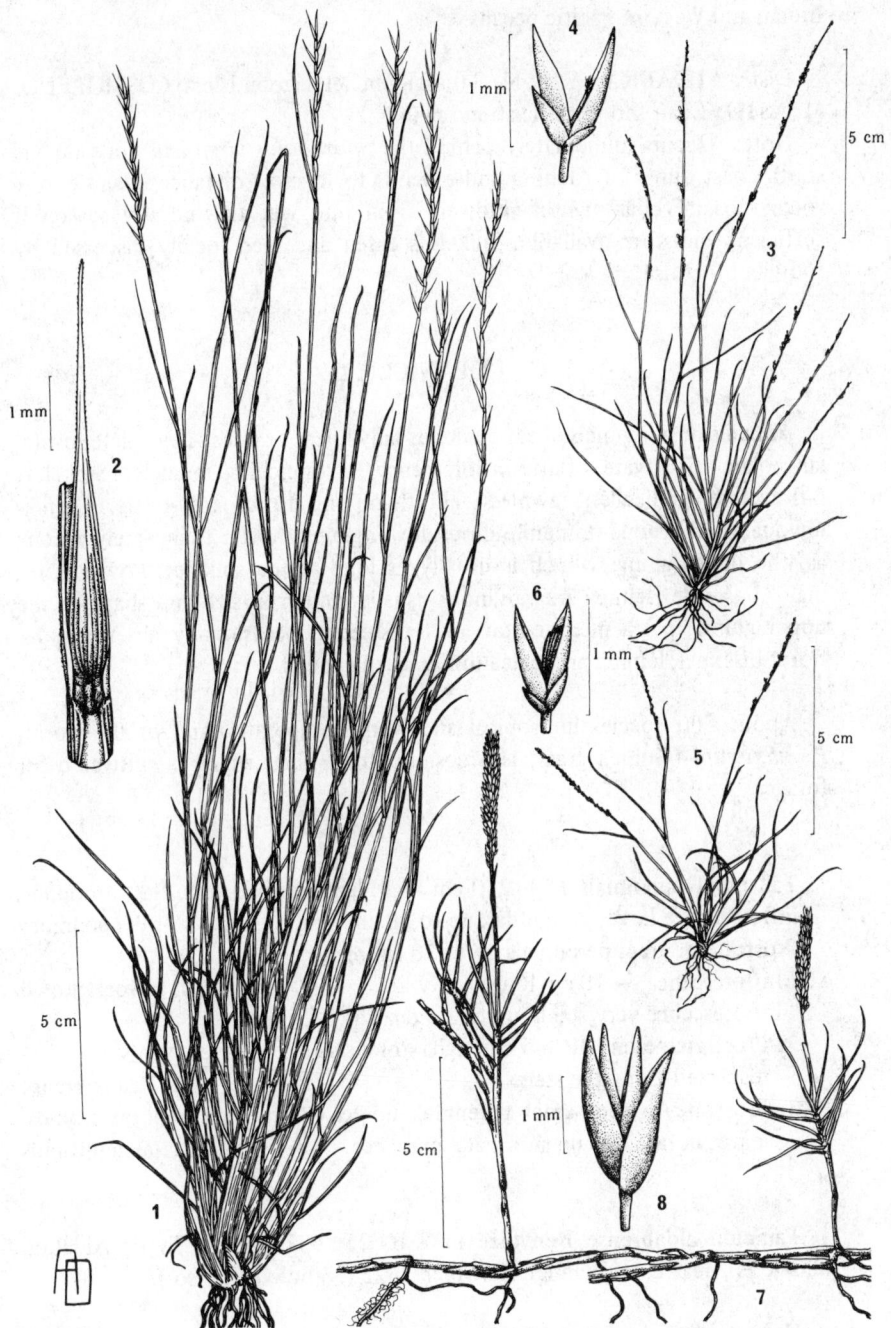

Fig. 55 1, *Lepturus repens*, habit; 2, spikelet. 3, *Sporobolus aldabrensis*, habit; 4, spikelet. 5, *Sporobolus testudinum*, habit; 6, spikelet. 7, *Sporobolus virginicus,* habit; 8, spikelet.

This is an extremely variable sea-shore species found throughout the Indian and Western Pacific oceans.

Distr. ALDABRA: W, P, M, S (east), Michel, lagoon islets; COSMOLEDO: M; ASTOVE; an Indo-Pacific strand plant.

Notes. Occurs immediately behind the *Sporobolus virginicus* zone on the south coast dunes of Aldabra and extends for a short distance inland on the rough coastal champignon. Flowers from mid wet to mid dry season if sufficient moisture available. Spikelets eaten and seed locally dispersed by fody and tortoise.

13. PANICUM L.

Annuals or perennials. Leaf-blades usually linear or linear-lanceolate, ovate-lanceolate or ovate, flat. Inflorescence a branching panicle. Spikelets 2-flowered, pedicelled, awnless, elliptic-oblong to broadly-ovate. Glumes unequal, herbaceous or membranous, lower glume shorter than, rarely as long as, the upper glume, which is usually as long as the spikelet. Lower floret male or sterile, lemma membranous usually similar in size and shape to the upper glume, palea membranous or translucent, occasionally absent; upper floret bisexual, lemma and palea thinly cartilaginous.

About 500 species in tropical and warm temperate parts of the world. *P. maximum* (Guinea grass) is a cosmopolitan plant which is cultivated for forage.

1. Tall erect perennials 150—250 cm high; leaf-blades broad, flat, drooping; inflorescence large, pyramidal, up to 70 cm long   **3. P. maximum**
1. Prostrate or erect perennials up to 30 cm high
  2. Inflorescence 5—10 cm long   **4. P. voeltzkowii**
  2. Inflorescence very small, up to 1.5 cm long
    3. Prostrate perennial 1—5 cm high; stolons slender; leaves evenly dispersed along the stems   **1. P. aldabrense**
    3. Prostrate or semi-erect perennial, up to 12 cm high; stouter stolons; leaves in fascicles on short side branches   **2. P. assumptionis**

1. Panicum aldabrense Renvoize in K.B. 25: 421 (1971). Type: Aldabra, Middle Is., near East Channel, *Renvoize* 1325 (K, holo., US, iso.).

Very small, prostrate, stoloniferous perennial; culms 1—5 cm high; stolons up to 0.5 mm diameter, bearing leaves all along their length. Leaf-blades linear-lanceolate, 0.5—2.0 × 0.05—0.1 cm. Inflorescence very small, 0.3—1 cm long, barely exceeding the leaf-blades. Spikelets very small, ovoid, 1—1.3 mm long. Fig. 54/4—5.

Distr. ALDABRA: W, P, M (east), S (east); endemic.

Notes. Known only from Aldabra where it forms small patches of turf, closely cropped by tortoises, most abundant in the platin area around Cinq Cases.

2. Panicum assumptionis Stapf in B.M.I.K. 1919: 140 (1919). Type: Assumption, *Dupont* 110 (K, holo.).

Small perennial grass, 7–12 cm high; stolons 0.5–0.75 mm in diameter, wiry, semi-decumbent, without leaves for most of their length except on short side branches. Leaves congested and generally enclosed at the base by an enlarged leaf-sheath; lower leaf-blades 2–5 × 0.15–0.3 cm, flat; upper leaf-blades up to 1 × 0.1 cm, folded or inrolled, tightly packed on short internodes. Inflorescence small, barely exceeding the leaf-blades. Spikelets very small, ovoid or obovoid-elliptic, 1–1.3 mm long, Fig. 51/6.

P. assumptionis is closely related to *P. aldabrense* from which it differs in having stouter, semi-erect stolons with short side branches bearing fascicles of leaves, with the lower leaves usually longer than the upper leaves.

Distr. ASSUMPTION; endemic.

Notes. Only known from 2 collections, *Dupont* 110 (K) and *Frazier* 610 (US).

3. Panicum maximum Jacq., Icon. Pl. Rar. 1: 2, fig. 13 (1781); F.M.S.: 436 (1877); Schinz in A.S.N.G. 21: 82 (1897); Voeltzkow in A.S.N.G. 26: 549 (1902); Dupont, Report: 41 (1907); Stapf in B.M.I.K. 1919: 134 (1919).

Perennial, 1.5–2.5 m high (in our area); leaf-blades narrowly linear-lanceolate, up to 70 × 2.5 cm. Inflorescence pyramidal, 40–70 cm long. Spikelets oblong, 3–3.3 mm long, apex obtuse, plump. Lower glume broadly ovate, 1 mm long, apex acute, membranous, 1–3-nerved; upper glume oblong, as long as the spikelet, apex acute, membranous, completely covering the fertile upper lemma which is conspicuously transversely wrinkled.

Distr. ALDABRA: W; COSMOLEDO: M; ASTOVE; widespread in the tropics and subtropics, often cultivated as a fodder crop.

Notes. Found near areas of settlement, growing on the fringes of coconut plantations. Flowering mid wet season.

Vernac. 'herbe de Guinée' or 'fataque'.

4. Panicum voeltzkowii Mez in Engler, Bot. Jahrb. 57: 187 (1921).

Tufted or stoloniferous perennials; culms delicate, geniculately ascending, up to 30 cm high. Basal leaf-sheaths densely pubescent; leaf-blades narrowly lanceolate, 4–8 × 0.3–0.4 cm, apex, tapering, glabrous or sparsely hairy;

ligule long ciliate. Inflorescence a finely branched, oblong or pyramidal panicle, 5—10 cm long. Spikelets small, ovoid, 1.5—1.75   mm long. Fig. 54/6—7.

Distr. COSMOLEDO: W; ASTOVE; Madagascar.

Notes. Found growing in low scrub near the lagoon beach on Wizard Is., *Fosberg* 49819 & 49827 (both K, US). One collection, *Ridgway* 115 (US) from Astove.

14. PASPALUM L.

Annuals or perennials. Leaf-blades flat, linear. Inflorescence of narrow, digitate or distant racemes on a common axis. Spikelets 2-flowered, 2- or 4-ranked on 1-side of a flattened rhachis, solitary or paired, ovate-lanceolate to orbicular, awnless, with the lower lemma abaxial i.e. away from the rhachis. Glumes dissimilar, lower glume a minute scale or absent, upper membranous, as long as the spikelet. Lower floret sterile, lemma similar to the upper glume; upper floret bisexual, lemma and palea leathery.

Over 250 tropical, sub-tropical and temperate species, mainly in South America.

Paspalum distichum L., Syst. Nat. ed. 10, 2: 855 (1759); F.M.S.: 431 (1877).*

P. vaginatum Swartz, Prodr. Veg. Ind. Occ.: 21 (1788).

Rhizomatous or stoloniferous perennial; culms erect or semi-decumbent, leafy, (10—)25—70 cm high. Leaf-sheaths often crowded on the culms and overlapping, conspicuously hairy at the mouth, otherwise glabrous; blades linear, 3—15 × 0.15—0.8 cm, apex acute, flat or folded. Inflorescence usually of 2 terminal racemes, rarely more, (1—)3.5—6 cm long. Spikelets solitary, arranged alternately in 2 rows on the rhachis, ovate, elliptic-ovate or oblong, 3—4.5 mm long. Lower glume absent; upper glume as long as the spikelet, thinly membranous. Lower lemmas as long as the spikelet; lower palea absent. Fertile floret 2.5 mm long, lemma and palea leathery. Fig. 54/8—9.

*The type of *Paspalum distichum* was not indicated in the original description but can be deduced from L., Amoen. Acad. 5: 389 (1760) as a specimen collected by Browne and acquired by Linnaeus in 1758. Unfortunately the type sheet bears pieces of two difference species: *P. paspalodes* (Michx.) Scribn. to which the name *P. distichum* has been universally applied as well as the true *P. distichum*, beside which is the inscription "Br."; indicating that it is the plant collected by Browne. This plant is widely known as *P. vaginatum*. The similarity of the two species has caused considerable confusion.

Distr. ALDABRA: P, M (west), S (west), Michel; COSMOLEDO: N.W. Is.; pantropical coastal plant.

Notes. Limited distribution on Aldabra, near Gionnet Channel, Anse Mais and Ile Michel.

15. PENNISETUM L.Rich.

Annuals or perennials. Leaf-blades usually linear, flat. Inflorescence a solitary, false spike. Spikelets 2-flowered, sessile or shortly pedicelled, solitary or in clusters of 2–5, surrounded by a whorl of free, glabrous or hairy, slender or flexuous, bristles, both the whorl of bristles and spikelets falling at maturity (in wild species only). Glumes membranous or translucent, equal or the lower smaller or suppressed. Lower floret male or sterile, lemma membranous; upper floret bisexual, lemma thinly leathery.

150 species throughout tropical and warm temperate parts of the world.

This genus intergrades with *Cenchrus L.* but can usually be distinguished by its free, slender whorl of bristles.

Pennisetum polystachion (L.) Schult., Mant. 2: 146 (1824); Schinz in A.S.N.G. 21: 82 (1897); Voeltzkow in A.S.N.G. 26: 549 (1902); Dupont Report: 41 (1907).

Panicum polystachion L., Syst. Nat. ed. 10, 2: 870 (1759).
Pennisetum setosum (Swartz) L. Rich. in Pers., Syn. Pl. 1: 72 (1805); F.M.S.: 441 (1877); Stapf in B.M.I.K. 1919: 134 (1919).

Erect or geniculately ascending, tufted annual or perennial, 50–150 cm high. Leaf-blades linear to linear-lanceolate, 12–30 × 0.5–0.75 cm, flat. Inflorescence 7–18 × 0.75–1 cm, excluding the bristles. Bristles 5–15 mm long, 1 often longer than the others, minutely rough to the touch towards the apex, ciliate towards the base. Spikelet solitary, enclosed by a whorl of bristles, oblong, 2–3 mm long. Fig. 51/7.

Distr. ALDABRA; ASSUMPTION; ASTOVE; pantropical.

Notes. Two collections from Aldabra; an unnumbered collection by Dupont cited by Stapf *l.c.* (1919) cannot be traced, the second, by Voeltzkow is without data. Further collections from Aldabra are required. On Assumption it is found growing in the sand near the shore.

Vernac. 'herbe ma tante' or 'ma tante'.

16. SCLERODACTYLON Stapf

Densely tufted perennials. Leaf-blades erect, rigid, circular in cross-section. Inflorescence of 1-several, terminal digitate spikes. Spikelets many-flowered, laterally compressed, arranged alternately in 2 rows on 1-side of the rhachis; Glumes unequal, keeled, 1-nerved, slightly shorter than the lemmas. Florets all bisexual. Lemmas keeled.

One species only, restricted to Tanzania, northern Madagascar, Aldabra and neighbouring islands.

Sclerodactylon macrostachyum (Benth.) A.Camus in Bull. Soc. Bot. Fr. 79: 38 (1932).

Eleusine macrostachya Benth. in J. Linn. Soc. Bot. 19: 107 (1881).
Sclerodactylon juncifolium Stapf in B.M.I.K. 1911: 318 (1911) & in B.M.I.K. 1919: 135 (1919), name superfluous.

Densely tufted perennial, 40–90 cm high, often producing wiry stolons bearing dense clumps of short leaves at regular intervals. Leaf-sheaths flattened, fan-like, keeled, glabrous, often yellowish and shining or powdery; blades 10–30 × 0.1–0.2 cm, circular in cross-section, glaucous. Inflorescence of (1–)2–3), rarely more, digitate spikes, up to 12 × 0.5–2.5 cm. Spikelets densely packed on the rhachis, 0.5–2.5 × 0.3 cm, 4–40-flowered. Glumes boat-shaped, hard, thin and brittle, 1-nerved; lower glume 2 mm long; upper 2.5–3 mm long. Lemmas 3–3.5 mm long. Florets ovate, 3 mm long. Fig. 53/9–10.

Distr. ALDABRA: W, P, M, S, Esprit, Michel; ASSUMPTION; Madagascar, S. Tanzania.
Notes. Usually restricted to coastal sites, colonising sandy deposits between rocks, on some dunes forming continuous pure stands, occasionally found inland. Flowering during the wet season, most florets shed by mid dry season. Seed locally dispersed by tortoises.

17. SPOROBOLUS R.Br.

Annuals or perennials. Leaf-blades linear, flat, folded or inrolled. Inflorescence an effuse, ovate or pyramidal panicle, or contracted and spike-like. Spikelets 1-flowered, pedicelled, generally small, spindle-shaped, awnless. Glumes membranous, 1-nerved or nerveless, equal or unequal, very short to as long as the spikelet. Florets bisexual. Lemma membranous, 1–3-nerved. Palea usually as long as the lemma; seed not firmly grown to fruit wall.

About 150 species in the tropics, sub-tropics and warm temperate regions of the world.

1. Tufted annual or short lived perennial
   2. Plants 12–30 cm high; spikelets 1.3–1.5 mm long; grain oblong,
      tetragonal, 0.7–0.9 mm long   **1. S. aldabrensis**

2. Plants 2–10 cm high; spikelets 1 mm long; grain elliptic, truncate,
   0.5–0.7 mm long   **2. S. testudinum**
1. Perennial; rhizomes long, slender   **3. S. virginicus**

1. Sporobolus aldabrensis Renvoize in K.B. 25: 417 (1971). Type: Aldabra,
South Is. near Flamingo Pool, *Renvoize* 987 (K, holo., US, iso.).

Loosely tufted annual; culms weakly ascending, usually branched just
above the base, 12–30 cm high, occasionally geniculate at the base and
rooting at the lower nodes. Leaf-sheaths glabrous, becoming papery; blades
wiry, (1–) 6–12 × 0.1 cm, glabrous, inrolled. Inflorescence a narrow,
interrupted spike (2.5–) 6–14 × 0.1–0.3 cm. Spikelets 1.3–1.5 mm long,
sessile or shortly pedicelled, on branches up to 2 cm long, usually appressed
to the main axis. Glumes translucent, nerveless, rarely faintly 1-nerved; lower
glume ovate, 0.5 mm long, apex truncate; upper ovate-lanceolate, 1 mm long,
apex acute or roughly 3-toothed. Lemma and palea sub-equal, both ovate-
lanceolate, 1.3–1.5 mm long, apex acute, translucent, nerveless or faintly
nerved. Anthers 0.5 mm long. Grain oblong, tetragonal, 0.7–0.9 mm long.
Fig. 55/3–4.

Distr. ALDABRA: S (east); endemic.
Notes. This species is only known from 2 collections, *Renvoize* 987 and
Fosberg 48867 (both K, US) from mixed scrub communities near Flamingo
Pool and Cinq Cases.

2. Sporobolus testudinum Renvoize in K.B. 25: 417 (1971). Type: Aldabra,
South Is., Cinq Cases, *Renvoize* 972 (K, holo., US, iso.).

Tufted annual or short lived perennial; culms unbranched, prostrate or
geniculately ascending, 2–10 cm high. Leaf-sheaths glabrous, persistent,
membranous, often purplish; blades wiry, 1–6 cm × 0.25–0.75 mm,
glabrous, inrolled. Inflorescence a narrow interrupted spike, 1.5–6 ×
0.1–0.2 cm. Spikelets 1 mm long, sessile or shortly pedicelled, on short,
appressed side branches. Glumes translucent, nerveless; lower glume ovate,
0.2–0.3 mm long, apex truncate or roughly toothed; upper ovate,
0.3–0.6 mm long, apex truncate or 3-toothed. Lemmas and paleas subequal,
both translucent, nerveless, ovate-lanceolate, 1 mm long, apex acute. Anthers
0.3–0.4 mm long. Grain elliptic, 0.5–0.7 mm long, apex truncate,
Fig. 55/5–6.

Distr. ALDABRA: S (east); endemic.
Notes. Occurs in small crevices in the region of fairly open mixed scrub
between the plateau area and the coast of the eastern end of South Is.
Flowering during the mid wet season. Seeds locally dispersed by tortoises.

3. Sporobolus virginicus (L.) Kunth, Rev. Gram. 1: 67 (1829); Stapf in B.M.I.K. 1919: 135 (1919); F.T.E.A. Gramin.: 370 (1974).

Agrostis virginicus L., Sp. Pl.: 63 (1753).
Vilfa virginica (L.) P.Beauv., Ess. Agrost.: 16, 182 (1812); F.M.S.: 449 (1877).

Perennial, spreading by long slender rhizomes; culms 10–30 cm high. Leaf-blades, 2–10 × 0.1–0.4 cm, apex stiff, pointed, 2-ranked, stiff, nearly always inrolled, with margins touching. Panicle spike-like, 2–10 × 0.2–0.8 cm. Spikelets 1.5–2.0 mm long. Lower glumes two thirds to three quarters the length of spikelet; upper glume as long as the spikelet. Fig. 55/7–8.

Distr. ALDABRA: W, P, M, S, Esprit, Moustique; ASSUMPTION; COSMOLEDO: W, M; ASTOVE; pantropical strand plant.
Notes. Found on sandy sea shores often fixing sand or dunes, rarely found away from the coast. Flowering during mid wet season but a few plants flower during the dry season. Seed locally dispersed by tortoises.

18. STENOTAPHRUM Trin.

Annuals or perennials. Leaf-blades flat, folded or inrolled. Inflorescence a terminal, bilaterally flattened or cylindrical, false spike. Spikelets 2-flowered, lanceolate to ovate-oblong, awnless, on a very short rhachis sunk into a depression on the main axis. Glumes both small, or the upper as long as the spikelet, membranous. Lower floret male or sterile, lemma papery or leathery, as long as the spikelet; upper floret bisexual, lemma and palea both leathery.

7 species, tropics and sub-tropics of the world, usually coastal.

False spike less than 1 cm long   **1. S. clavigerum**
False spike 5–15 cm long   **2. S. micranthum**

1. Stenotaphrum clavigerum Stapf in B.M.I.K. 1919: 142 (1919). Type: Assumption, *Dupont* 255 (K, holo.).

Delicate tufted annual 25 cm high; stems very slender. Leaf-blades linear-lanceolate, 1.5–6 × 0.1–0.3 cm, flat. False spikes 5–8 × 1.5–2 mm. Spikelets 5, rarely 6, 1.8–2 mm long, sunk in depressions on opposite sides of the main axis. Fig. 54/10–11.

Distr. ALDABRA: W, M, S; ASSUMPTION; endemic.
Notes. Frequently found growing in crevices in mixed scrub coastal communities. Flowering late wet to mid dry season.

2. Stenotaphrum micranthum (Desv.) Hubbard in Hubbard & Vaughan, Grasses Maurit. & Rodrigues.: 73 (1940).

Ophiurinella micrantha Desv., Opusc.: 75, fig. 5/4 (1831).
Stenotaphrum subulatum Trin. in Mem. Acad. St. Petersb. 6, Sci. Nat. 1: 190 (1834); F.M.S.: 440 (1877); Stapf in B.M.I.K. 1919: 134 (1919).

Erect or geniculately ascending annual, up to 50 cm high. Leaf-blades lanceolate, 3–12 × (0.3–) 0.5–1 cm, apex narrowly acute. False spikes usually cylindrical, 5–15 × 0.2–0.25 cm. Spikelets many, 3–3.5 mm long, arranged in groups of 1–6 on short racemes sunk into depressions on opposite sides of the main axis. Fig. 51/8–9.

Distr. COSMOLDEO: M; ASTOVE; Indo-Pacific strand plant.
Notes. Found growing in coconut plantations.
Vernac. 'herbe la seine'.

19. ZEA L.

Robust annuals. Leaf-blades broad, flat, drooping. Inflorescence of 2 different kinds. Male inflorescence: several, subdigitately arranged, spreading, terminal racemes; spikelets 2-flowered, arranged in pairs, 1 sessile, the other pedicelled. Female inflorescence: a single, stout, axillary spike borne in the axils of the upper leaves, enclosed by the sheaths; spikelets 2-flowered, lower floret sterile, upper floret female.

A genus of 1 species only, originally from Central America, now widely distributed throughout the warmer parts of the world as a crop plant for both human consumption and forage.

Zea mays L., Sp. Pl.: 971 (1753); F.M.S.: 458 (1877).

Tall plants up to 5 m high, (2.5 m in our area), leaf-blades up to 90 × 10 cm, flat, drooping. Male racemes: 10–20 cm long. Female spike: tightly enclosed by leaf-sheaths when young with only the long purple styles drooping out at the apex in a conspicuous tuft.

Distr. ALDABRA: W; COSMOLEDO: M; Central America, now widely distributed throughout the world.
Notes. Cultivated in a small plot behind the dunes at the northern end of Menai Is., also at Settlement on Aldabra; formerly cultivated on Assumption and Astove.
Vernac. 'mais', 'sweet corn' or 'maize'.

MOSSES

C.C. Townsend, Herbarium, Royal Botanic Gardens, Kew

1. Creeping mosses; stems adhering to the substratum by rootlets (pleuro-
   carpous mosses)
   2. Main stem with minute, 2–4-celled organs (paraphyllia) among the
      leaves; nerve of leaf single, reaching above the middle; upper cells
      scarcely longer than broad, with several small papillae on the lumen
                                                              5. Thuidium sp.
   2. Main stem without paraphyllia; nerve double, not reaching the centre of
      the leaf; upper cells distinctly longer than broad, papillose only through
      the extruded upper end of the cell   **6. Trachyphyllum inflexum**
1. Erect (acrocarpous mosses)
   3. Leaves 2-ranked, blade appearing double in the basal half on the upper
      side of the nerve   **3. Fissidens Sect. Crenularia**
   3. Leaves arranged all round the stem, lamina nowhere appearing double
      4. Cells in upper part of leaf smooth, obviously longer than wide,
         the ± pointed ends intermeshing with those of adjacent cells
         (prosenchymatous)   **1. Bryum sp.**
      4. Cells in upper part of leaf papillose, ± isodiametric, quadrate or some-
         what rounded with the ends not intermeshing (parenchymatous)
         5. Basal cells of leaves large, clear, abruptly demarcated from the lamina
            cells; apex of upper leaves with entire margins, with a rounded termi-
            nal group of gemmae   **2. Calymperes tenerum**
         5. Basal cells of leaves not abruptly demarcated, passing gradually into
            the lumina cells; margin of upper leaves ± toothed for some distance
            below the nongemmiferous apex   **4. Hyophila potieri**

1. Bryum sp.

Distr. ALDABRA: W.
   Notes. Sterile and indeterminable specimens; *Stoddart* 989 (K) from path-
way in coconut grove and *Hnatiuk* 731728 (K) from rock crevice in Basin
Cabri.

2. Calymperes tenerum C. Muell. in Linnaea 37: 174 (1872).

C. sancta-mariae Besch. in Ann. Sci. Nat. Bot. VI, 9: 346 (1880); Renauld &
Cardot, Hist. Nat. Pl. Madagasc., Mousses: 187, fig 43/2 (1915).

Distr. ALDABRA: W, P, M, S, Esprit; Indo-Pacific.
Notes. Found growing on tree trunks, farily common.

3. Fissidens sp.

Distr. ALDABRA: W.
Notes. Single collection from rock crevices in Basin Cabri, *Hnatiuk* 731739.

4. Hyophila potieri Besch. in Ann. Sci. Nat. Bot. VI 9: 341 (1880) & in Rev. Bryol. 1880: 21 (1880); Renault & Cardot, Hist. Nat. Pl. Madagasc., Mousses 208, fig. 41/1 (1915).

Distr. ALDABRA: W, P, S, Esprit; Madagascar, Mascarenes, Seychelles.
Notes. Growing above *Thuidium* sp., on rocks in depressions in dense scrub.

5. Thuidium sp.

Distr. ALDABRA: P.
Notes. Growing below *Hyophila potieri*, on rocks in depressions in dense scrub. Sole record *Hnatiuk* 730950 (K), sterile and indeterminable.

6. Trachyphyllum inflexum (Harv.) Gepp in Hiern, Cat. Welw. 2, 2: 299 (1901).

Hypnum inflexum Harv. in Hook., Icones 1: 24, fig. 6 (1836).

Distr. ALDABRA: Esprit; southern Asia, Oceania and tropical Africa.
Notes. Found growing on rock and tree trunks in dense scrub.

INDEX TO BOTANICAL NAMES

Names accepted for the flora or used in the discussions are in roman, synonyms in italics.

INDEX TO VERNACULAR NAMES

The Seychellois vernacular names have been compiled from Baker (1877), 'Flora of Mauritius and the Seychelles'; Dupont (1907), 'Report on a visit to St Pierre, Astove, Cosmoledo, Assumption and the Aldabra Group of the Seychelles Islands'; Bailey (1961), 'List of flowering plants and ferns of Seychelles'; labels on herbarium sheets and from the transient Seychellois labour on Aldabra. No attempt has been made to correct the more obvious corruptions of the French language.

Printed in England for Her Majesty's Stationery Office by Hobbs the Printers of Southampton
(63) Dd0597195 500 3/80 G327